Carcinogens and Mutagens in the Environment

Volume I
Food Products

Editor
Hans F. Stich, Ph.D.
Head
Environmental Carcinogenesis Unit
British Columbia Cancer Research Centre
Professor of Zoology
University of British Columbia
Vancouver, British Columbia, Canada

CRC Press, Inc.
Boca Raton, Florida

Library of Congress Cataloging in Publication Data

Main entry under title:

Carcinogens and mutagens in the environment.

Includes bibliographies and indexes.
Contents: v. 1. Food products.
1. Carcinogens. 2. Mutagens. 3. Environmentally induced diseases. I. Stich, H. F. (Hans F.), 1927- . [DNLM: 1. Carcinogens, Environmental. 2. Mutagens. 3. Environmental pollutants. QZ 202 C2651]
RC268.6.C365 616.99'4071 81-21764
ISBN 0-8493-5881-7 AACR2

Direct all inquiries to CRC Press, Inc., 2000 Corporate Blvd., N.W., Boca Raton, Florida, 33431.

International Standard Book Number 0-8493-5881-7

Library of Congress Card Number 81-21764
Printed in the United States

PREFACE

Daily each member of the human population ingests, inhales, or comes into contact in other ways with thousands of chemicals, including carcinogens, mutagens, and teratogens. The individual is constantly bombarded by news that one or more of the most cherished food items is hazardous to health. Regulatory agencies face the difficulty each day of making decisions on the basis of totally inadequate scientific data. It is therefore not surprising to find a frightened public, irritated industrialists, pressed civil servants, and underpaid scientists clamoring for more data on human populations and a larger influence in the decision-making processes that may affect everyone's lives.

Our lack of understanding of environmental toxicology becomes only too evident from the observation that we cannot even guess about the relative contribution of man-made agents and naturally occurring chemicals to the total carcinogenic or mutagenic load in a particular environment. The information available to assess a particular hazard to human health is shockingly inadequate when we consider the millions of chemicals in our environment, the few thousand chemicals tested for genotoxicity, and the paltry hundred or so compounds that have been properly evaluated for their carcinogenicity and teratogenicity. The difficulties involved in trying to assess the carcinogenic and genotoxic load will be further compounded when testing programs are extended to complex mixtures. The carcinogenicity and genotoxicity of a complex mixture does not seem to be a simple sum of the activities of the individual compounds because of a multitude of synergistic, antagonistic, co-carcinogenic, and co-mutagenic interactions between its various components.

Moreover, the composition of any mixture when ingested or inhaled will be subject to continuous change within the body that will in turn affect its carcinogenic and genotoxic potential. Finally, the testing of combinations and permutations between only a few compounds would rapidly exhaust the total available manpower and funding of an entire nation. A simple calculation reveals that the testing of all permutations between 10 different chemicals would lead to about 10 million possible permutations, 1½ billion dishes, approximately 79 million working hr and 9.8×10^{9} U.S. dollars when the Ames' Salmonella mutagenicity test is used on 5 strains, at 5 doses, and with or without S9 activation. With 20 different compounds, the number of man years would amount to approximately 3.3×10^{16} and the cost to 6.5×10^{21} U.S. dollars. These are staggering figures considering that 20 compounds is a relatively small number when we realize that 2000 or more chemicals are found, for example, in such widely consumed beverages as coffee and tea. In spite of these and other apparently insurmountable difficulties, the search for solutions will and must continue. The papers in this volume are an example of this effort.

H. F. Stich

THE EDITOR

Hans F. Stich, Ph.D., is Head, Environmental Carcinogenesis Unit, British Columbia Cancer Research Centre and Professor in the Department of Zoology and in the Department of Pathology, University of British Columbia, Vancouver, B. C., Canada.

Dr. Stich received his Ph.D. in Developmental Biology in 1949 from the Faculty of Arts and Science at the University of Wurzburg, Germany. He also studied tropical medicine at the same university.

He is a member of the Environmental Contaminants Advisory Committee on Mutagenesis, Health and Welfare Canada; a Director of the British Columbia Cancer Foundation; Professional Consultant of the B.C. Cancer Control Agency, Councilor of the Environment Mutagen Society, on the editorial board of Mutation Research and Cancer Research, was Chairman of the Grant Panel on Environment Toxicology, and Chairman of the Grant Panel on Cell Biology and Genetics of the Natural Science and Engineering Research Council Canada.

Dr. Stich has presented numerous invited lectures at international and national meetings as well as guest lectures at various universities and institutes. He has published over 180 papers to date. His major research interests include the etiology of oral and esophageal carcinomas among betel nut and tobacco chewers, the use of genotoxicity tests to detect naturally occurring carcinogens in food products and beverages, and the assessment of a mutagenic load by examining the synergistic and antagonistic interactions between mutagens, co-mutagens, and antimutagens.

CONTRIBUTORS

Richard J. Albertini, M.D., Ph.D.
Professor of Medicine and Medical Microbiology
University of Vermont
Burlington, Vermont

Elizabeth F. Allen, Ph.D.
Research Assistant Professor of Pathology
University of Vermont
Burlington, Vermont

L. F. Bjeldanes, Ph.D.
Associate Professor of Food Toxicology
Department of Nutritional Sciences
University of California
Berkeley, California

George T. Bryan, M.D., Ph. D.
Professor of Human Oncology
University of Wisconsin Center for Health Sciences
Madison, Wisconsin

Hai-Ying Cai, Ph.D.
Pathologist, Experimental Oncologist
Cancer Research Institute
Chinese Academy of Medical Sciences
Beijing, People's Republic of China

K. K. Carroll, Ph.D.
Professor of Biochemistry
University of Western Ontario
London, Ontario, Canada

Frank Cordle, Ph.D., M.P.H.
Chief
Epidemiology and Clinical Toxicology Unit
Bureau of Foods
Food and Drug Administration
Washington, D.C.

Bruce D. Dannenberg
Technologist
University of Vermont
Burlington, Vermont

R. Dietrich
Food Research Institute
Department of Food Microbiology and Toxicology
University of Wisconsin
Madison, Wisconsin

Bruce P. Dunn
Environmental Carcinogenesis Unit
British Columbia Cancer Research Centre
Vancouver, British Columbia, Canada

G. Eisenbrand, Dr. rer. nat.
Research Scientist
Institute of Technology and Chemotherapie
German Cancer Research Center
Heidelberg, West Germany

P. S. Elias, M.D.
Director
International Food Irradiation Project
Federal Research Centre for Nutrition
Karlsruhe, West Germany

G. A. H. Elton, D. Sc.
Chief Scientist (Fisheries and Food)
Ministry of Agriculture, Fisheries, and Food
London, England

J. S. Felton, Ph.D.
Senior Biomedical Scientist
Biomedical Sciences Division
Lawrence Livermore National Laboratory
University of California
Livermore, California

W. Gary Flamm, Ph.D.
Deputy Associate Commissioner for Health Affairs
Food and Drug Administration
Washington, D.C.

Walter A. Hargraves
Assistant Scientist
Food Research Institute
Department of Food Microbiology and Toxicology
University of Wisconsin
Madison, Wisconsin

F. T. Hatch, M.D., Ph.D.
Assistant Associate Director
Biomedical and Environmental Research Program
Lawrence Livermore National Laboratory
University of California
Livermore, California

Takeshi Hirayama, M.D.
Chief
Epidemiology Division
National Cancer Center Research Institute
Tokyo, Japan

Wayne T. Iwaoka, Ph.D.
Associate Professor
Institute for Food Science and Technology
University of Washington
Seattle, Washington

R. D. Klassen, B. Sc.
Chemist
Food Directorate
Health Protection Branch
Health and Welfare Canada
Ottawa, Ontario, Canada

Ib Knudsen, D.V.M.
Deputy Director
Institute of Toxicology
National Food Institute
Soborg, Denmark

Cheryl A. Krone
Research Associate
Institute for Food Science and Technology
University of Washington
Seattle, Washington

Elizabeth C. Miller, Ph.D.
Wisconsin Alumni Research Foundation Professor of Oncology
McArdle Laboratory for Cancer Research
Center for Health Sciences
University of Wisconsin
Madison, Wisconsin

James A. Miller, M.D.
Wisconsin Alumni Research Foundation Professor of Oncology
McArdle Laboratory for Cancer Research
Center for Health Sciences
University of Wisconsin
Madison, Wisconsin

Sanford A. Miller, Ph.D.
Director
Bureau of Foods
Food and Drug Administration
Washington, D.C.

T. Muise, B.Sc.
Graduate Student
Department of Biochemistry
University of Ottawa
Ottawa, Ontario, Canada

A. Mahir Pamukcu, D.V.M., Ph.D.
Adjunct Professor of Human Oncology
University of Wisconsin Center for Health Sciences
Madison, Wisconsin

M. W. Pariza
Associate Professor and Chairman
Food Research Institute
Department of Food Microbiology and Toxicology
University of Wisconsin
Madison, Wisconsin

David H. Phillips
Chester Beatty Research Institute
Institute of Cancer Research
Royal Cancer Hospital
London, England

Emil Poulsen, D.V.M., D.V.Sc.
Director
Institute of Toxicology
National Food Institute
Soborg, Denmark

W. D. Powrie
Professor and Head
Department of Food Science
University of British Columbia
Vancouver, British Columbia, Canada

R. Preusmann
Professor, Dr. rer. nat.
Institute of Toxicology and Chemotherapie
German Cancer Research Center
Heidelberg, West Germany

Miriam P. Rosin, Ph.D.
Assistant Professor
Department of Pathology
University of British Columbia
Staff Member
Environmental Carcinogenesis Unit
British Columbia Cancer Research Center
Vancouver, British Columbia, Canada

Ronald C. Shank, Ph.D.
Associate Professor of Toxicology
Community & Environmental Medicine
University of California
Irvine, California

B. Spiegelhalder, Dr. rer. nat.
Research Scientist
Institute of Toxicology and Chemotherapie
German Cancer Research Center
Heidelberg, West Germany

B. Stavric, Ph.D.
Research Scientist
Food Directorate
Health Protection Branch
Health and Welfare Canada
Ottawa, Ontario, Canada

Hans F. Stich, Ph.D.
Head
Environmental Carcinogenesis Unit
British Columbia Cancer Research Centre
Vancouver, British Columbia, Canada

Leonard Stoloff
Bureau of Foods
Food and Drug Administration
Washington, D.C.

D. R. Stoltz, D.V.M.
Research Scientist
Toxicology Research Division
Food Directorate
Health and Welfare Canada
Ottawa, Ontario, Canada

David L. Sylwester, Ph.D.
Professor of Mathematics
University of Vermont
Burlington, Vermont

John H. Weisburger, Ph.D.
Vice President for Research
Naylor Dana Institute for Disease Prevention
American Health Foundation
Valhalla, New York

Gary M. Williams., M.D.
Associate Director
Naylor Dana Institute for Disease Prevention
American Health Foundation
Valhalla, New York

C. H. Wu
Research Associate
Department of Food Science
University of British Columbia
Vancouver, British Columbia, Canada

Ernst L. Wynder, M.D.
President
American Health Foundation
New York, New York

TABLE OF CONTENTS

Part I

Epidemiological Evidence

Chapter 1

THE USE AND LIMITATIONS OF EPIDEMIOLOGIC AND ANIMAL DATA TO DESCRIBE CURRENT CANCER RISK IN THE U.S.

Frank Cordle and Sanford A. Miller

TABLE OF CONTENTS

I. INTRODUCTION

A recent report from the Office of Technology Assessment[1] concerned with improving methods to better determine cancer risks, has indicated that despite the current antiregulatory mood in the U.S., the public fear and dread of cancer are not likely to decrease and Americans still favor regulations that protect public health and the environment. Surveys taken by the Yankelovich, Skelly, and White polling firm show that public support for vigorous food and drug regulation remains extremely high, almost the only exception to the general appeal of deregulation.

The continuing concern about cancer is likely to provide impetus for continuing efforts to reduce the incidence and to improve the treatment. However, while efforts to improve treatment are seen as highly desirable and excite little controversy, the efforts to reduce cancer incidence by regulatory intervention generate great debate about whether the expected benefits from the regulations justify their costs. Underlying the problem with cost-benefit decisions is the lack of adequate data to make such a determination, especially in the area of cancer incidence and mortality, and any improvement thereof resulting from regulatory or program intervention.

Quantification of risks and benefits is likely never to be so precise that estimates of these variables can be plugged into a formula to produce a number that dictates a decision. Instead, after all the analyses a decision about whether a risk is acceptable or unacceptable and reasonable or unreasonable will be made by a few individuals. Those judgments reflect societal values and would, ideally, be made by citizens as a whole. In our form of government, elected representatives are responsible for expressing societal values, and this responsibility is sometimes delegated by elected representatives to executive officials.

Unfortunately, there appears to be a growing tendency on the part of some high-level administrative officials toward the fast, short, and expedient approach to the decision-making process in public health activities. This kind of decision-making approach appears to be somewhat analogous to the trends observed in the last decade in the automobile and steel industries in the U.S., where the management approach has been totally dedicated to the immediate profit with little regard for investment in the future. Currently, too many administrators fault epidemiology for the lack of immediate answers to problems in the public health regulatory area, disregarding the fact that if the answers are unavailable today, and if no investment is made in adequate epidemiological resources for the future, the answers will always be unavailable.

The principal data deficiencies for the epidemiological assessment of cancer risks, in particular in the regulatory area to determine risk of either cancer incidence or mortality, are inadequate information about exposures and lifestyle characteristics. Since cancer has a long latency period, relating cancer in the present population to particular exposures might require information from 20 or more years ago. Even when information was collected, records may have been destroyed before they could become useful in cancer epidemiology studies. As a preliminary to a discussion of the present controversy over cancer trends in the U.S. and the equally heated debate over animal tests vs. epidemiological studies to yield data describing cancer risk (increases or decreases in such risks), a description of some of the problems associated with each of the data sources might be useful.

Although studies of trends over time in the incidence of and mortality from cancer in defined populations may provide the data for clues to the etiology of the disease (provided the data are adequate and valid) there is, unfortunately, no national reporting system available to define the incidence of cancer in the entire U.S. As a result, a considerable amount of discussion and debate has taken place over the last several years and continues today, about the use of incidence data collected by the various surveys conducted by the National Cancer Institute (NCI) as well as mortality data collected by the National Center for Health Statistics (NCHS), in the interpretation of cancer trends in the U.S.

II. NCI INCIDENCE SURVEYS

Perhaps the three most frequently analyzed surveys are the somewhat mislabeled National Cancer Surveys carried out during three periods: the first (FNCS), 1937 to 1939; the second (SNCS), 1947 to 1948; and the third (TNCS), 1969 to 1971. Devesa and Silverman[2] have described in detail the methods used in these surveys. For a variety of reasons, each of the three surveys could not be conducted in the same geographic areas. As a result, only seven metropolitan areas are common in the three surveys (SCA). For the TNCS, several new areas were added, including two entire states, so that the cancer incidence in rural areas could be described.

In addition, due to the growth in urban areas since the SNCS, data for the entire Standard Metropolitan Statistical Area (SMSA) for each area defined in the 1970 census were included in the TNCS rather than restricting the area to those counties included in the earlier surveys.

Although the population surveyed in each of the three studies was reported to represent 10% samples of the population of the U.S. at the time of the survey, this is not entirely accurate. In fact, although the populations from the various areas in each of the three studies may have represented 10% of the total U.S. population at the time of each study, no attempt was made to design a valid 10% probability sample of the U.S. for purposes of the studies. One can appreciate the need for a pragmatic approach in such studies, but one must also recognize the potential problems associated with any inferences directed toward the country as a whole.

It must also be pointed out that when the seven common areas of the three studies are used for comparision, population bases become 5.0, 5.2, and 7.2% of the total U.S. population at the time of each respective survey, rather than the often cited 10%.

The NCI started the Surveillance, Epidemiology, and End Results (SEER) program in 1973 to obtain annual cancer incidence and patient survival data on a population base[3]. When the SEER program was established in 1973, there were eight geographic participants including Puerto Rico, Detroit, San Francisco, Connecticut, Hawaii, New Mexico, Utah, and Iowa. By 1976 the SEER program included 11 areas: 5 entire states, 5 large metropolitan areas, and the entire commonwealth of Puerto Rico.

The purpose of the SEER program is to measure cancer incidence and patient survival on a continuing basis, so that a new survey does not have to be undertaken every few years to measure cancer incidence. The participating organizations in the SEER program were selected on the basis of their demonstrated ability to operate a population-based cancer registry system, and for the unique population subgroups that each of them offered.

Unfortunately, by the nature of the selection process, distortions in population distribution were introduced. Thus, although the participants as a whole came from areas that represented 10% of the total U.S. population (again, not a 10% random sample of the U.S. population), they represented only about 9% of the black population but a larger proportion of other ethnic groups, e.g., 47% of the American Japanese, 36% of the American Chinese, and 15% of the American Indians.[3] As seems evident, with each new study the number of geographic areas that might be compared are reduced, with seven in the three NCI surveys and only four from the NCI survey comparable with the SEER program.

III. HUMAN MORTALITY DATA

Although mortality data by specific cause, age, sex, race, and a variety of demographic variables have been available since 1935, variations in the reporting methods by various states, as well as the changes that have occurred over time in assigning International Classification of Diseases (ICDA) codes to cause of death, present some difficulties in assessing cancer trends over time on the basis of mortality.

For example, data on mortality before 1949 are not available for several cancer sites. Trends in mortality from cancer of the uterine cervix and corpus prior to 1949 cannot be evaluated separately because the subsite of the uterus was not coded until the sixth revision of the ICDA. In the fourth and fifth revisions of the ICDA, primary and secondary liver cancers were assigned the same ICDA code, and were not coded separately until 1949.

In addition to the numerous classification changes that have occurred in the past 45 years, the rules for assigning cause of death were substantially revised in 1949, and could introduce artifacts in any observed trends. Perhaps the most important factors affecting error and bias in both mortality and incidence data over time are

1. Improper diagnosis
2. Improvements in case ascertainment
3. Inclusion of prevalent cases
4. Incomplete or incorrect information on death certificates
5. Increased access to medical care
6. Changes in diagnosis over time, among others

Now that these comments have provided some background on survey methodology, observations on comparability between data sets and the validity of observed trends in incidence and mortality are in order.

Devesa and Silverman[2] have published the results of their analysis of the data from the three NCI surveys. There they describe trends in both the incidence and mortality from cancer during the period 1935 to 1974. They reported that between two survey points, 1947 and 1970, the "overall age-adjusted incidence rate for all sites combined decreased 3.9 percent." This summary figure is the result of ups and downs in race-sex-site groups, including an overall striking decrease for women and an overall increase for men. Lung cancer was by far the most active site, increasing more than 3.5 times between the FNCS, when it ranked eighth in incidence as a primary site, and the TNCS, when it was ranked first. Uterine cancer, which had the highest incidence rate in FNCS, had decreased 40%. Even more dramatic was the overall 72% decrease in stomach cancer by the time of the TNCS.

However, Devesa and Silverman believe that some of the apparent changes may be artifactual, notably part of the dramatic increase in cancers of the lung and prostate in nonwhite males. In addition, the lower rates for nonwhites in the earlier surveys may reflect underdiagnosis. However, they conclude:[2]

> "Changes are likely to have occurred in the prevalence of carcinogenic agents either in the general or personal environment, since the shifts in trends, especially when considered by race and sex, are greater than those that could be explained by the problems discussed earlier, changes in access to medical care, changes in diagnosis, or changes in a variety of areas."

Some of the incidence data from the first, second, and third cancer surveys are presented in Table 1. For all sites for all race-sex groups, the rates per 100,000 are 259.3 in the first survey, 288.9 in the second survey, and 277.7 in the third. All rates

Table 1
CANCER INCIDENCE CHANGES 1937—1939, 1947—1948, 1969—1971 (rate/100,000)[a]

	WM	WF	NWM	NWF	Total
All sites					
1937—1939	247.1	286.1	153.8	243.3	259.3
1947—1948	283.7	305.0	225.6	273.2	288.9
1969—1971	309.0	256.8	330.2	231.5	277.7
Stomach					
1937—1939	41.8	25.4	37.3	18.4	32.9
1947—1948	32.4	17.8	38.6	18.9	25.2
1969—1971	12.1	5.8	18.6	7.9	9.1
Esophagus					
1937—1939	5.9	1.5	4.1	1.0	3.6
1947—1949	6.3	1.8	6.6	1.5	4.0
1969—1971	4.4	1.3	14.3	3.4	3.5
Digestive system					
1937—1939	102.9	77.2	72.6	44.7	86.7
1947—1949	102.7	75.3	89.0	57.0	87.2
1969—1971	79.6	54.0	93.7	56.4	66.6
Colon					
1937—1939	NA	NA	NA	NA	NA
1947—1949	23.8	26.0	13.7	11.9	23.8
1969—1971	29.0	24.8	22.9	23.6	26.4
Liver, gallbladder, & biliary tract					
1937—1939	8.9	10.6	6.6	6.7	9.4
1947—1949	7.0	7.8	5.9	5.9	7.4
1969—1971	5.1	3.9	8.1	3.8	4.6
Liver					
1937—1939	NA	NA	NA	NA	NA
1947—1949	5.1	4.1	5.1	3.0	4.2
1969—1971	2.9	1.2	6.3	1.6	2.2
Gallbladder & biliary tract					
1937—1939	NA	NA	NA	NA	NA
1947—1949	1.9	3.7	0.8	2.9	2.8
1969—1971	2.2	2.7	1.8	2.2	2.4
Lung, bronchus & trachea					
1937—1939	13.1	4.1	7.7	2.9	8.5
1947—1949	29.5	6.7	23.3	4.9	17.6
1969—1971	68.0	14.9	77.9	13.5	39.2
Urinary system					
1937—1939	18.7	10.0	7.3	5.1	13.5
1947—1949	22.8	10.6	10.1	8.8	15.8
1969—1971	29.8	9.6	17.0	7.1	17.7
Bladder					
1937—1939	14.1	6.6	3.8	3.9	9.7
1947—1949	17.2	7.1	4.8	5.6	11.3
1969—1971	21.3	5.6	9.8	3.5	11.7
Breast					
1937—1939	0.8	67.1	1.0	46.2	32.7
1947—1949	0.9	73.6	0.2	50.4	36.9
1969—1971	0.8	73.3	0.7	53.7	38.8
Pancreas					
1937—1939	6.7	5.0	4.3	2.2	5.5
1947—1949	8.9	5.6	9.9	4.1	7.1
1969—1971	10.7	6.5	14.1	8.2	8.7

Note: NA = not available.

[a] Age adjusted to 1950 standard U.S. population.

From Devesa, S. S. and Silverman, D. T., *J. Natl. Cancer Inst.*, 60, 545, 1978. With permission.

Table 2
CANCER MORTALITY RATES[a] FOR SELECTED SITES, BY RACE AND SEX, U.S. 1935—1974[b]

	WM	WF	NWM	NWF
All sites				
1935	130.6	146.6	77.4	127.8
1945	144.5	139.8	105.0	127.7
1959—61	159.1	120.9	173.1	134.8
1969—71	173.2	117.9	205.7	131.2
1973—74	177.3	118.2	219.7	135.3
Esophagus				
1935	3.4	1.1	1.5	0.7
1945	3.7	0.9	4.2	0.9
1959—61	3.8	0.9	9.1	2.2
1969—71	3.7	1.0	10.8	2.6
1973—74	3.8	1.0	12.1	3.3
Stomach				
1935	31.5	21.7	24.6	20.8
1945	24.7	14.6	24.5	14.7
1959—61	13.6	6.9	23.5	10.2
1969—71	8.7	4.3	16.6	6.9
1973—74	7.8	3.7	15.1	7.0
Pancreas				
1935	4.5	3.6	2.5	1.7
1945	5.9	4.4	3.9	2.1
1959—61	9.0	5.3	10.0	6.0
1969—71	9.6	5.7	11.3	6.9
1973—74	9.5	5.7	11.0	7.4
Lung, bronchus, trachea, pleura				
1935	6.4	3.1	3.5	1.2
1945	14.2	4.3	8.9	3.2
1959—61	35.4	5.3	34.3	5.5
1969—71	52.2	10.4	57.5	10.7
1973—74	56.5	13.0	67.0	13.1

[a] Average annual death rate/100,000.
[b] Age adjusted to 1950 U.S. population.

From Devesa, S. S. and Silverman, D. T., *J. Natl. Cancer Inst.*, 60, 545, 1978. With permission.

are standardized to the 1950 U.S. population. For all sites for males, both white and nonwhite, the rates appear to show an increase over the time span, while rates for females, both white and nonwhite, indicate fluctuations in the middle-study time frame.

Overall, and for whites, the incidence rates for cancer of the stomach for the period 1937 to 1971 show a steady decline, whereas for nonwhites the decline appears only in the TNCS. Whether or not this reflects changes in the nonwhites' access to medical care or differences in the addition of geographic areas to the TNCS is unclear.

Incidence rates for cancer of the esophagus show an increase for nonwhites and fluctuations for whites. Incidence rates for bladder cancer increase for all males, whereas the trend for females is unclear.

These particular sites were selected for comparision with mortality trends for the same sites. Table 2 presents mortality data for selected intervals between 1935 and 1974.

A comparison of mortality data from Table 2 with incidence data from Table 1 indicates that the most serious problem in attempting to describe trends between the time periods of the NCI surveys lies with the first time period (FNCS). This is understandable, inasmuch as most of the factors that involve errors and bias in the incidence data appear to have been operating most strongly in the early survey.

If one excludes data from the FNCS and then compares the trends in incidence data from the SNCS with the data from the TNCS, and these trends, in turn, with those from the mortality data for 1945 to 1971, some reasonable similarities appear. For all sites there was an increase for white and nonwhite males, a decrease for white females, and a slight increase for nonwhite females in both incidence and mortality. Cancer of the stomach shows a clear decrease in the rates for each race-sex group in both incidence and mortality, while cancer of the pancreas shows an increase for each race-sex group in both incidence and mortality. Cancer of the lung, bronchus, trachea, and pleura shows an increase for each race-sex group in both incidence and mortality.

It seems reasonably clear that incidence trends from the SNCS to the TNCS for most sites are also reflected in the mortality trends, giving some assurance that the trends may be real. Whether or not these incidence trends can be extended into comparisons with the SEER program remains a question.

Pollack and Horm[3] have described trends in cancer incidence and mortality in the U.S. during the period 1969 to 1976, with comparisons of the SEER program results with those of the TNCS. The paper, according to the authors, had three objectives:

1. To assess the comparability of the cancer incidence rates across the total SEER program over the period 1973 to 1976.
2. To assess the validity of use of TNCS incidence rates for 1969 to 1971 and SEER rates for 1973 to 1976 to analyze incidence trends over the period 1969 to 1976.
3. To present trends in cancer incidence and mortality for 1969 to 1976, where data are sufficiently comparable, for some of the major forms of cancer and to summarize these trends by use of a convenient set of measures.

These investigators report that for whites, rates for all SEER areas combined, for individual cancer sites, and for all cancer sites combined, were found to be "reasonably comparable." However, rates for blacks over the four-year period were not comparable, and therefore analysis of incidence data was carried out only for whites. The authors concluded with regard to the second stated purpose, that "the use of incidence rates for TNCS areas for 1969 to 1971 and for SEER areas for 1973 to 1976 appears to provide a good approximation to trends over that total time period for the white population."[3] Pollack[4] also reached the same conclusion after comparing Connecticut tumor registry data, the data from the three NCS, and the SEER program. Others have disagreed, notably Doll and Peto.[5]

As the data reveal, a variety of interpretations can be attached to changes in either incidence or mortality data. For example, incidence data collected in the SEER program (1973 to 1976), in combination with data from the TNCS, carried out from 1969 through 1971, have been interpreted as showing an increase of more than 10% in cancer incidence during the last decade, while some of the major changes seen in the incidence data parallel those seen in mortality data, e.g., increases in lung cancer for both males and females and decreases in stomach and uterine cancers, while other sites vary.

However, publication of this analysis sparked a controversy about the true nature of incidence trends, since only two years earlier an analysis of data from the three NCS had shown an overall decrease of about 4% between 1947 and 1970. Some observers are concerned about the possibility that, after at least half a century of stable or declining rates, cancer incidence has increased, and the increase might result from newly introduced chemical carcinogens. Those who dispute the importance of the observed increase contend that it reflects changes in the reporting of cancer incidence between TNCS and SEER (1973 through 1976), rather than real changes in cancer incidence.

Doll and Peto,[5] for example, consider the TNCS/SEER comparison "completely unreliable." These authors compared incidence from the SNCS/TNCS/SEER series with registered incidence from Connecticut and upstate New York. They also looked at U.S. mortality for the concurrent period, and found that the TNCS/SEER portion yielded "fantastic and irregular variations in incidence . . . ten times greater than can plausibly be attributed to chance, and a hundred times greater than the corresponding annual changes in mortality over the past few decades."

In addition to their discussion concerning the difficulties of epidemiological interpretation of the NCI survey data and data from the SEER program to determine cancer incidence and mortality trends over time, Doll and Peto[5] have also described some general points of consideration concerning the current status of the use of epidemiology to determine risk in a variety of settings. This description seems particularly appropriate for those in the regulatory process, and their expectations from epidemiology. They reported:[5]

> When positive epidemiologic results show an association between exposure or some other variable and cancer, based on a valid study, they tend to dominate any decision to be made on carcinogenicity. However, even risks that will ultimately kill, for instance, 1 percent or more of the exposed population, may be overlooked or attributed to chance unless a very large-scale investigation is undertaken. In these circumstances, too, when the cancer rates among exposed people are only a moderate multiple of those among the unexposed, problems of interpretation may become acute, and it may be extremely difficult to disentangle the various contributions of biased information, confounding of two or more factors, and cause and effect.

In other words, unless epidemiological studies have been carried out in reasonably large, well-defined groups of people who have been heavily exposed to a particular substance for two or three decades without apparent effect, they can offer no guarantee that continued exposure to moderate levels will, in the long run, be without some increase in risk.

This is not to say that human evidence can ever be totally dispensed with. It is always relevant, but the weight that can be given to it varies greatly with the duration and intensity of the exposure experienced by individuals. Positive evidence is always important. Negative human evidence may mean very little, unless it relates to prolonged and heavy exposure. If, however, it does, and is consistent in a variety of studies (correlation studies over time, cohort studies of exposed individuals, and case-control studies of affected patients), and if the laboratory evidence is also limited in its scope, (for instance, to a particular type of tumor in a few species), negative human evidence may justify the conclusion that for practical purposes the agent need not be treated as a human carcinogen.[5]

Epidemiology has, at present, an undeservedly poor reputation among administrators and regulatory decision makers who first have artificially limited themselves by wondering which chemicals to restrict, and then find that almost none of the few thousand chemicals they are worried about have been adequately studied by epidemiologists.

However, this is to condemn epidemiology for failing to achieve ends that it does not have either the resources or data to meet. Epidemiology starts, not with a series of rodents under carefully controlled laboratory conditions for time, exposure, and outcome over a relatively short lifetime, but rather with people in an uncontrolled environment and with little possibility of controlling lifetime exposure to a variety of chemical substances. Epidemiology is, admittedly, more likely to overlook many undetectably small effects of various chemicals than laboratory studies, but it is much less likely to overlook the large determinants of contemporary cancer rates and trends, especially if these are not simple environmental pollutants or dietary contaminants.

IV. ANIMAL DATA

In most cases, no adequate human data are available for estimating risks, and it is necessary to make both numeric and biologic extrapolation from measured responses in animals to estimate human risk. Less uncertainty attends making numeric extrapolations from observation of human responses at high-exposure levels than such extrapolations from animal data. However, extrapolations from human experience are often complicated by poor exposure data.

The more troubling and fundamental problem with biologic extrapolation concerns questions about how closely the test animal resembles humans. This problem is partially related to differences in the greater genetic variability of human populations. Populations of test animals are highly inbred and are genetically uniform. Populations of humans are outbred and include greatly differing genotypes. There is no way to deal with the problem of the human who may differ in sensitivity because there is seldom, if ever, a way to associate sensitivities with the individual. An additional component of the problem concerns differences in metabolism between test animals and humans. Few are well-understood, and many may be unidentified.

Although extrapolation techniques are used to estimate the probability of human cancer from study-derived data, opinions differ about the extent to which, and how, extrapolation methods should be used in estimating the amount of human cancer that might be caused by exposure to a carcinogen. The disagreements among the groups that hold different opinions about the use of such extrapolations are vocal and current.[1] A particular problem in quantitative extrapolation arises from the fact that different extrapolation models produce estimates of cancer incidence that differ by factors of 1000 or more at levels of human exposure. Given such uncertainty, many individuals refuse to choose one model over another for estimating the impact of a carcinogen on humans, and oppose the use of quantitative extrapolation. Fewer objections are raised against choosing a model to establish priorities for further evaluation of carcinogens on the basis of their likelihood of causing cancer. Regardless of which particular model is chosen, it should produce approximately the same relative ranking as any other.

Proponents of quantitative extrapolation argue that careful attention to the available data aids in choosing the correct model and reduces chances for error. Arguments about the applicability of these techniques will continue, especially because efforts to apply cost-benefit analysis to making decisions about carcinogens will require quantitative estimates of cancer incidence.

There are now no convincing data to dictate which extrapolation model is best for estimating human cancer incidence, whether from epidemiological data or animal data, or even that one model will be consistently better than all others. However, one particular model for estimating human incidence from animal data (linear, no threshold extrapolation; relating animals and humans on the basis of total lifetime exposure divided by body weight) has been reported to estimate human cancer incidence within a factor of 10 to 100 when compared to incidence measured by epidemiological studies. While this agreement may be gratifying, good data exist to make these comparisons for fewer than 20 substances.

V. SUMMARY

In summary, it seems apparent that for some time to come, and perhaps permanently, the determination of changes in cancer incidence and mortality will depend upon less than satisfactory data sources. At the same time the determination of human risk will involve the use of both animal and human epidemiological data. However, given the current state of risk assessment using animal models and/or epidemiological data derived from human experience, it becomes mandatory that the conclusions reached from these efforts be approached with caution. As a framework for determining priorities, or for identifying a threshold for requiring regulatory action, they can prove valuable. With a better knowledge of mechanisms, more sophisticated interpretation can take place. We must, however, recognize the limitation of the current methodology, avoid over-interpretation of the data and making promises to the public that cannot be fulfilled.

REFERENCES

1. Research staff, *Assessment of Technologies for Determining Cancer Risks from the Environment,* Office of Technology Assessment, U.S. Government Printing Office, Washington, D.C., 1981.
2. **Devesa, S. S. and Silverman, D. T.**, Cancer incidences and mortality trends in the United States: 1935—74, *J. Natl. Cancer Inst.*, 60, 545, 1978.
3. **Pollack, E. S. and Horm, J. W.**, Trends in cancer incidence and mortality in the United States: 1969—76, *J. Natl. Cancer Inst.*, 64, 1091, 1978.
4. **Pollack, E. S.**, Cancer Incidence Trends in the U.S.: Some Methodological Problems, paper presented at UICC Symp. Trends in Cancer Incidence, Oslo, 1980.
5. **Doll, R. and Peto, R.**, The causes of cancer: quantitative estimates of avoidable risks of cancer in the U.S. today, *J. Natl. Cancer Inst.*, 66, 1192, 1981.

Chapter 2

EPIDEMIOLOGY OF HUMAN CARCINOGENESIS: A REVIEW OF FOOD-RELATED DISEASES

Takeshi Hirayama

TABLE OF CONTENTS

I. INTRODUCTION

Undoubtedly, the best way to prevent cancer is to reduce exposure to carcinogens to a minimum. Cessation of cigarette smoking and avoidance of occupational or other environmental exposure to carcinogens must be considered as being of primary importance in cancer prevention.[1] Alcohol has also been found to elevate the risk of selected cancers, especially when combined with cigarette smoking.[2] In addition, food-related carcinogens are of unique importance because of the large size of the exposed population and also the high frequency of intake from habit or custom, plus the possible influence on cancer of multiple sites, if not all.[3] At present, numerous initiators and promoters have been identified in food additives, foods themselves, or have been formulated during processing, storage, cooking, or even during the process of metabolism.

The prevailing lifestyle in each country is of particular importance. Excessive fat intake, common in Western countries, is now considered to be a promoter of cancers of the breast[4] and pancreas,[5] and the frequent intake of highly salted food common in Japan[6,7] is thought to be the probable promoter of stomach cancer.

On the other hand, selected host factors must also be considered as significant risk modifiers. Nutritional status is one of the most probable determinants of cancer risk. Certain nutritional elements have been found to have promising risk-lowering effects. Laboratory and epidemiological studies thus far conducted have revealed that selected nutritional elements are of particular importance in this respect: ascorbic acid or vitamin C as an initiator inhibitor, β-carotene and/or vitamin A as a promoter inhibitor.[8] Dietary fiber is also considered to be an important risk modifier for colon cancer in man.

Evidence on the effect of diet and nutrition on human cancer has been obtained by different types of epidemiological studies (Table 1).

II. DESCRIPTIVE EPIDEMIOLOGY

The distribution and change in the patterns of diet and nutrition in relation to those of cancer incidence should first be studied by taking various demographic variables into consideration. For instance, national nutritional surveys have been conducted and reported annually in Japan for randomly selected households (6000 in number, see Figure 1). Information on the amount of intake of nutritional elements and each food item has been collected according to districts, socioeconomic levels, and urban-rural differences. Height, weight, and other physiological measurements also have been recorded on each occasion.

For cancer frequency, detailed mortality statistics are available annually based on vital statistics (Figure 2). Morbidity statistics have also been collected from the standard population-based cancer registries covering almost 43% of the population of Japan. Nationwide statistics of subclinical cancer cases also are available for selected cancers, such as stomach and cervix, utilizing the results of mass screening throughout the country. Results of every autopsy conducted in Japan have been compiled as a yearly monograph, providing high quality information on cancer frequency by each cell type. These data are being fully utilized as sources for the descriptive epidemiology on nutrition and cancer.

The two types of information listed above, one on diet and nutritional intake, the other on cancer frequency, have been compared by making maps of each and also by conducting the correlation studies described below.

Table 1
SELECTED APPROACHES IN EPIDEMIOLOGIC STUDIES ON DIET AND CANCER

Descriptive epidemiology	Demographic studies	Geographical pathology (observation of cancer patterns and their changes); ethnic, socioeconomic comparisons; migrant studies
Analytic epidemiology	Correlation studies	Nutrition survey data × cancer incidence data by site
	Case-control studies	Survey of dietary history (quantity, frequency) by interview; comparison between cases and controls
	Prospective studies	First dietary survey of healthy population and follow-up; ambush method
Metabolic epidemiology	Laboratory studies	Study of mechanisms, e.g., intestinal flora
Experimental epidemiology	Intervention studies	Observation of response by altering diet
Operational epidemiology	Dietary strategy	Dietary engineering; integration into whole scheme of cancer control programs; built-in evaluation programs needed

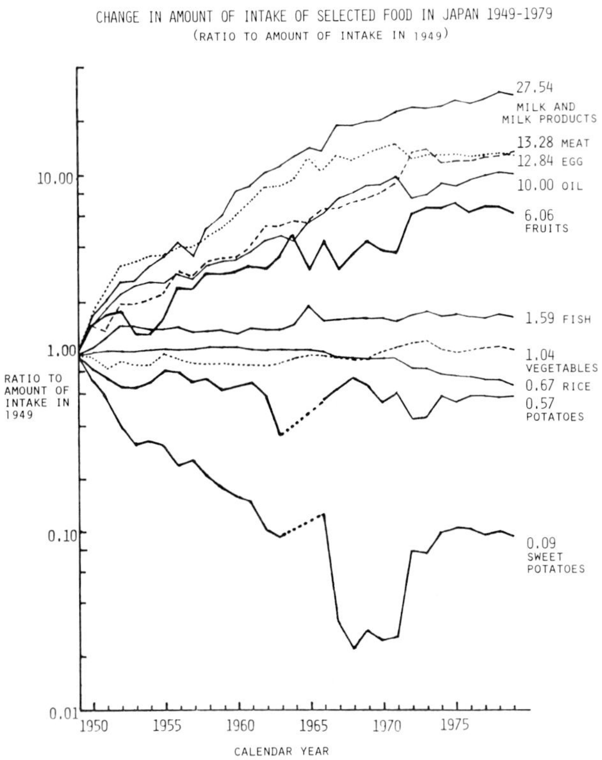

FIGURE 1. Change in amount of intake of selected food in Japan, 1949 to 1979 (ratio to amount of intake in 1949).

III. ANALYTICAL EPIDEMIOLOGY

A. Correlation Studies

Some of the risk determinants are closely related to geography, which makes the geographical distribution of certain cancers quite unique. Examples include lung can-

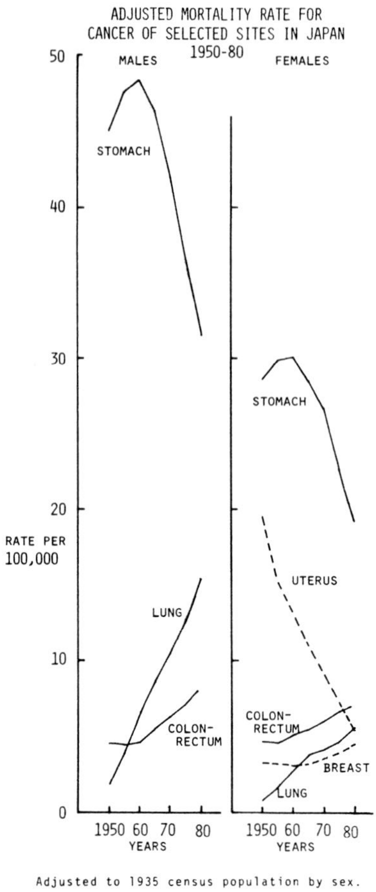

FIGURE 2. Adjusted mortality rate for cancer of selected sites in Japan, 1950 to 1980 (adjusted to 1935 census population by sex).

cer in industrialized urban areas in most countries, Burkitt's lymphoma in endemic malaria areas in Africa and Papua New Guinea, nasopharyngeal cancer in southern China, stomach cancer in Tohoku-Hokuriku districts in Japan facing the Japan Sea, esophageal cancer in selected areas of the world such as the Caspian Sea littoral, Linxian county in Honan Province, China, and certain mountainous areas of Japan, and liver cancer in Asia and Africa.

Also, some of these types are closely associated with the distribution of known carcinogenic risk factors (e.g., higher frequency of cigarette smoking with indoor[9] and outdoor[10] air pollution in urban areas, higher frequency and extent of hepatitis B virus carrier rate[11] and/or exposure to aflatoxin in endemic liver cancer areas, and higher frequency and amount of bracken fern intake in endemic areas of esophageal cancer in Japan[12]).

Several of the geographical peculiarities in cancer occurrence were found to be quite similar to the distribution of promoters or promoter-like substances (e.g., *Croton tiglium,* the parental plant of croton oil that is widely used as a herb drug for nasopharyngeal cancer[6,7] and hot food and drinks for esophageal cancer[14]). In the case of stomach cancer, a significant negative association was also found geographically with the

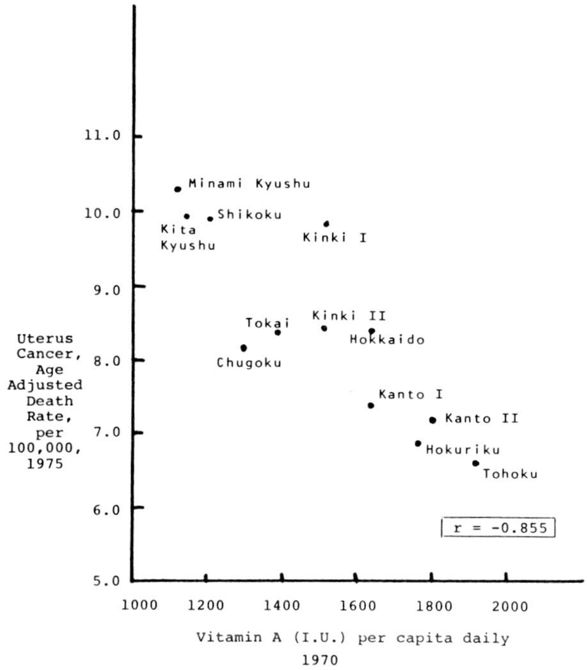

FIGURE 3. Vitamin A intake and adjusted death rate for cancer of the uterus, Japan.

frequency of intake of green-yellow vegetables[5] that are rich in β-carotene, a precursor of vitamin A, a known, potent, promoter-inhibitor. A significant negative correlation was observed between per capita vitamin A intake and the cervical cancer adjusted death rate (Figure 3).

Some groups of risk factors vary together. Careful analysis, such as the computation of partial correlation coefficients, might become necessary.[4,15,16]

B. Case-Control Studies and Cohort Studies

When such geographical associations are observed, case-control studies and/or large-scale cohort studies should follow to confirm the associations and also to clarify the nature and extent of the associations.

Up to now, the influence of diet and nutrition on cancer has been examined in detail by a series of case-control studies conducted in many countries. For example, dietary patterns were found to be principal determinants of cancer of the stomach[6,7] and esophagus[14] by means of case-control studies, taking the influence of confounding variables into consideration.

The risk-enhancing effect of daily intake of salty food and the risk-lowering effect of daily milk drinking were observed in a case-control study in Japan (Table 2).

A series of cohort studies of groups of individuals of defined dietary status were also conducted. They clearly revealed the importance of diet and nutrition in altering the risk of cancer of all sites and/or selected sites.[5] Examples include the effect of daily intake of yellow-green vegetables in reducing the risk of cancers of the stomach (Table 3, Figure 4) and lung (Table 4, Figure 5); the effect of daily intake of soybean paste soup in reducing the risk of cancer of the stomach while daily intake of hot tea gruel enhanced the risk (Tables 5 and 6, Figure 6); the effect of daily intake of meat

Table 2
DIET AND STOMACH CANCER — COMPARISON OF 454 CASES AND MATCHED CONTROLS (KANAGAWA PREF. 1960-61)

	Age									
	−49		50-59		60 +		Total			Standardized
	S	C	S	C	S	C	S	C	Mantel extension χ	ratio rate
Male										
Milk									−8.141 (p<0.0001)	
None	55	28	68	28	76	40	199	96		1.000
Occasional	8	16	12	25	17	22	37	63		0.293
Daily	20	39	18	45	26	57	64	141		0.220
Total	83	83	98	98	119	119	300	300		
Salty food									6.093 (p<0.0001)	
None	43	52	45	66	48	85	136	206		1.000
Occasional	18	22	26	21	43	25	87	68		1.989
Daily	22	9	27	11	28	9	77	29		4.070
Total	83	83	98	98	119	119	300	300		
Female										
Milk									−5.583 (p<0.0001)	
None	39	8	31	14	24	21	94	43		1.000
Occasional	9	17	9	17	4	3	22	37		0.422
Daily	8	31	18	27	12	16	38	74		0.289
Total	56	56	58	58	40	40	154	154		
Salty food									5.886(p<0.0001)	
None	30	38	19	43	13	34	62	115		1.000
Occasional	18	11	16	10	20	5	54	26		4.306
Daily	8	7	23	5	7	1	38	13		7.729
Total	56	56	58	58	40	40	154	154		

Note: S: stomach cancer; C: control group.

in enhancing the risk of cancers of the breast (Table 7, Figure 7), pancreas, liver, esophagus, and lung (Table 4); and the effect of daily intake of alcoholic drinks in elevating the risk of cancers of the mouth, pharynx, esophagus, and liver.[2]

These associations were confirmed by controlling confounding variables such as sex, age, socioeconomic status and cigarette smoking.

C. Diet/Nutrition and Cancer of Each Site

The available information with regard to diet/nutrition and cancer sites can be summarized as follows:

Lip, oral cavity, and pharynx[17] — Vitamin A deficiency appears to enhance the risk. Synergistic effects of alcohol and tobacco chewing and/or smoking exist.

Nasopharynx[13,16,18] — The risk is high in those showing high antibody titer against Epstein-Barr virus and who are frequent users of herbal drugs, the genome of the former possibly being induced by the herbal drugs or other related promoters. A possible association with salted fish was reported in Hong Kong with supporting evidence observed in Japan. The influence of nitrosamines is suspected. Protein deficiency may also exist.

Esophagus[12,14,19-22] — Iron deficiency (sideropenic dysphagia) is associated with upper esophageal cancer. The combined intake of hot tea gruel and bracken fern (BF) enhances the risk in mountainous areas of Japan. A synergistic effect of alcohol and tobacco was observed in many countries, particularly in Brittany, France. Opium and

Table 3
STANDARDIZED MORTALITY RATE FOR CANCER OF THE STOMACH BY SMOKING AND GREEN-YELLOW VEGETABLE INTAKE
(Prospective study, 1966-78, Japan)

Cigarette smoking	Green-yellow vegetables	Observed person (Yr)	Standardized mortality rate (per 100,000)
Male			
Daily	Daily	732,340	194.7
	Others	297,056	213.4
None	Daily	185,433	135.5
	Others	61,380	141.0
Female			
Daily	Daily	119,544	94.7
	Others	55,534	101.0
None	Daily	1,071,275	73.1
	Others	353,607	89.6

	Smoking	Green-yellow vegetables
Mantel-Haenszel χ	6.848	−4.927
One-tail p-value	3.7666×10^{-12}	4.1730×10^{-7}
Rate ratio	1.388	0.838
Confidence limits	1.283	0.790
	1.501	0.889

Table 4
STANDARDIZED MORTALITY RATE FOR CANCER OF THE LUNG BY SMOKING AND GREEN-YELLOW VEGETABLE INTAKE
(Prospective study, 1966-78, Japan)

Cigarette smoking	Green-yellow vegetables	Observed person (Yr)	Standardized mortality rate (per 100,000)
Male			
Daily	Daily	732,340	84.0
	Others	297,056	89.6
None	Daily	185,433	18.3
	Others	61,380	28.4
Female			
Daily	Daily	119,544	30.3
	Others	55,534	37.3
None	Daily	1,071,275	15.1
	Others	353,607	18.1

	Smoking	Green-yellow vegetables
Mantel-Haenszel χ	11.734	−1.720
One-tail p-value	4.4094×10^{-32}	0.043
Rate ratio	3.330	0.898
Confidence limits	2.813	0.811
	3.942	0.995

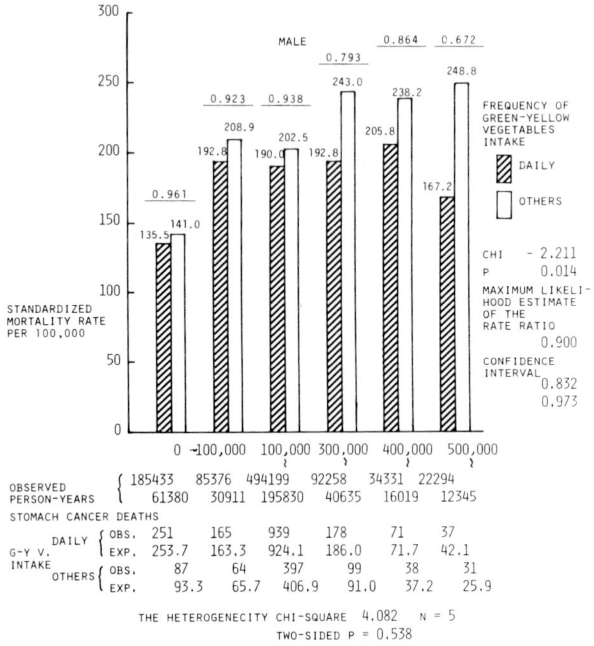

FIGURE 4. Standardized mortality rate for stomach cancer by total number of cigarettes smoked and by frequency of green-yellow vegetable intake (prospective study, 1966 to 1978, Japan).

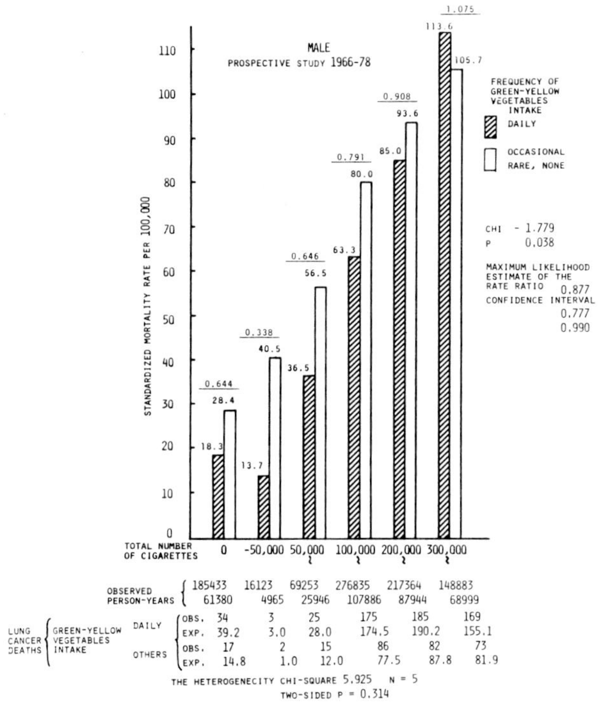

FIGURE 5. Standardized mortality rate for lung cancer by total number of cigarettes smoked and by frequency of green-yellow vegetable intake (prospective study, 1966 to 1978, Japan).

Table 5
CANCER OF THE STOMACH IN DAILY INTAKERS OF SOYBEAN PASTE SOUP

(Prospective Study 1966-78, Tondabayashi Health Center Districts, Osaka, Japan)

		Male Hot tea gruel		Female Hot tea gruel	
	Total	Daily	Others	Daily	Others
Daily					
Ca. stomach	72	15	33	9	15
Population	3,942	442	1,373	602	1,525
Rate/100	1.8	3.4	2.4	1.5	1.0
Others					
Ca. stomach	391	155	107	78	51
Population	14,033	2,837	3,265	4,007	3,924
Rate/100	2.8	5.5	3.3	2.0	1.3

Table 6
CANCER OF THE STOMACH IN DAILY INTAKERS OF SOYBEAN PASTE SOUP[a]

(Observed and expected frequency based on the maximum likelihood pooled point estimate of odds ratio)

Soybean paste soup	Male Hot tea gruel		Female Hot tea gruel		
	Daily	Others	Daily	Others	Total
Daily					
Obs.	15	33	9	15	72
Exp.	16.9	32.3	8.4	14.3	71.9
Others					
Obs.	155	107	78	51	391
Exp.	153.1	107.7	78.6	51.7	391.1

Mantel-Haenszel χ	−2.574; one-sided p = 0.00503
Odds Ratio:	
Standardized rate ratio	0.715
Mantel-Haenszel pooled point estimate	0.713
Maximum likelihood pooled point estimate	0.713
Confidence limits (90%) around the maximum likelihood estimate	0.57
Likelihood ratio test of heterogeneity	0.89
$X^2 = 0.363$ n = 3	two-sided p = 0.948

[a] Prospective study, 1966-78, Tondabayashi Health Center Districts, Osaka, Japan

Table 7
STANDARDIZED MORTALITY RATE FOR STOMACH CANCER AND BREAST CANCER BY HABITS OF SMOKING, GREEN-YELLOW VEGETABLE INTAKE, AND MEAT INTAKE

Smoking	Green-yellow vegetables	Meat	Observed person (Yr)	Standardized mortality rate: Stomach cancer	Standardized mortality rate: Breast cancer
Daily	Daily	Daily	11,612	65.6	33.3
		Others	107,091	97.8	10.1
	Others	Daily	3,050	152.6	0
		Others	52,039	100.7	11.2
None	Daily	Daily	75,157	89.1	16.2
		Others	990,529	73.3	10.8
	Others	Daily	13,064	28.2	31.5
		Others	338,633	92.3	7.5

	Smoking	Green-yellow vegetables	Meat
Stomach cancer			
Mantel-Haenszel χ	2,551	−2.870	0.239
One tail p-value	0.005	0.002	0.405
Rate ratio	*1.234*	*0.838*	1.027
90% confidence limit	1.077	0.757	0.853
	1.413	0.927	1.238
Breast cancer			
Mantel-Haenszel χ	0.572	1.181	2.682
One tail p-value	0.284	0.119	0.004
Rate ratio	1.143	1.248	*1.876*
90% confidence limit	0.778	0.917	1.275
	1.679	1.698	2.768

nutritional deficiencies are suspected in northern Iran; fungus (*Fusarium poae*), nitrosamines, and molybdenum deficiency are suspected in northern China, and consumption of beer made from maize together with molybdenum deficiency, in Africa.

Stomach[6,7,13,23-25] — The risk observed is higher in cigarette smokers and in groups taking salted fish and pickles, and in groups not taking green-yellow vegetables, soybean diet, and dairy products daily. An association with smoked and fried food, heated fats, and proteins is also suspected. A positive correlation with acidic soil and zinc in the garden soil has been observed in Britain and Japan. Nitrates are also suspected to be precursors of nitrosamines. Deficiencies of vitamin C, β-carotene and retinol may also exist.

Colon and rectum[5,26-28] — High beef/high animal fat diets are suspected, but a large-scale cohort study in Japan showed a striking risk-reducing effect of daily meat intake (Table 8, Figure 8). High dietary fiber intake may also reduce the risk. The risk was also observed to be higher in subjects with a low blood cholesterol.

Liver[11,29,30] — The risk is quite high in hepatitis B virus carriers with liver cirrhosis. Dietary exposure to aflatoxin is considered to be quite important in Africa and Asia. Daily meat intake also enhances the risk and the risk is higher in smokers and drinkers.

Gallbladder[31] — The risk is higher in obese women. Dietary fats and total calories have been suggested as promoting factors.

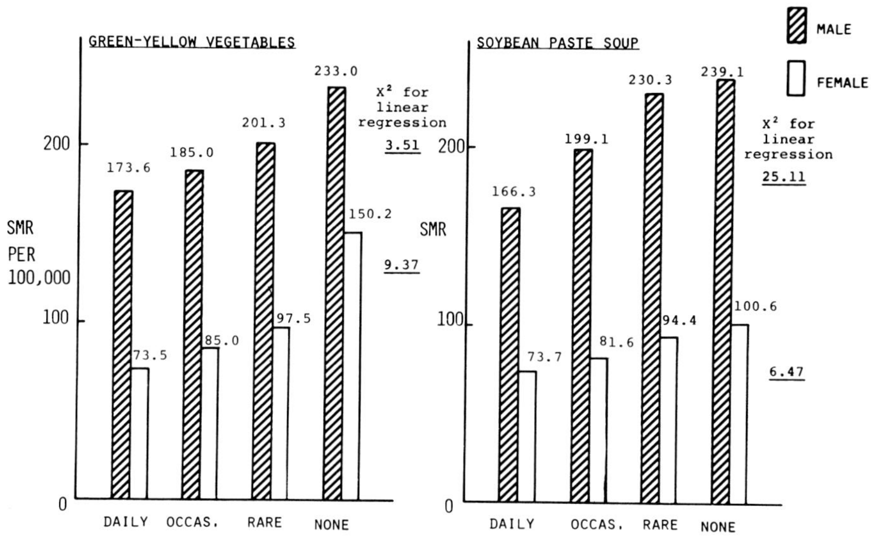

FIGURE 6. Standardized mortality rate for stomach cancer by habit of eating green-yellow vegetables and soybean paste soup (prospective study, 1966 to 1978, Japan).

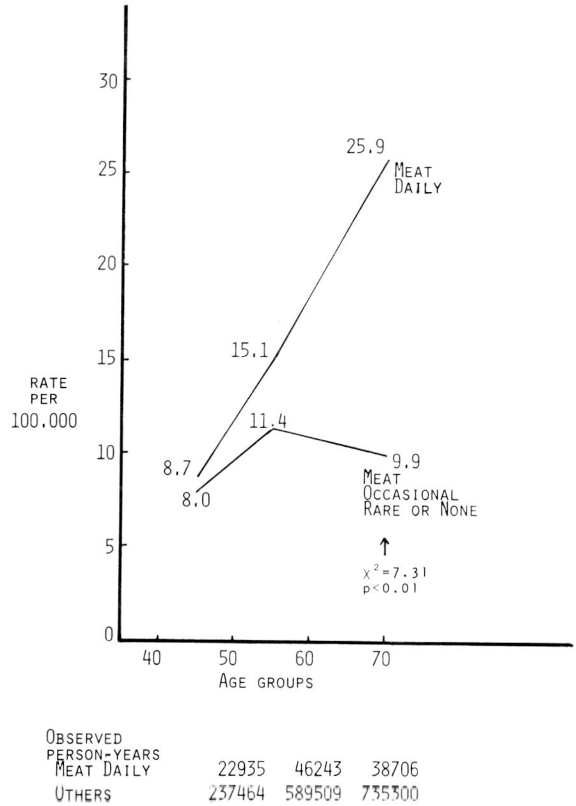

FIGURE 7. Standardized mortality rate for breast cancer by frequency of meat intake (prospective study, 1966 to 1978, Japan).

Table 8
STANDARDIZED MORTALITY RATE BY SMOKING, GREEN-YELLOW VEGETABLE INTAKE, MEAT INTAKE, AND SOCIOECONOMIC STATUS

MALES																
CIGARETTE SMOKING	DAILY								NONE							
GREEN-YELLOW VEGETABLES	DAILY				OTHERS				DAILY				OTHERS			
MEAT	DAILY		OTHERS		DAILY		OTHERS		DAILY		OTHERS		DAILY		OTHERS	
SOCIO-ECONOMIC STATUS	HIGH	LOW	HIGH	LOW	HIGH	LOW	HIGH	LOW	HIGH	LOW	HIGH	LOW	HIGH	LOW	HIGH	LOW
OBSERVED PERSON-YEARS	37334	30975	154896	465988	10133	8980	78866	178013	8126	6892	36176	120731	1425	1255	14542	38663
	STANDARDIZED MORTALITY RATE (PER 100,000)								STANDARDIZED MORTALITY RATE (PER 100,000)							
CANCER OF ALL SITES	456.61	496.61	478.99	501.22	652.82	551.26	518.13	490.05	368.06	363.75	295.00	307.76	777.07	165.65	286.44	277.75
BUCCAL CAVITY & PHARYNX	9.44	2.31	5.68	6.08	0	20.11	3.99	2.52	0	0	2.85	1.43	0	0	6.95	0
ESOPHAGUS	29.96	44.01	28.89	26.60	41.75	32.35	16.37	18.04	14.06	19.83	5.39	10.27	0	0	19.78	9.00
STOMACH	161.92	180.16	182.02	201.27	200.35	220.71	239.76	204.90	164.83	148.15	124.19	138.65	362.59	165.65	127.31	130.69
SMALL INTESTINE & COLON	7.57	0	16.62	11.94	0	0	7.42	10.60	0	0	9.95	8.94	75.76	0	6.95	12.24
RECTUM	6.80	9.06	14.58	17.54	10.28	0	15.29	19.45	0	0	5.31	17.88	0	0	19.25	5.52
GALL BLADDER	16.41	14.97	11.56	10.85	15.18	0	8.54	9.05	12.98	0	16.95	8.61	0	0	20.84	5.93
LIVER	56.24	52.81	46.57	38.80	151.41	10.09	52.13	43.93	14.06	77.73	33.19	25.50	183.98	0	17.85	24.71
PANCREAS	26.02	33.09	16.26	19.95	18.97	31.31	13.80	21.51	29.00	0	20.93	12.19	0	0	5.06	7.50
PERITONEUM	2.81	3.22	3.80	6.16	24.35	19.83	1.17	2.36	0	10.71	7.85	7.22	0	0	0	11.99
NOSE	0	0	1.32	4.95	0	7.95	1.17	6.30	0	16.81	2.70	3.18	78.81	0	6.38	0
LARYNX	0	2.88	8.36	5.53	7.04	0	5.26	3.60	0	0	0	0.79	0	0	0	0
LUNG	96.54	107.30	82.80	86.03	108.87	111.44	90.40	87.56	25.30	47.13	17.98	17.27	0	0	29.55	37.23
PROSTATE	6.71	6.56	6.81	8.84	39.60	0	6.43	11.23	46.37	0	5.47	9.73	0	0	5.17	6.05
KIDNEY	10.17	0	2.51	3.23	0	0	4.28	2.82	14.94	0	5.70	4.60	0	0	0	0
BLADDER	8.53	6.09	9.55	7.74	0	30.24	14.53	10.73	0	0	5.70	8.06	0	0	9.24	0
SKIN	0	0	5.68	2.69	0	0	1.50	1.89	0	0	0	3.36	0	0	0	3.89
BRAIN	3.82	0	7.67	4.85	0	11.43	2.42	2.86	20.26	0	2.54	1.60	0	0	0	0
BONE	0	6.30	0.72	4.25	0	0	3.26	2.12	0	0	2.39	0.48	0	0	5.35	0
LYMPHOMA	2.81	5.58	8.94	14.65	0	7.95	10.94	7.94	0	16.81	4.12	13.98	0	0	0	6.90
LEUKEMIA	0	0	4.56	6.70	0	19.83	5.61	5.16	25.96	0	7.50	5.37	0	0	6.38	1.50

Mantel-Haenszel χ and Maximum Likelihood Estimate of the Rate Ratio

			MALES		
MEAT	MANTEL-HAENSZEL X	ONE-TAIL P-VALUE	RATE TATIO	CONFIDENCE LIMITS	
CANCER OF ALL SITES	0.974	0.165	1.047	0.969	1.132
BUCCAL CAVITY & PHARYNX	0.769	0.221	1.376	0.695	2.723
ESOPHAGUS	2.000	0.023	1.445	1.068	1.957
STOMACH	- 0.768	0.221	0.941	0.826	1.072
SMALL INTESTINE, COLON & RECTUM	- 3.394	3.451 x 10⁻	0.360	0.220	0.591
GALL BLADDER	0.586	0.279	1.186	0.734	1.916
LIVER	2.734	0.003	1.453	1.161	1.819
PANCREAS	1.814	0.035	1.464	1.036	2.067
LARYNX	- 1.480	0.069	0.369	0.122	1.118
LUNG	1.796	0.036	1.219	1.017	1.462
PROSTATE	1.299	0.097	1.503	0.897	2.517
BLADDER	- 0.604	0.273	0.801	0.438	1.466
LYMPHOMA	- 1.738	0.041	0.465	0.225	0.960
LEUKEMIA	- 0.734	0.232	0.684	0.291	1.604

Note: Prospective study, 1966-78, Japan.

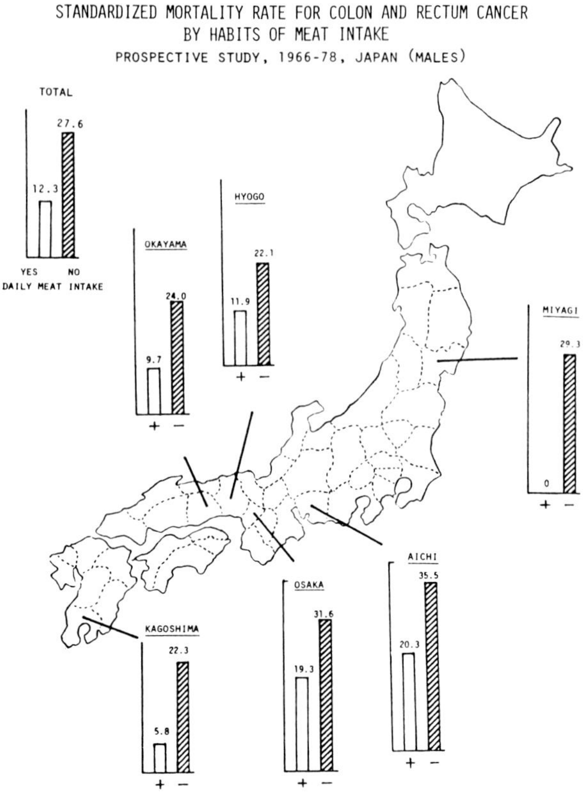

FIGURE 8. Standardized mortality rate for colon and rectum cancer by frequency of meat intake (prospective study, 1966 to 1978, Japan, males).

Pancreas[5,32] — A positive association with meat intake (Table 8) and per capita fat consumption has been observed. The risk increases when cigarette smoking is combined. An association with coffee has been reported.

Larynx[33] — The risk is quite high in daily smokers. A synergistic effect of alcohol and tobacco has been reported, and vitamin A deficiency is also suspected.

Lung[9,10,34-36] — The risk is quite high in daily cigarette smokers. A dose-response relationship clearly exists with the number of cigarettes smoked, the extent of inhalation, and the start of smoking at an early age. Frequent intake of green-yellow vegetables, β-carotene, or vitamin A may reduce the risk.

Breast[4,37-40] — Excessive fat intake, probably of animal origin, enhances the risk. A clear-cut geographical and trend association has also been observed. Large body weight and height appear to be associated with postmenopausal breast cancer.

Cervix[41] — Vitamin A deficiency may enhance the risk in addition to other high risk conditions. The risk has been observed to be higher in cigarette smokers.

Choriocarcinoma[42] — Deficient protein and folic acid content are suspected. The influence of maternal age is apparent.

Corpus[43] — High fat consumption and obesity are suspected.

Ovary[44] — High fat consumption is suspected.

Prostate[45,46] — The risk is lower in daily consumers of green-yellow vegetables. The risk tends to be higher in groups of high fat consumption.

Kidney[47,48] — A nephrotoxic strain of *Penicillium verrucosum* from stored maize (Balkan nephropathy) along with renal pelvis cancer is suspected.

Bladder[49,50] — An association with coffee and vitamin A deficiency has been reported. The risk appears to be lower in daily consumers of green-yellow vegetables.

Thyroid[51] — There is a possible relationship to iodine deficiency.

Non Hodgkin's lymphoma[52] — Magnesium deficiency is suspected.

There are thus many instances in which diet could influence the risk of cancer at various sites. Mechanisms by which diet may affect the incidence of cancer have been summarized below (modified table from Doll[15]). Each of these should be carefully considered when studying the associations listed above.

1. As a source of carcinogens or precarcinogens:
   - Natural components of plants (bracken fern)
   - Products of chemical, bacterial, or fungal action during processing or storage (aflatoxin)
   - Products of cooking (amino acid pyrolysate, Trp-P-1, Glu-P-2, IQ, MeIQ)
   - Food additives (AF2)
   - Contaminants (products of fuel combustion, pesticide residues)

2. By affecting the formation of carcinogens:
   - Provision of substances for the formation of nitrosamines (secondary amines, nitrates, nitrites)
   - Inhibition of formation of nitrosamines, as in the stomach (vitamin C)
   - Alteration of excretion of bile salts and cholesterol into the large bowel (fat)
   - Alteration of bacterial flora of the gut (fiber, meat)
   - Alteration of metabolism of carcinogens (enzyme induction by meat, fat, indoles in vegetables, antioxidants)
   - Alteration of enzyme formation (trace elements)
   - Effect on formation of estrogen (fats, total calories)

3. By modifying the effects of carcinogens:
   - Through transport (alcohol, fiber)
   - Through effect on concentration in the bowel (fiber)
   - Through inhibition of promotion (vitamin A, β-carotene)

D. Intervention Studies

History provides valuable natural experiments in intervention. The improvement of diet and nutrition has been shown to reduce the risk of selected cancers, such as stomach cancer, as seen in Japanese migrants to the U.S.[53] or in Japan itself in recent years.[54] The extent of risk reduction appears to be much less in Japanese migrants to Brazil. A reduction of stomach cancer risk by dietary improvement was also shown in the cohort study in Japan by following groups with recently changed dietary status. Those who increased the frequency of green-yellow vegetable intake during the past five years showed a significant risk reduction during several years of observation after such a diet change[12] (Figure 9).

Cessation of smoking was found to reduce the risk of lung cancer, not only in individuals, but also in professional groups (e.g., British doctors) and communities (e.g., Ina City, Nagano Prefecture, Japan). Daily intake of green-yellow vegetables combined with cessation of smoking appears to expedite the process of lung cancer risk reduction.

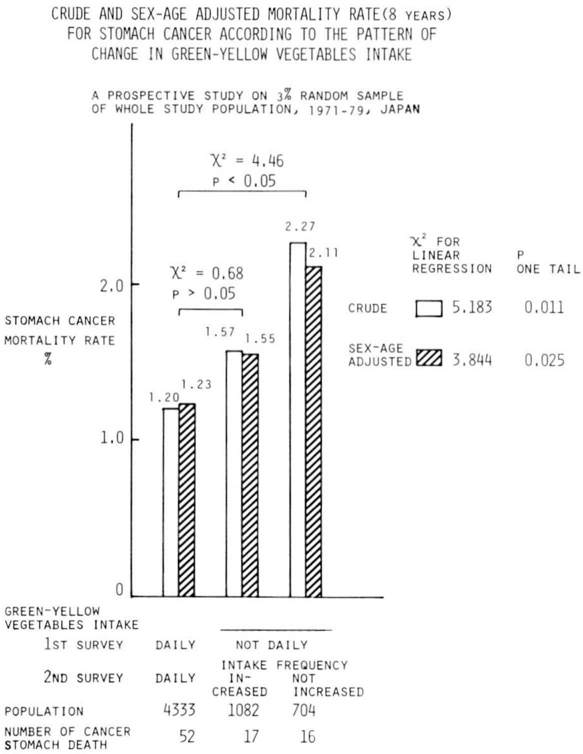

FIGURE 9. Crude and sex-age adjusted mortality rate (8 years) for stomach cancer according to the pattern of change in green-yellow vegetable intake (prospective study on 3% random sample of whole study population, 1971 to 1979, Japan).

E. Strategy for Cancer Control

Primary prevention is, without doubt, the most effective method of cancer control. Measures must be directed to limit the exposure to carcinogens from cigarette smoking, occupational contamination, and environmental pollution, including dietary carcinogens. Exposure to the medical use of carcinogenic drugs or radiation should also be reduced to a minimum. Environmental monitoring and management is necessary, not only for initiators but also for promoters. Measures should also be taken to improve nutrition, such as increased intake of β-carotene (promoter-inhibitor), vitamin C (initiator-inhibitor) and dietary fiber, and the avoidance of an excess intake of animal fat (probable promoter). It seems to be quite important to understand that the influence of diet covers multiple cancer sites. Just as the cessation of cigarette smoking can reduce cancer incidence at different sites at the same time, selected dietary change, such as increased intake of green-yellow vegetables and decreased intake of meat and alcohol, appears to be quite promising in lowering the risk of cancer at many sites at the same time. Such comprehensive primary prevention is presently believed to be the most promising way to control cancer in man in modern societies.

REFERENCES

1. **Hirayama, T.,** Operational epidemiology of cancer, *J. Cancer Res. Clin. Oncol.,* 99, 15, 1981.
2. **Hirayama, T.,** Prospective studies on cancer epidemiology based on census population in Japan, in *Prevention and Detection of Cancer, Part 1: Prevention,* Nieburgs, H. E., Ed., Marcel Dekker, New York, 1977, 1139.
3. **Hirayama, T.,** Epidemiological evaluation of the role of naturally occurring carcinogens and modulators of carcinogenesis, in *Naturally Occurring Carcinogens-Mutagens and Modulators of Carcinogenesis,* Miller, E. C., Miller, J. A., Hirono, I., Sugimura, T., and Takayama, S., Eds., University Park Press, Baltimore, 1979, 359.
4. **Hirayama, T.,** Epidemiology of breast cancer with special reference to the role of diet, *Prev. Med.,* 7, 173, 1978.
5. **Hirayama, T.,** A large-scale cohort study on the relationship between diet and selected cancers of digestive organs, in *Gastrointestinal Cancer: Endogenous Factors,* Banbury Rep. 7, Bruce, W. R., Correa, P., Lipkin, M., Tannenbaum, S. R., and Wilkins, T. D., Eds., Cold Spring Harbor Laboratory, Cold Spring Harbor, N.Y., 1981, 409.
6. **Hirayama, T.,** The epidemiology of cancer of the stomach in Japan with special reference to the role of diet, in Proc. 9th Int. Cancer Congr., L'Union International Contre le Cancer Monogr. Ser. 10, UICC, Geneva, 1967.
7. **Hirayama, T.,** The epidemiology of cancer of the stomach in Japan, with special reference to the role of diet, *Gann Monogr.,* 3, 15, 1968.
8. **Hirayama T.,** Diet and cancer, *Nutr. Cancer,* 1, 67, 1979.
9. **Hirayama, T.,** Non-smoking wives of heavy smokers have a higher risk of lung cancer: a study from Japan, *Br. Med. J.,* 282, 183, 1981.
10. **Hirayama, T.,** Epidemiology of lung cancer based on population studies, in *Clinical Implications of Air Pollution Research,* American Medical Association, Chicago, Ill., 1976, 69.
11. **Hirayama, T.,** Epidemiological consideration on the natural history of hepatitis B virus infection, *Asian Med. J.,* 19, 77, 1976.
12. **Kamon, S. and Hirayama, T.,** Epidemiology of cancer of the esophagus in Miye, Nara and Wakayama prefectures, with special reference to the role of bracken fern, *Proc. Jpn. Cancer Assoc.,* 34, 211, 1975.
13. **Hirayama, T. and Ito, Y.,** A new view of the etiology of nasopharyngeal carcinoma, *Prev. Med.,* 10, 1981.
14. **Hirayama, T.,** An epidemiological study of cancer of the esophagus in Japan, with special reference to the combined effect of selected environmental factors, in *Proc. Int. Semin. Epidemiol. Oesophageal Cancer (India),* Monogr. 1, 1971.
15. **Doll, R.,** The epidemiology of cancer, in *Accomplishments in Cancer Research,* Fortner, J. G. et al., Eds., General Motors Cancer Research Foundation, Detroit, Mich., 1979.
16. **Hirayama, T.,** Descriptive and analytic epidemiology of nasopharyngeal cancer, in *Nasopharyngeal Carcinoma, Etiology and Control,* de Thé, G. and Ito, Y., Eds., IARC Sci. Publ. 20, International Agency for Research on Cancer, Lyon, 1978, 167.
17. **Hirayama, T.,** An epidemiological study of oral and pharyngeal cancer in Central and South-East Asia, *Bull. WHO,* 34, 41, 1966.
18. **Lin, T. M., Yang, C. S., Tu, S. M., Chen, C. J., Kuo, K. C., and Hirayama, T.,** Interaction of factors associated with cancer of the nasopharynx, *Cancer (Philadelphia),* 44, 1419, 1979.
19. **Burrell, R. J. W., Roach, W. A., and Sadwell, A.,** Esophageal cancer in the Bantu of the Transkei associated with mineral deficiency in garden plants, *J. Natl. Cancer Inst.,* 36, 201, 1966.
20. **The Committee of Principal Investigators,** A cooperative trial in the primary prevention of ischemic heart disease using clofibrate, *Br. Heart J.,* 40, 1069, 1978.
21. **Hirayama, T.,** Smoking and drinking. Is there a connection? *Smoke Signals,* 16, 1, 1970.
22. **Tuyns, A. J., Péquignot, G., and Jensen, O. M.,** Les cancers de l'oesophage en Ille-et-Vilaine en fonction des niveaux de consommation d'alcool et de tabac. Des risques qui se multiplient, *Bull. Cancer,* 64, 45, 1977.
23. **Haenszel, W., Kurihara, M., Segi, M., and Lee, R. K. C.,** Stomach cancer among Japanese in Hawaii, *J. Natl. Cancer Inst.,* 49, 969, 1972.
24. **Hirayama, T.,** A study of epidemiology of stomach cancer, with special reference to the effect of diet factor, *Bull. Inst. Publ. Health,* 12, 85, 1963.
25. **Stocks, P. and Davies, R. L.,** Zinc and copper content of soils associated with the incidence of cancer of the stomach and other organs, *Br. J. Cancer,* 18, 14, 1964.
26. **Burkitt, D.,** Related disease-related cause? *Lancet,* 2, 1229, 1969.
27. **Haenszel, W., Berg, J. W., Kurihara, M., and Locke, F. B.,** Large bowel cancer in Hawaiian Japanese, *J. Natl. Cancer Inst.,* 51, 1765, 1973.

28. **Rose, G., Blackburn, H., Keys, A., Taylor, H. I., Kannel, W. B., Paul, O., Reid, D. D., and Stamler, J.,** Colon cancer and blood cholesterol, *Lancet,* 1, 181, 1974.
29. **Hirayama, T.,** Epidemiological patterns and risk factors of liver cancer in Japan, in *Liver Cancer: Cancer Problems in Asian Countries,* Shanmugaratnam, K. et al., Eds., Singapore Cancer Society, 1976, 3.
30. **Linsell, C. A. and Peers, F. G.,** Field studies on liver cell cancer, in *Origins of Human Cancer,* Hiatt, H. H., Watson, J. D., and Winsten, J. A., Eds., Cold Spring Harbor Laboratory, Cold Spring Harbor, N.Y., 1977, 549.
31. **Bernstein, R. A., Werner, L. H., and Rimm, A. A.,** Relationship of gallbladder disease to parity, obesity, and age, *Health Serv. Rep.,* 88, 925, 1973.
32. **MacMahon, B., Yen, S., Trichopoulos, D., Warren, K., and Nardi, G.,** Coffee and cancer of the pancreas, *N. Engl. J. Med.,* 12, 630, 1981.
33. **Wynder, E. L., Bross, I. J., and Day, E.,** Epidemiological approach to the etiology of cancer of the larynx, *JAMA,* 160, 1384, 1956.
34. **Bjelke, E.,** Dietary vitamin A and human lung cancer, *Int. J. Cancer,* 15, 561, 1975.
35. **Hirayama, T. and Hamano, Y.,** The significance of the smoking factor in the epidemiology of cancer of the lung, *Gann,* 46, 418, 1955.
36. **Hirayama, T.,** Metal-material workers and lung cancer in Japan. Occupational carcinogenesis, *Ann. N.Y. Acad. Sci.,* 271, 269, 1976.
37. **De Waard, F. and Baanders-Van, H. E. A.,** A prospective study in general practice of breast cancer risk in postmenopausal women, *Int. J. Cancer,* 14, 153, 1974.
38. **Hirayama, T. and Wynder, E. L.,** A study of the epidemiology of cancer of the breast. II. The influence of hysterectomy, *Cancer (Philadelphia),* 15, 28, 1962.
39. **Miller, A. B., Kelly, A., Choi, N. W., Matthew, V., Morgan, R. W., Munan, L., Burch, J. D., Feather, J., Howe, G. R., and Jain, M.,** A study of diet and breast cancer, *Am. J. Epidemiol.,* 107, 599, 1978.
40. **Wynder, E. L., Bross, I. D. J., and Hirayama, T.,** A study of the epidemiology of cancer of the breast, *Cancer (Philadelphia),* 13, 559, 1960.
41. **Hirayama T.,** Cancer epidemiology in Japan, *Environ. Health Perspect.,* 33, 14, 1979.
42. **Acosta-Sison, H.,** Trophoblastic or chorionic tumors as observed in the Philippines, in *Choriocarcinoma,* Holland, J. F. and Hreshchyshyn, M. M., Eds., UICC Monogr. Ser. No. 3, Springer-Verlag, New York, 1967, 33.
43. **Armstrong, B. K.,** Diet and hormones in the epidemiology of breast and endometrial cancer, *Nutr. Cancer,* 1, 91, 1979.
44. **Carroll, K. K. and Khor, H. T.,** Lipids and tumors, in *Progress in Biochemical Pharmacology,* Vol. 10, Carroll, K. K., Ed., S. Karger, Basel, 1975, 308.
45. **Hirayama, T.,** Epidemiology of prostatic cancer with special reference to the risk lowering effect of green-yellow vegetables intake, *Proc. Jpn. Cancer Assoc.,* 37, 280, 1978.
46. **Hirayama, T.,** Epidemiology of prostate cancer with special reference to the role of diet, *Natl. Cancer Inst. Monogr.,* 53, 149, 1979.
47. **Barnes, J. M., Austwick, P. K. C., Carter, R. L., Flynn, F. V., Periastianis, G. C., and Alredge, W. N.,** Balkan (endemic) nephropathy and a toxin producing strain of *Penicillium verrucosum* var. *cyclopium:* an experimental model in rats, *Lancet,* 1, 671, 1977.
48. **Chernozemsky, I. N., Stoyanov, I. S., Petkova-Bocharova, T. K., Nicolov, I. G., Dragonov, I. V., Stoichev, I. I., Tanchev, Y., Naidenov, D., and Kaicheva, N. D.,** Geographic correlation between the occurrence of endemic nephropathy and urinary tract tumors in Vratza district, Bulgaria, *Int. J. Cancer,* 19, 1, 1977.
49. **Cole, P.,** A population-based study of bladder cancer, in *Host Environment Interactions in the Etiology of Cancer in Man,* IARC Sci. Publ. 7, International Agency for Research on Cancer, Lyon, 1973, 83.
50. **Mettlin, C. and Graham, S.,** Dietary risk factors in human bladder cancer, *Am. J. Epidemiol.,* 110, 255, 1979.
51. **Wahner, H. W., Cuello, C., Correa, P. et al.,** Thyroid cancer in an endemic goiter area, Cali, Colombia, *Am. J. Med.,* 40, 58, 1966.
52. **Bois, P., Sandborn, E. B., and Messier, P. E.,** A study of thymic lymphosarcoma developing in magnesium-deficient rats, *Cancer Res.,* 29, 763, 1969.
53. **Fukazawa, T., Hirayama, T., and Kimura, M.,** Various computation methods of age and occupation standardized mortality ratios and its application to the analysis of mortality by ethnic group in Hawaii, *Bull. Inst. Publ. Health,* 16, 152, 1967.
54. **Hirayama, T.,** Changing patterns in the incidence of gastric cancer, in *Advances in the Biosciences,* Vol. 32, Fielding, J. W. et al., Eds., Pergamon Press, New York, 1981, 1.

Chapter 3

DIETARY FAT AND ITS RELATIONSHIP TO HUMAN CANCER

K. K. Carroll

TABLE OF CONTENTS

I. INTRODUCTION

Epidemiological data have shown that cancer incidence and mortality vary greatly in different parts of the world.[1,2] In countries such as the U.S. and Canada, cancers of the breast, colon, prostate, rectum, ovary, and pancreas account for a large proportion of the total mortality from cancer. There is good reason to think that diet plays an important role in the genesis of these types of cancer. In particular, mortality in each case, shows a strong positive correlation with the amount of dietary fat available for consumption in different countries (Table 1).

Cancers of the lung, stomach, and uterus also account for relatively large numbers of deaths in the U.S. and Canada. Cancer of the lung began to increase in American males about 1930 and is now by far the most common single cause of cancer-related deaths in males. In females, the increase in mortality from lung cancer began about 1960, and is showing an upward trend similar to that in males.[4] Experience in Canada has paralleled that in the U.S. (Figure 1). Reports of the U.S. Surgeon General have provided extensive documentation of a clear relationship between smoking habits and lung cancer.[7,8]

Cancer of the stomach has been decreasing in the U.S. and Canada, as well as in most other countries. Nitrosamines and nitrosamides have been implicated as causative factors in stomach cancer and salt may act as a modifier.[4] Epidemiological data show that stomach cancer and stroke are decreasing in parallel and it has been proposed that this is related to decreasing salt intake.[9] Mortality from stomach cancer does not appear to be related to dietary fat (see Table 1).

Mortality from uterine cancer has also been decreasing in the U.S. and Canada in recent years[4] (Figure 1). This is probably due to improved treatment for this type of cancer. It is interesting to note that incidence of uterine cancer is positively correlated with dietary fat,[10] whereas mortality shows no such correlation.[11]

II. EVIDENCE RELATING DIETARY FAT TO CARCINOGENESIS

As indicated above, dietary fat shows a strong positive correlation with mortality from some of the more common types of cancer (Table 1). This correlation is illustrated for cancers of the breast and prostate in Figure 2. Studies on migrating populations have provided evidence that the differing cancer mortalities in different countries are related to environmental rather than hereditary factors.[4,11] This is also supported by time-course studies within countries. For example, a recent marked increase in fat intake in Japan has been accompanied by an increase in the death rate from breast cancer.[13]

High-fat diets tend to be high in caloric content and in animal protein as well. It is therefore not surprising to find that mortality from breast cancer also shows a strong positive correlation with both caloric intake and dietary animal protein.[3] Since these dietary variables cannot be readily dissociated in diets eaten by large populations, it is difficult to assess their relative importance in carcinogenesis on the basis of epidemiological data alone. Experiments with animals have thus played an important role in focusing attention on dietary fat.

Numerous studies over the past four decades have consistently shown that rats and mice fed high-fat diets develop mammary tumors more readily than their counterparts on low-fat diets.[11] During the past 15 years, similar observations have been made with respect to effects of dietary fat on colon tumors in rats.[4,14] It has also been reported within the past year that high-fat diets enhance the development of pancreatic tumors in rats,[15] and hamsters.[16]

Table 1
CORRELATION COEFFICIENTS BETWEEN DIETARY FAT AND AGE-ADJUSTED CANCER MORTALITY IN DIFFERENT COUNTRIES

Site	Correlation coefficient	
	Male	Female
Breast		+0.935
Intestine (except rectum)	+0.928	+0.911
Prostate	+0.892	
Leukemia, aleukemia	+0.857	+0.838
Rectum	+0.834	+0.786
Ovary, fallopian tube, broad ligament		+0.726
Lung, bronchus, trachea	+0.710	+0.252
Pancreas	+0.666	+0.501
Skin	+0.634	+0.550
Bladder, other urinary organs	+0.523	+0.345
Buccal cavity, pharynx	+0.260	−0.245
Larynx	+0.170	−0.276
Esophagus	+0.088	−0.014
Uterus		−0.064
Stomach	+0.010	−0.112
Liver, biliary passages	−0.676	−0.487

Restriction of caloric intake causes a general inhibition of carcinogenesis in animals, but in most of the experiments in which inhibition has been observed, the caloric intake has been restricted to less than 80% of normal.[17] The effect of dietary fat on mammary tumorigenesis in animals appears to be independent of caloric intake.[3]

Experiments with animals have failed to provide strong evidence that dietary protein has much effect on carcinogenesis at intakes in the normal range.[14,17] In fact, the results of experiments by Clinton et al.[18] indicated that initiation of mammary tumors occurred more readily in rats fed a low-protein diet.

III. EFFECTS OF DIFFERENT KINDS OF DIETARY FAT

Studies in the author's laboratory on mammary tumors induced in female Sprague-Dawley rats by 7,12-dimethylbenz[*a*]anthracene provided evidence that polyunsaturated fats stimulate mammary tumorigenesis more effectively than saturated fats.[19] This seemed at odds with data on human populations, since the epidemiological evidence indicated that breast cancer mortality correlated better with total dietary fat than with either animal or vegetal fat, although vegetal fat tends to be more unsaturated than animal fat.[17] An explanation for this apparent lack of agreement was provided by further experiments in the laboratory that showed that high fat diets promoted mammary tumorigenesis only when the dietary fat contained a certain amount of polyunsaturated fat.[20] This is apparently related to a requirement for essential fatty acids,[21] but the amount involved is so small that it would probably be provided by most human diets.

The effects of different types of dietary fat have been studied less systematically in

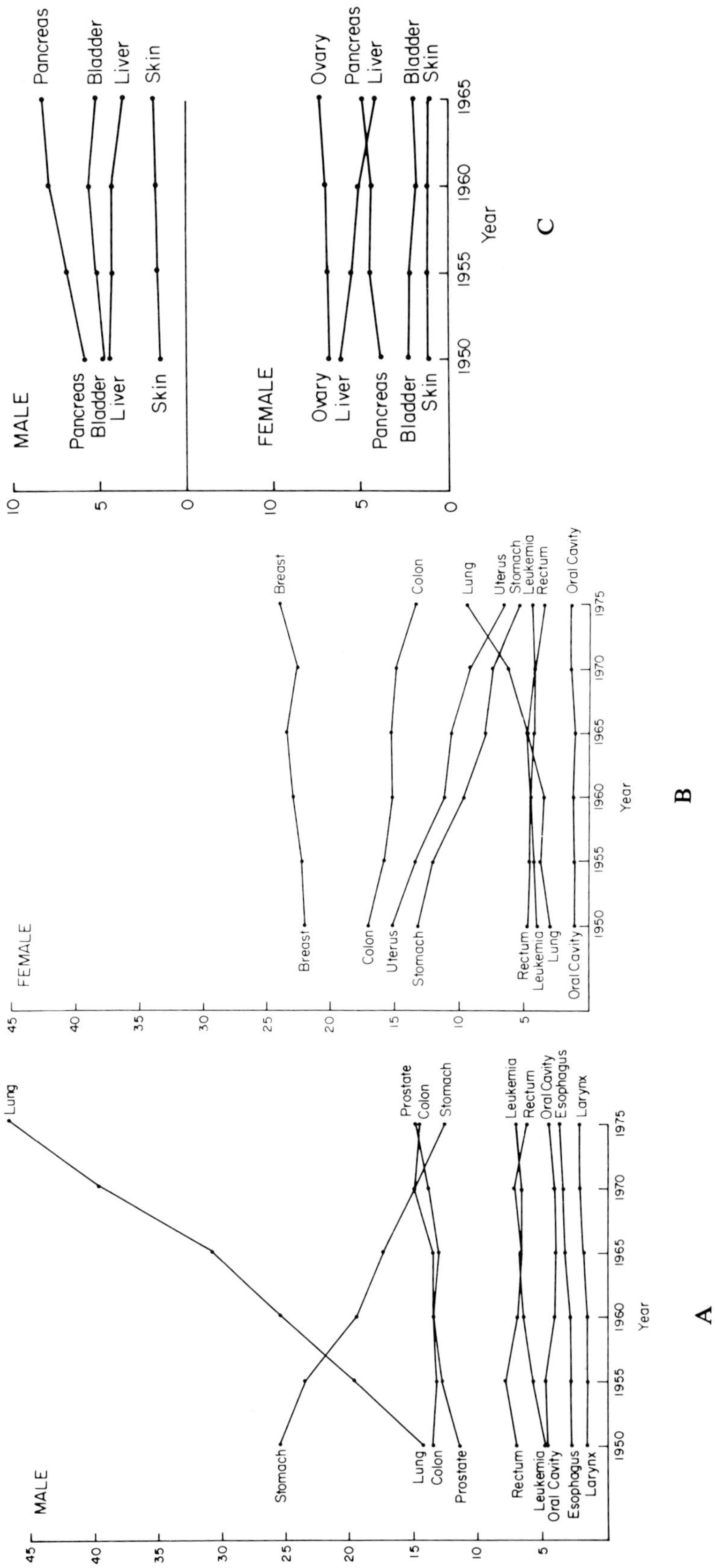

FIGURE 1. Trends in cancer mortality in male (A) and female (B) populations, from data collated by Segi et al.[2,5,6] Data for tumors at some sites (C) were only available for the years 1950 to 1965. Left axis is age-adjusted mortality rates per 100,000 population.

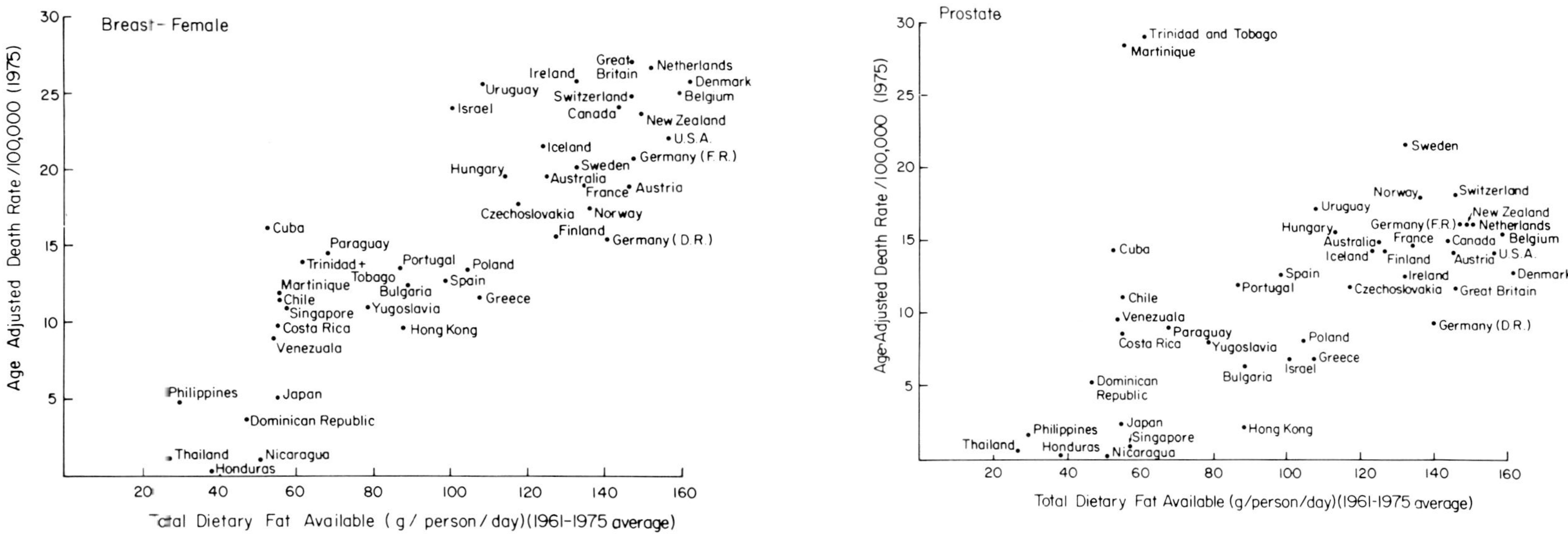

FIGURE 2. Correlation between per capita availability of dietary fat,[12] and mortality from cancers of the breast (A) and prostate (B) in different countries.[2] The values for dietary fat are averages over the 15 years prior to the year for which the death rates were recorded. The high death rates for prostatic cancer in Trinidad and Tobago, and in Martinique, may be due in part to yearly variations in small populations. Thus, the age-adjusted death rate for Trinidad and Tobago in 1973 was only 13.4/100,000.

the case of colon cancer. The type of fat does not seem to matter at high levels of dietary fat, but there is some evidence that polyunsaturated fat is more effective at low levels of dietary fat.[14] Pancreatic cancer was found to be enhanced by a diet containing 20% corn oil, but not by one containing 18% hydrogenated coconut oil and 2% corn oil.[15]

IV. MECHANISM OF ACTION OF DIETARY FAT

Dietary fat appears to act mainly at the promotional stage of carcinogenesis.[3,11,14] Its effect can be observed with spontaneous tumors as well as with tumors induced by a variety of carcinogenic agents. With induced tumors, fat appears to be most effective when fed after the animals have been exposed to carcinogen. There is also evidence that dietary fat can affect the growth and development of transplantable mammary tumors.[22,23]

There are a number of possible mechanisms by which dietary fat may promote mammary tumorigenesis. These include alterations in the hormonal environment or changes in the composition of cellular membranes.[24,25] Essential fatty acids are known to inhibit certain immune responses (Ir) and may influence tumorigenesis in this way. They also serve as precursors of prostaglandins, which can act as tumor promoters.[26] Dietary fat may influence colon carcinogenesis by stimulating the production of bile acids, some of which have been shown to act as promoters.[14]

V. LIMITING FACTORS IN CARCINOGENESIS

Every cancer presumably starts from a transformed cell that is thought to be formed by some initiating event. This may be due to interaction with a chemical, a virus, or some form of radiation. Initiation is thought to involve a mutational change in the genome of the cell. It occurs within a short period of time and is considered to be essentially irreversible.[27]

However, the mere presence of a transformed cell does not necessarily mean that a tumor will inevitably develop. Such cells may be eliminated by the body's defense mechanisms, but even if they persist, their potential for growth may not be expressed immediately. It appears that they are quite susceptible to environmental influences, particularly during the early stages of tumorigenesis. This phase, known as the promotional stage, may last for a considerable period of time so that there is often a relatively long latency period between initiation and demonstrable cancer.

For any given type of cancer in a particular population, it is of interest to consider what may be the limiting factor or factors that determine cancer incidence and mortality. In some cases, the degree of exposure to a carcinogenic agent may be the limiting factor. In other situations where the agent is not a direct-acting carcinogen, the conditions that determine its rate of conversion to the active form, or that affect its distribution and length of life in the body, may be limiting. The existence and number of cells susceptible to transformation by a particular agent can also be a limiting factor. Once initiation has taken place, subsequent proliferation of the transformed cell may be markedly influenced by various environmental factors, as indicated above.

Until recently, most of the emphasis in studies on carcinogenesis had been concentrated on identification of carcinogenic or mutagenic agents and other factors involved in tumor initiation. These are many and varied, and a substantial number of them may contribute to the overall total of initiated cells. Even if the more significant carcinogens or co-carcinogens could be identified, there is no certainty that they can be eliminated from the environment or otherwise avoided. Furthermore, it is uncertain whether ex-

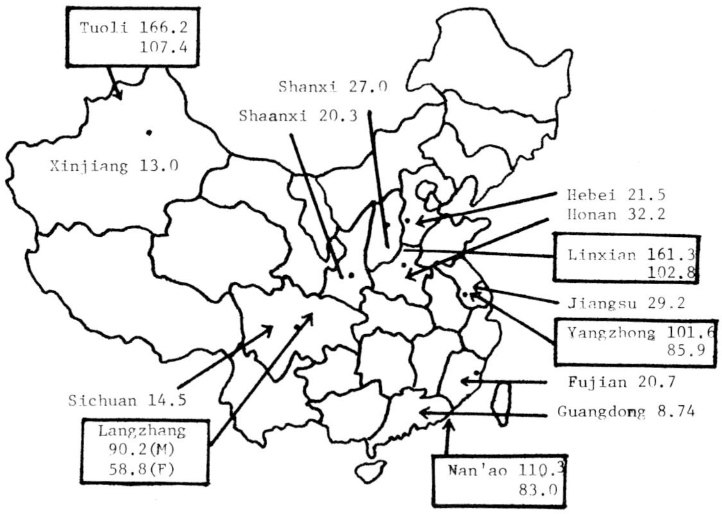

FIGURE 1. Mortality rate of esophageal cancer in some high-incidence areas in China.

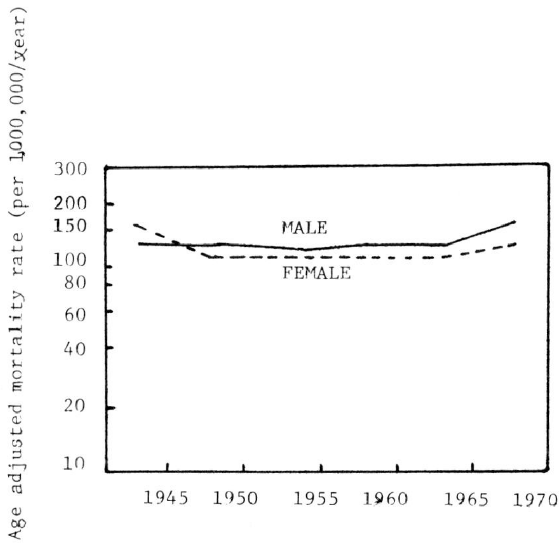

FIGURE 2. Changes in mortality rates of carcinoma of the esophagus in 76 Linxian county brigades between 1941 and 1970.

of Linxian county. Nitrates and nitrites were found in most water samples and were especially high in summer, with an average of 12.65 and 0.052 mg/ℓ, respectively. The average concentration of nitrates and nitrites was 10.55 mg/ℓ (range 0 to 75 mg/ℓ) and 0.03 mg/ℓ (range 0 to 2.63 mg/ℓ), respectively, over the four seasons.

The average concentration of nitrates and nitrites in the water of most wells in the 49 brigades showed a positive correlation with the average incidence rates of EC and marked epithelial hyperplasia, a precancerous lesion of the esophagus, for the years 1969 to 1976.[3]

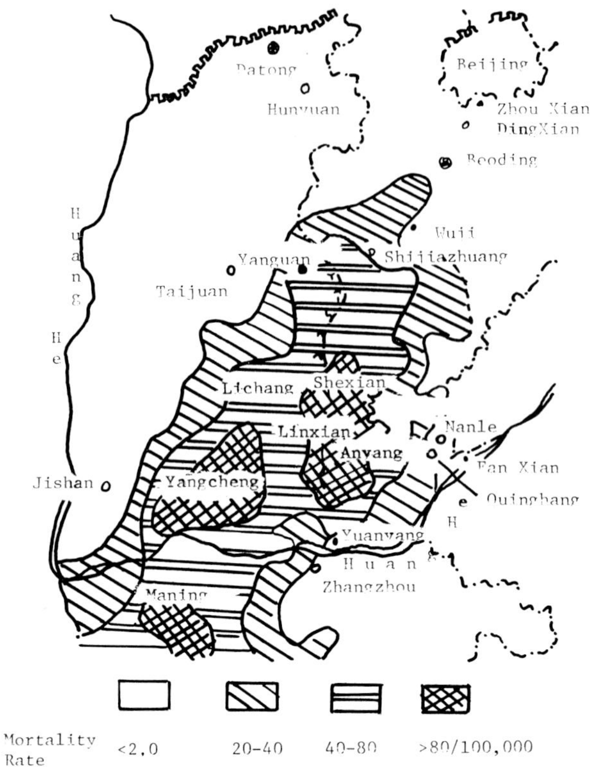

FIGURE 3. Distribution of esophageal cancer mortality rates in areas of the Taihang Mountains.

Nitrites in the drinking water used in peasant homes, in well water, water stored in jars, warm water in cast-iron pots, and liquid food (gruel) were found in increasing concentrations, respectively. The average concentrations of nitrites in samples taken from warm water pots and gruel were 0.512 and 0.696 mg/ℓ, respectively, at values 10 to 20 times higher than those taken from water jars.[3]

Nitrates and nitrites were recovered in most gastric juice samples from Linxian peasants, with an average amount of 31.67 and 1.37 mg/ℓ, respectively. A comparative study of nitrites in saliva carried out on peasants in Linxian county and Xinyangxian county (a low-risk area for EC) showed an average of 4.5 mg/ℓ (216 samples) compared to 5.74 mg/ℓ (51 samples). There were no other significant differences. However, a significant difference in the content of salivary nitrites was observed between patients with marked epithelial hyperplasia or EC and normal controls (Table 1).[7-9]

2. Nitrosamines and Their Precursors in Food Samples

Wheat, corn, millet, millet bran, dried sweet potato, and pickled vegetables from high and low EC incidence areas in Honan province were analyzed for nitrosamines by thin-layer chromatography (TLC). Of the 124 food samples from Linxian county 29 were "positive" in the nitrosamine assay, whereas only 1 of 86 samples from Fanxian county (low-incidence area) was positive. The presence of diethylnitrosamine (DENA) and dimethylnitrosamine (DMNA) in some of the food samples was confirmed by gas chromatography.[8]

Table 1
AVERAGE CONTENT OF NITRITE IN SALIVA OF LINXIAN PEASANTS WITH LESIONS OF THE ESOPHAGUS

Esophageal lesions	No. of persons	Nitrites (mg/ℓ±SD)	*p*
Normal epithelium	45	3.17 ± 3.23	
Mild epithelial hyperplasia	43	4.50 ± 4.20	>0.05
Marked epithelial hyperplasia	33	6.00 ± 6.60	=0.01
Cancer	38	5.06 ± 5.50	<0.05

B. Enhancement of Nitrosamine Formation in Food by Fungi

In high-incidence EC areas, dietary habits and improper preservation of food usually cause contamination of foodstuffs by fungi. Li et al.[10,11] and Lu et al.[12] showed that some common species of fungus belonging to *Fusarium, Geotrichum, Aspergillus* and other genera not only could reduce nitrates to nitrites but could also decompose proteins, thus increasing the amount of amines in food and promoting the formation of nitrosamines.

After inoculation of cornbread with common fungi frequently encountered in Linxian county *(Fusarium monili forme, Geotrichum candidum, Cladosporium herbarum, Penicillium chrysogenum* and *Aspergillus flavus),* the content of secondary amines increased during a four-day incubation. A 17-fold increase in secondary amines was noted in *F. monili forme*-incubated cornbread. When a small amount of $NaNO_2$ was added to moldy cornbread, followed by an eight-day incubation, gas chromatography/mass spectrometry (GC-MS) analysis showed the presence of four *N*-nitroso compounds: DMNA, DENA, methylbenzylnitrosamine, and a new volatile nitrosamine, *N*-1-methylacetonyl-*N*-3-methylbutylnitrosamine (MAMBNA), with a molecular weight of 186.1. This new compound was synthesized and confirmed by GC-MS, and its concentration in contaminated cornbread was estimated to be 0.2 to 0.3 ppm.[9,11]

Li et al.[10] reported that MAMBNA exerts a mutagenic effect on *Salmonella typhimurium* after activation by an S9 mix in the Ames test. Papillomas of the forestomach are induced in mice fed MAMBNA together with $NaNO_2$ for 5 months.

In another experiment, in rats, cornbread was disinfected before inoculation of the fungus and to each 400 g sample, 400 mg $NaNO_2$ was added after an 8-day incubation. The presence of MAMBNA in the cornbread was determined regularly by TLC. During the course of the experiment, 10 papillomas, 4 early carcinomas, and 1 squamous-cell carcinoma developed in the forestomach among the 35 rats that were fed the moldy food for 238 to 883 days. Most carcinoma lesions were observed after 775 days of treatment. In addition, there were papillary hyperplastic growth and other lesions in the forestomach of treated rats. A second group of female rats was fed the fungus-inoculated cornbread alone, and 2 papillomas and 2 early carcinomas developed in the forestomach among the 31 rats after 554 to 701 days. Epithelial hyperplasia of the esophagus, forestomach, and glandular stomach was observed in some animals of this group. Another group of 10 females, serving as controls, was fed conventional cornbread. No epithelial lesions were noted in their forestomachs. The experimental results demonstrate the presence of carcinogenic mycotoxins in the *F. monili forme*-contaminated cornbread and show that MAMBNA is formed following the addition of $NaNO_2$ into the moldy cornbread, that in turn, produced more papillomas and carcinomas in the forestomach of treated rats.

Recently, Wang and Zhang[13] reported 2 papillomas in the esophagus and 3 papillomas in the forestomach of rats fed *F. monili forme*-inoculated food for 168 to 467 days, but no carcinomas developed in the experimental rats.

C. Mutagenicity and Carcinogenicity of Pickled Vegetables

Pickled vegetables are popular dishes served daily in high-incidence areas of EC, and the mortality rate appears to be correlated with the amount of pickled vegetables consumed.[14] Among 39 rats of the Wistar strain fed with extracted or concentrated fluid from pickled vegetables for 333 to 730 days, 1 developed an adenocarcinoma in the glandular stomach, 4 had fibrosarcoma of the liver, and another had angioendothelioma of the thoracic wall in addition to epithelial dysplasia lesions in the esophagus and forestomach. No tumors were noted in control rats.

Geotrichum candidum link is the predominant species of fungi occurring in pickled vegetables from Linxian county. Experimental studies showed that daily gastric inoculation of 0.5 mℓ *G. candidum* culture fluid to mice induced various hyperplastic lesions of the forestomach, and the fungus culture fluid enhanced the carcinogenic action of methylbenzylnitrosamine (0.25 mg/kg per day) in the forestomach when administered simultaneously.[15] In another experiment, however, this promoting effect was not observed in mice.[16]

Concerning the mutagenicity of pickled vegetables from Linxian county, Takahashi et al.[17] found mutagens in the methanol extract when tested with *Salmonella typhimurium* TA100 and TA98 after treatment with S9 mix. A systematic investigation was recently carried out by Chen et al.[18] in the Cancer Research Institute of the Centre National de la Recherche Scientifique (CNRS) in Paris. They found that extracts of Linxian pickles increased sister chromatid exchange in V79 cells, induced mutations in *Salmonella,* and induced transformed foci in cultures of Syrian hamster embryonic cells. Moreover, the pickle also exerted a promoting effect in 3-methylcholanthrene-initiated C3H/10T½ cell cultures.

A dichloromethane or other extract of pickled vegetables from Linxian county was found to contain a nitroso compound — Roussin red methyl ester (di-μ-methanethiolate-tetranitrosodiiron) — as indicated by TLC analysis and confirmed by GC-MS.[19] The compound was first synthesized in 1858 by Roussin. This chemical may provide NO^+ ion, which would react readily with secondary amines to form nitrosamines. In experiments with mice, the gastric intubation of Roussin red alone, or with sarcosine ethyl ester, induced epithelial hyperplasia of the esophagus and forestomach and papilloma of the forestomach within 194 to 269 days.[10]

D. Effects of Dietary Factors and Trace Elements

In the past, cornbread and millet were the staple foods in Linxian county, although the consumption of wheat has increased in recent years. There was a low intake of vitamins A and C, riboflavin, animal protein, fat, fresh vegetables, and fruit.

Zheng et al.[20] studied the urinary excretion of a test dose of riboflavin (5 mg) in peasants of Linxian county. The majority of the subjects excreted over 1000 μg of riboflavin in the urine within 4 hr, an amount below the adequate level. About 38.8% of the test subjects excreted less than 400 μg. During winter the riboflavin deficiency was severe; only 4.6% excreted over 1000 μg and 52.2% less than 400 μg.

The DCEC[3] carried out a preliminary analysis of chemical elements in drinking water (222 samples) and grains and vegetables (216 samples) and showed that trace elements, such as molybdenum, zinc, copper, magnesium, manganese, and cobalt are deficient in the drinking water of high risk areas. It is known that the trace element molybdenum is an important constituent of certain enzymes, such as nitrate reductase and other oxidases, and a deficiency of molybdenum may lead to the accumulation of

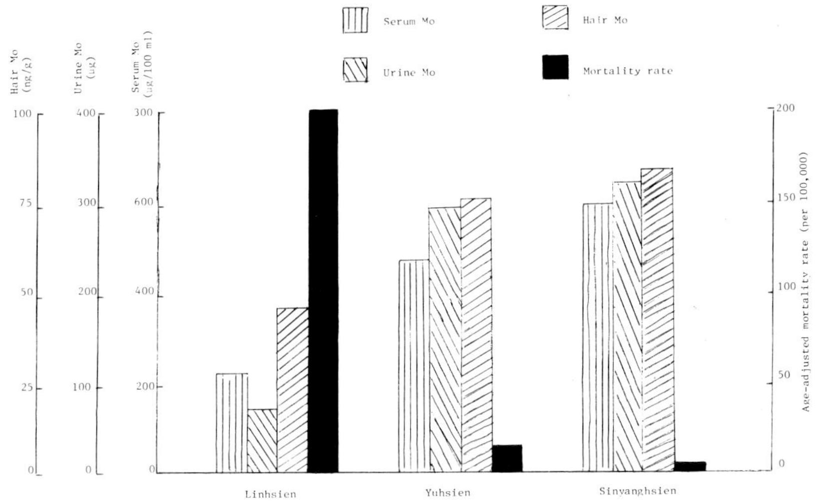

FIGURE 4. Molybdenum contents in the sera, urine, and hair of normal subjects in three counties with different mortality rates of esophageal cancer.

nitrates in plants. Burrell et al.[21] reported a relative deficiency of molybdenum, copper, zinc, magnesium, and boron in the soil of Transkei, South Africa, where cancer of the esophagus is prevalent. Nemenko et al.[22] also noted a low content of molybdenum in the Guriev region of Kazakhstan, U.S.S.R. Furthermore, the molybdenum content in 360 hair samples, 126 serum samples, and 276 overnight urine samples from inhabitants of Linxian county, and from the counties of Yuxian and Xinyangxian (low-risk areas) have been analyzed using catalytic polarography. The results indicated that the molybdenum level in the human body was much lower in Linxian county than in the low-risk areas (Figure 4).[3]

E. Etiology of Esophageal Cancer: Relationship Between Animals and Man

A preliminary study in Linxian county revealed 8 hens with upper digestive tract cancer from a total of 152 sick chickens. In another Linxian county study of 18,774 chickens over 6 months of age, it was demonstrated that 33 birds suffered from pharyngeal and esophageal cancer — a morbidity rate of 175.78/100,000 chickens. A comparative examination was also carried out in high- and low-incidence areas. In a commune in Fanxian county, Honan province, a low-incidence area, 2 cases of pharyngoesophageal cancer were detected among 11,399 chickens over 6 months of age, showing an incidence rate of 17.55/100,000. Moreover, in two communes in Hunyan county, Shanxi province, 9420 chickens were examined in the same manner, revealing only 2 cases of cancer, i.e., an incidence rate of 21.23/100,000.This comparative study revealed that the incidence rate of upper digestive tract cancer among chickens parallels the human EC rate in the high- and low-incidence areas.[23,24]

There is a relationship between the higher incidence rates for the Honan migrants and for the indigenous residents in Zhongxiang county. A survey of chickens 2 years of age or more, carried out at the end of 1975, showed that of the 5484 birds fed by Honan migrants, 12 were proved histopathologically to have squamous-cell carcinoma, while not a single pharyngoesophageal carcinoma was found among the 2371 chickens cared for by the indigenous residents. This suggests that there is a correlation between the incidence rates for the chickens and humans in the migrant communities, possibly based on the human dietary habits.[25]

Sarcosine ethyl ester and $NaNO_2$ were used in attempts to induce pharyngeal and esophageal cancer in domestic fowl of Linxian county by in vivo or in vitro formation of nitrososarcosine ethyl ester (NSEE). Pathological changes of the pharynx, esophagus, and forestomach in chickens were induced by NSEE or its precursor compounds within 300 days in 67 chickens. In these 67 chickens, varying degrees of epithelial hyperplasia and atypical hyperplastic lesions appeared in the pharynx, esophagus, and glandular stomach, and 1 chicken showed neoplastic changes of the glandular stomach.[26] In another long-term experiment in which methylbenzylnitrosamine was administered to 72 domestic fowl for as long as 998 days, there was no occurrence of pharyngoesophageal cancer.[27]

In another study, a 7.5-year old goat in Linxian county was found with EC, showing lesions similar to those in domestic chickens of the locality.[28] This may have some relationship to the local environment, especially to the carcinogenic content of the food.

F. Clastogenicity of Fish Juice

In the Shantou region of Guangdong province, a high incidence of EC was also reported in recent investigations. A survey from 1974 to 1976 showed that the adjusted rate was 8.74, which is 18.56% of the total deaths from malignant tumors. However, mortality was rather high in the eastern central areas. For example, the death rate was as high as 60/100,000 in the Yuhu commune in Jieuang county and Nan'ao Island, on the border of Guangdong and Fujian provinces, had a rate of 110.38 for males and 63.08 for females. The occupants of this county are mostly fishermen who spend over ⅓ of their time at sea. Their diet, therefore, consists mainly of fish, and they employ a special fish juice as seasoning made from decomposed fish that have been aged for over two years. Chinese hamster ovary (CHO) cells were used to examine the clastogenic action of the fish juice and the frequency of chromosome breaks and exchanges in cells arrested at metaphase was analyzed. The results showed that the fish juice, at a final concentration of 62.5 $\mu\ell/m\ell$, induced chromosome aberrations. At pH 9, the frequency of chromosome aberrations was similar to that seen under neutral conditions, but the clastogenic activity was increased at pH 4.2. The experiment was repeated and similar results were obtained in the second trial. Under neutral or alkaline conditions, the frequency of chromosome aberrations was enhanced 3 to 4.5 times by S9 and this was increased even further under acidic conditions The frequency of cells with chromosome aberrations was 21.9% at a nitrosated fish juice dose of 83.3 $\mu\ell/m\ell$ without activation. The average number of exchanges per metaphase plate was 0.42. Upon activation by S9, the frequency of cells with chromosome aberrations was increased to 62.13%, the majority of which were translocations. The average number of exchanges per metaphase plate was 1.582, which was approximately 4 times that observed in the absence of S9 (Tables 2 and 3).[29]

III. PREVENTION

Prevention is always being emphasized in China even though the cause of EC is not completely clear. Several preventive measures have been undertaken.

A. Preventive Measures: Five Measures of Prevention

In Linxian county, the prevention work being carried out can be described using the common motto, "three, five, seven":

"Three Earlies": Early discovery, early diagnosis, early treatment

Table 2
MICROSOMAL ACTIVATION OF FISH JUICE (FJ)

(% Metaphase Plates with Chromosome Aberrations)

FJ concentration (µl/ml)	A[a]		B[a]		C[a]	
	−S9	+S9	−S9	+S9	−S9	+S9
83.3	T[b]	T	T	T	T	T
62.5	MI[c]	T	7.7[d] (0.077)[e]	24.0(0.400)	MI	MI
41.6	22.5(0.709)	MI	2.7(0.027)	13.6(0.309)	7.3(0.073)	33.6(0.445)
31.2	0.0	MI	0.0	8.1(0.100)	0.0	12.7(0.227)
20.8	0.0	31.7(0.634)	0.0	6.0(0.095)	0.0	1.8(0.036)
15.6	0.0	14.7(0.234)	0.0	2.7(0.036)	0.0	0.0
10.4	0.0	11.8(0.118)	0.0	0.9(0.009)	0.0	0.0

[a] Original pH of FJ = 5.1—5.5. A: Final pH = 4.2; B. Final pH = 7.0; C: Final pH = 9.0.
[b] T = toxic, no cells.
[c] MI = mitotic inhibition; less than one metaphase plate per 6000 cells.
[d] Percentage metaphase plates with chromosome aberrations.
[e] Average number of chromosome exchanges per metaphase plate. CHO cells not exposed to FJ, Control = 0.0.

Table 3
ENHANCEMENT OF THE CLASTOGENIC ACTION OF NITROSATED FISH JUICE (FJ) BY S9

(% METAPHASE PLATES WITH CHROMOSOME ABERRATIONS)

FJ concentration (µl/ml)	A[a]		B[a]		C[a]	
	−S9	+S9	−S9	+S9	−S9	+S9
83.3	T[b]	T	21.9[c](0.402)[d]	62.1(1.582)	MI[e]	MI
62.5	MI	T	0.0	41.4(1.008)	4.5(0.045)	24.7(0.522)
41.3	5.9(0.401)	MI	0.0	15.9(0.235)	0.0	3.7(0.060)
31.2	0.0	23.2(0.309)	0.0	10.8(0.150)	0.0	1.8(0.027)
20.8	0.0	23.6(0.372)	0.0	1.8(0.027)	0.0	0.0
15.6	0.0	0.9(0.090)	0.0	0.0	0.0	0.0

[a] Original pH of FJ = 5.1—5.5. A: Final pH = 3.8; B: Final pH = 6.0; C: Final pH = 8.5
[b] T = toxic, no cells.
[c] Percentage metaphase plates with chromosome aberrations.
[d] Average number of chromosome exchanges per metaphase plate.
[e] MI = mitotic inhibition; less than one metaphase plate per 6000 cells. CHO cells not exposed to FJ, Control = 0.0

"Five Preventions": Prevention of mildew, elimination of amines, use of molybdenum fertilizer, treatment of marked dysplasia, changing of bad living habits

"Seven Warning Signs": Obstructive feeling on swallowing, pain behind breast bone, feeling of something stuck in the gullet, stomach-ache, dry throat, belching when taking food, soreness in the back

Table 4
THE RELATIONSHIP BETWEEN INCIDENCE OF DYSPLASIA AND CARCINOMA OF THE ESOPHAGUS

Lesion by cytological mass survey	Malignant changes	
	No.	%
Marked dysplasia (79 cases)	21	26.6
Normal (11,011 cases)	13	0.12

The five preventive measures have been practiced in Linxian county since 1974. This paper introduces as the main point, the treatment of marked dysplasia.

B. Relationship Between Epithelial Dysplasia and Carcinoma of the Esophagus

In EC high-incidence areas, the incidence of esophageal dysplasia as well as the incidence of esophageal carcinoma is very high. The transformation of dysplasia to carcinoma is higher than that of the normal group, and the marked dysplasia progressing to carcinoma has been increasing every year. It was reported[30] that out of the 79 cases of marked dysplasia followed for 2 to 7 years, 21 cases of marked dysplasia patients (26.6%) had progressed to cancer. In one group of 11,011 cases of normal smears, only 13 cases (0.12%) progressed to cancer during the years of observation (Table 4). According to the report of our Cancer Institute in 1975, follow-up examination, after 1 to 12 years, of 530 marked dysplasia cases previously discovered in Yao Can commune, Linxian county, Honan, revealed a cancer rate of 14.8%. Marked dysplasia of the esophageal epithelium is considered to be a precancerous lesion, and thus by blocking the transformation of marked (severe) dysplasia to carcinoma it may be possible to reduce the incidence rate of EC. During the past few years the Chinese Academy of Medical Sciences has carried out research into the treatment of marked dysplasia of the esophagus.

C. Prevention and Treatment of Marked Epithelial Dysplasia (Precancerous Lesions)

1. Establishment of the Precancerous Model in Animals

Since no available precancerous animal model of the esophagus or forestomach has been reported in the literature, presented herein is a favorable method for inducing precancerous lesions and cancer of the esophagus and forestomach in rats and mice, using NSEE. The results prove that in rats, large doses of NSEE (0.5 g/kg bw) given 9 times within 31 days, can induce precancerous lesions in the epithelium of the esophagus of all animals. Precancerous lesions developed in the epithelium of the forestomach of all mice within 35 days after about 10 doses of NSEE or its precursor compounds were given, e.g., 2 g/kg bw sarcosine ethyl ester HCl and 0.3 g/kg bw sodium nitrite, mixed at a suitable pH. After cessation of the feeding of these carcinogens, about 50% of the treated mice developed carcinomas in 77 days. These experiments were repeated three times and the same results were obtained. This is an ideal precancerous model, since the side effects of NSEE are few, the latency period is short, and the incidence rate of the lesions is high.[31]

2. Experimental Therapy with Animals

The precancerous model in animals has been used in experimental drug therapy, etiology, and carcinogenesis studies:

Table 5
THE INHIBITORY EFFECT OF TILORONE ON DYSPLASIA AND MALIGNANT CHANGES OF THE ESOPHAGUS INDUCED BY NSEE IN RATS

Group	Days in experiment	No. of rats	Carcinogenesis (%) Dysplasia	Carcinoma
Control	112	10	100	80
Therapy	112	10	100	40
Control	112	20	100	60
Prevention	112	19	89	16 ($p < 0.02$)

Note: Control: NSEE was used alone; therapy: Treated with 0.02% Tilorone solution after formation of dysplasia; prevention: Tilorone solution was given for 1 week prior to the carcinogen.

The antitumor B and antitumor B3 drugs to inhibit carcinogenesis in mice — When the control group was given the carcinogen alone, the incidence of precancerous lesions was 100% and that of cancer 46.6%. After the formation of dysplasia, the second group was given antitumor B, the precancerous lesions were reduced to 68% and the cancer to 22.7%. In the third group, given antitumor B3, precancerous lesions were reduced to 65.2% and cancer to 8.6%. The difference is significant.[32]

The inhibitory effect of Tilorone on dysplasia, and malignant changes of the esophagus by nitrosamine in rats — After the formation of dysplasia, Tilorone solution was given liberally to the group for therapy and the animals were sacrificed at 112 days. The incidence of esophageal carcinomas was 80% in the control group and after Tilorone therapy the incidence was reduced to 40%. In a second experiment, Tilorone solution was given for one week before use of the carcinogen. The incidence of carcinomas was 60% in the control group and was reduced to 16% in the prevention group. This difference is significant (Table 5).[33]

3. Treatment of Marked Dysplasia Patients in Linxian County

The marked dysplasia of esophagus patients was classified according to the five-class typing for epithelial cells in the esophagus (i.e., normal, mild dysplasia, marked dysplasia grade I, marked dysplasia grade II [suspected cancer], and carcinoma) and then were given therapy. Patients with marked epithelial dysplasia of the esophagus were treated with antitumor B in 72 cases, with antitumor B3 in 162 cases, and by Tilorone in 43 cases. The regression rates were 68.1%, 48.1%, and 48.8%, respectively. In 215 cases of an untreated control group with a similar dysplasia, the natural regression rate was 29.8%. These differences were statistically significant. The incidence of malignancies developing in the therapy groups were 1.4%, 2.5%, and 2.3%, respectively, and that of the untreated patients 7.4%.[32] However, the effect of therapy on suspected cancer was not significant. Therefore, greater attention should be paid to these patients for prevention of esophageal carcinoma.

4. Microspectrophotometric Studies on the DNA Content in Epithelial Cells of the Esophagus from Patients with Marked Esophageal Dysplasia Treated with Antitumor B3 or Tilorone

A group of 85 patients with marked dysplasia of the esophageal epithelium was treated with antitumor B3 or Tilorone. The microspectrophotometer was used to measure the DNA content in nuclei of exfoliated epithelial cells of the esophagus after

Table 6
THE INHIBITORY EFFECT OF RETINOIC ACID ON EPITHELIAL DYSPLASIA AND MALIGNANT CHANGES OF THE FORESTOMACH INDUCED BY NSEE IN MICE

Group	Days in experiment	No. of mice	Carcinogenesis (%) Dysplasia	Carcinoma
Control[a]	77	14	100	50
Retinoic acid[b]	77	15	100	13.2
Control	115	15	100	73.2
Retinoic acid	115	16	68	12.5 ($p < 0.01$)

[a] NSEE was used alone.
[b] After formation of dysplasia, retinoic acid (10 mg/kg bw) was given.

therapy. After the drug therapy, the dysplasia characteristics of the DNA content of nuclei as measured by microspectrophotometry, were decreased.[34]

D. Retinoids and Esophageal Cancer

It is the impression of several investigators that the incidence of EC and the precancerous dysplasia state are higher in families with nutritionally inadequate diets than in those with better diets in the same area. Nutritional deficiencies are due to the lack of animal proteins and fats and vitamins A, B_2 and C.

Retinoids can inhibit the development of carcinoma and can cure precancerous lesions.[35] Furthermore, retinoic acid has been used in the prevention and treatment of carcinogenic changes of the epithelium of the forestomach of mice. An inhibitory effect was obtained when retinoic acid was given either in large doses (200 mg/kg bw) or in small doses (10 mg/kg bw) (Table 6).[36] Thus the experimental results indicate that retinoic acid may activate an antipromoting agent to prevent the further progression of preneoplastic lesions. However, use of this chemical in man is limited because of its undesirable side effects, and it is thus not possible to use it in Linxian. In recent years we have, therefore, modified the polar terminal group of the retinoic acid molecule in order to diminish the toxicity without lessening its biological activity.

We have tested nine new retinoic acid derivatives; five have an inhibitory effect on the carcinogenesis of the forestomach of mice, especially the RI.[37] At the same time, the RI (one form of the retinoid) can inhibit carcinogenesis of EC and precancerous lesions in rats.

In a recent study, indications were that RI can partially inhibit the clastogenic action of fish juice.[29]

IV. CONCLUSIONS

Following the investigation into the etiology of EC in the high-incidence areas, the following contributing factors related to foods, dietary, and eating habits were impressive:

1. Some mutagenic or nitrogenous compounds in the food and water
2. Fungi contamination in food and water
3. High nitrate and nitrite content in the water and foods, and in vivo in some people

4. Low molybdenum content in the soil
5. Nutritional deficiencies
6. Undesirable eating habits

It is believed that the implementation of the "five preventive measures" and the gradual rise in living standards will help eliminate cancer-causing factors. In addition, attention given to treatment of the marked dysplasia patients will reduce the incidence of EC in high-incidence areas of China.

ACKNOWLEDGMENTS

The author wishes to thank the colleagues of the Department of Epidemiology and the Department of Chemical Etiology of the Cancer Research Institute, Chinese Academy of Medical Sciences, for making their published research results available to her, Dr. H. Stich for arrangement of additions, Dr. R. H. C. San for valuable help during the preparation of this manuscript, and Miss R. Johnson for typing of this paper.

REFERENCES

1. **Liu, B. G.**, in *Studies on Mortality Surveys of Malignant Tumors in China*, 1981, in press (in Chinese).
2. **Kaplan, H. S. and Tsuchitani, P. J.**, *Cancer in China*, Alan R. Liss, New York, 1978.
3. **Li, M., Li, P., and Li, B.**, Recent progress in research on esophageal cancer in China, *Adv. Cancer Res.*, 33, 173, 1980.
4. **Magee, P. N. and Barnes, J. M.**, Carcinogenic nitroso compounds, *Adv. Cancer Res.*, 10, 163, 1967.
5. **Druckrey, H., Preussmann, R., Ivankovic, S., and Schmähl, D.**, Organotrophe carcinogene Wirkungen bei 65 verschiedenen *N*-Nitroso-Verbindungen an BD-Ratten, *Z. Krebsforsch.*, 69, 103, 1967.
6. **Wang, Y. L., Lu, S. H., and Li, M. M.**, Determination of nitrates and nitrites in well water from Yaocun commune, Linxian county, Honan province, *Chin. J. Oncol.*, 1, 201, 1979 (in Chinese).
7. Cancer Institute, Department of Chemical Etiology and Carcinogenesis, Chinese Academy of Medical Sciences, *Res. Cancer Treat. Control*, 4, 19, 1978.
8. Cancer Institute Research staff, Collected Papers, Chinese Academy of Medical Sciences, 1978, 38 and 67 (in Chinese).
9. **Anon.**, Further studies on the etiology of esophageal cancer, *Chin. Med. J.*, 58, 593, 1978 (in Chinese).
10. **Li, M. H., Lu, S. H., Ji, C., Wang, Y. L., Wang, M. Y., Cheng, S., and Tian, G. Z.**, Proc. 10th Int. Symp. Princess Takamatsu Cancer Res. Fund, (abstr.), Tokyo, 1979, 27.
11. **Li, M. H., Lu, S. H., Ji, C., Wang, M. Y., Cheng, S. J., and Jin, C. L.**, A new nitrosamine in moldy food: *N*-1-methylacetonyl-*N*-3-methylbutylnitrosamine (MAMBNA), *Sci. Sin.*, 22, 401, 1979.
12. **Lu, S. H., Wang, Y. L., and Li, M. H.**, Effects of fungi on formation of carcinogenic nitrosamines and their precursors in food, *Acta Acad. Med. Sin.*, 2, 24, 1980, (in Chinese).
13. **Wang, F. R. and Zhang, S. D.**, Experimental studies of esophageal cancer on rats induced by *Fusarium monili forme*, *Res. Cancer Treat. Control*, 3, 12, 1979 (in Chinese).
14. Cancer Institute, Department of Epidemiology, Chinese Academy of Medical Sciences, Preliminary surveys on the epidemiological factor of esophageal cancer in China, *Acad. Sin. Res. Cancer Treat. Control*, 2, 85, 1977 (in Chinese).
15. **Yang, J., Gao, J., Zhu, L. P., Gao, Z. J., Li, B. G., Wang, G. X., Wang, S. H., and Zhang, J. C.**, Promoting action of *Geotrichum candidum* link in pickle on carcinogenesis of esophageal cancer in Linxian, *Res. Cancer Treat. Control*, 3, 20, 1974 (in Chinese).
16. **Luo, D. Y., Wang, C. J., Bai, S. H., Dong, S. L., Xu, Y. Q., and Ceng, B. X.**, The promoting effect of *Geotrichum candidum* of the pickles on carcinogenesis of cancer of the lung and forestomach in mice in EC high incidence area of Yanting county, *Res. Cancer Treat. Control*, 3, 67, 1978 (in Chinese).

17. **Takahashi, Y., Nagao, M., Fujino, T., Yamaizumi, Z., and Sugimura, T.,** Mutagens in Japanese pickle identified as flavonoids, *Mutat. Res.*, 68, 117, 1979.
18. **Cheng, S. J., Sala, M., Li, M. H., Wang, M. Y., Pot-Deprun, J., and Chouroulinkov, I.,** Mutagenic, transforming, and promoting effect of pickled vegetables from Linxian county, China, *Carcinogenesis*, 1, 685, 1980.
19. **Wang, G. H., Zhang, W. X., and Chai, W. G.** The identification of natural Roussin red methyl ester, *Acta Chim. Sin.*, 38, 95, 1980.
20. **Zheng, S. F., Qin, Q. S., Li, Q. L., and Liu, X. F.,** *Chin. J. Prev. Med.*, 1980 (in Chinese).
21. **Burrell, R. J. W., Roach, W. A., and Shadwell, A.,** Esophageal cancer in the Bantu of the Transkei associated with mineral deficiency in garden plants, *J. Natl. Cancer Inst.*, 36, 201, 1966.
22. **Nemenko, B. A., Moldakylosa, M. M., and Zoruna, C. H.,** Distribution of esophageal cancer in Guriev region in relation with the mineral contents of the drinking water, *Vop. Onkol.*, 9, 75, 1976 (in Russian).
23. Cancer Institute, Chinese Academy of Medical Sciences, Epidemiology and pathological morphology of pharyngeal and esophageal cancers in domestic fowls, *Acta Zool. Sin.*, 19, 309, 1973.
24. Cancer Institute, Chinese Academy of Medical Sciences, Epidemiology and pathology of pharyngo-esophageal cancers in domestic fowls from Honan migrant communities and native inhabitants in Zhongxian county, Hubei province, *Acta Zool. Sin.*, 22, 319, 1976.
25. Cancer Institute Chinese Academy of Medical Sciences, and Cancer Hospital, Hubei province, *Acta Zool. Sin.*, 22, 314, 1976.
26. **Cai, H. Y., Lin, P. Z., and Li, Q. L.,** in *Collection of Papers for the 20th Anniversary of the Cancer and Ritan Hospital,* (245 papers in Chinese), Chinese Academy of Medical Sciences, Beijing, 1978, 146.
27. **Li, M. H.,** Esophageal carcinoma in a goat from Linxian, Honan province, unpublished data, 1980.
28. Cancer Institute Research Staff, Chinese Academy of Medical Sciences, *Acta Zool. Sin.*, 22, 327, 1976.
29. **Cai, H. Y. and Stich, H. F.,** Clastogenic action of fish juice of high incidence areas of esophageal cancer in China, *Chin. J. Oncol.*, in press.
30. Coordinating Groups for the Research of Esophageal Carcinoma, Honan Province and Chinese Academy of Medical Sciences, Studies on relationship between epithelial dysplasia and carcinoma of the esophagus, *Chin. Med. J.*, 1, 110, 1975.
31. **Cai, H. Y., Lin, Q. Z., and Guan, X. J.,** Studies on cancer and precancerous lesions of the esophagus and forestomach induced by *N*-nitrososarcosine ethyl ester in rats and mice, *Chin. Med. J.*, 60, 87, 1980.
32. **Cai, H. Y., Shu, Y. J., Wang, G. Q., Wu, E. R., Lin, P. Z., Liu, S. F., and Li, M. X.,** Therapy of severe epithelial dysplasia of esophagus. I. Therapeutic effects of antitumor B, antitumor B3, and tilorone, *Chin. J. Oncol.*, 2, 92, 1980.
33. **Cai, H. Y., Han, J., Lin, P. Z., Zhu, T. X., Guan, X. J., and Gu, Q. M.,** Influence of tilorone on carcinogenesis of the esophagus by *N*-nitrososarcosine ethyl ester in rats, *Chin. J. Oncol.*, 1, 186, 1979.
34. **Lin, P. Z., Cai, H. Y. et al.,** Therapy of severe epithelial dysplasia of esophagus. II. Microspectrophotometric studies on the DNA content in epithelial cells of the esophagus from patients with severe esophageal dysplasia, *Chin. J. Oncol.*, in press.
35. **Bollag, W.,** Retinoids and cancer, *Cancer Chemother. Pharmacol.*, 3, 207, 1979.
36. **Cai, H. Y., Guo, Z. R., Lin, P. Z., Huang, L., and Guan, X. J.,** Inhibitory effect of retinoic acid on carcinogenesis of the forestomach in mice, *Res. Cancer Prev. Treat.*, 14, 12, 1978.
37. **Cai, H. Y., Zhang, J. S., and Cui, X. Y.,** Inhibitory effect of some new retinoids on carcinogenesis. I. Inhibitory effect on carcinogenesis of the squamous cell epithelium of forestomach in mice, *Chin. J. Pharmacol.*, in press.

Chapter 5

LIFESTYLE AND CANCER ETIOLOGY

Gary M. Williams, John H. Weisburger, and Ernst L. Wynder

TABLE OF CONTENTS

I. MECHANISMS OF CARCINOGENESIS

During the last 30 years much knowledge has been acquired through research on the mechanisms of carcinogenesis. Evidence has been developed indicating that neoplasia may stem from a somatic mutation involving a change in normal cells at the level of the genetic material.[1-4] This had been quite a controversial area, but an adequate scientific basis now exists for this concept.

A change in the genetic material can be visualized as arising through several mechanisms. In one, the genetic material can be changed through a direct attack by radiation, by chemicals, or by viruses. Radiation or chemicals can damage the genetic material at a number of sites along the DNA chain, and these are converted to permanent alterations by mispairing of bases during replication of the damaged regions. With viruses, a more specific insertion of DNA segments or, through reverse transcriptase of RNA sequences, a change takes place through the operation of specific polymerases and hence yields an abnormal DNA containing new information.

Other mechanisms can produce abnormal DNA. One involves faulty DNA polymerase operation during DNA synthesis resulting in inaccurate transcription of the parent DNA segment. Loeb and co-workers[5] have demonstrated that certain carcinogenic metal ions may operate through this scheme. Abnormal DNA can also ensue, especially during postreplicative DNA synthesis, through the errors introduced by specific DNA polymerases concerned with DNA repair.[6-8] Furthermore, the infidelity of DNA polymerases may lead to production of further abnormalities in DNA during replication of early tumor cells, and thus may represent the means whereby tumor cells progress to less differentiated, more malignant cancer types during their growth and development.

The production of abnormal DNA by any of the above mechanisms is only the first step in a long sequence of events terminating in a malignant invasive neoplasm. An important element is the ability of an abnormal cell population to achieve a selective growth advantage in the presence of surrounding normal cells. The cell duplication process is highly dependent on a number of endogenous and exogenous controlling elements. Two of the key controlling elements are the promoters and inhibitors that either enhance or retard the growth process.

As numerous experiments documenting this phenomenon indicate, promoters do not lead to the production of an invasive cancer in the absence of an antecedent cell change.[4,9-11] Thus, in exploring the causes of any specific human cancer, consideration must be given to the agents leading to an abnormal genome as well as to any other agents possibly involved in the growth and development into neoplasms of the resulting abnormal cells and their further progression to malignancy.

Chemical carcinogens have been classified by Weisburger and Williams[4,12] into eight classes that, in turn, belong to two main groups: genotoxic carcinogens and agents (including promoters) operating by epigenetic pathways (Table 1). This classification, in relation to an understanding of the relevant mechanisms, is important in dissecting the complex causes of diverse kinds of cancer and arriving at a specification of the role of each agent, genotoxic carcinogen, co-carcinogen, or promoter in the overall carcinogenic process for each kind of cancer.

II. ENVIRONMENT AND CANCER

The majority of human cancers (70 to 90%) have been suggested to relate to environmental causes.[13-17] The environment is complex, and is often misunderstood as consisting primarily of ubiquitous chemicals, more specifically, those due to modern tech-

Table 1
CLASSIFICATION OF CARCINOGENIC CHEMICALS

Type	Mode of action	Example
Genotoxic		
Direct-acting	Electrophile, organic compound genotoxic, interacts with DNA	Ethylenimine, bis(chloromethyl)ether
Procarcinogen	Requires conversion through metabolic activation by host or in vitro to type 1	Vinyl chloride, benzo(α)pyrene, 2-naphthylamine, DMNA
Inorganic carcinogen	Not directly genotoxic, leads to changes in DNA by selective alteration in fidelity of DNA replication	Nickel, chromium
Epigenetic		
Solid-state carcinogen	Exact mechanism unknown; usually affects only mesenchymal cells and tissues; physical form vital	Polymer or metal foils, asbestos
Hormone	Mainly alters endocrine system balance and differentiation; often acts as promoter	Estradiol, diethylstilbestrol
Immunosuppressor	May stimulate "virally induced," transplanted, or metastatic neoplasm	Azathioprine
Cocarcinogen	Enhances effect of type 1 or type 2 agent when given at the same time; may modify conversion of type 2 to type 1	Phorbol esters, pyrene catechol, ethanol, *n*-dodecane, SO_2
Promoter	Enhances effect of type 1 or type 2 agent when given subsequently; may stimulate progression of "spontaneously" occurring preneoplastic lesions	Phorbol esters, phenol anthralin, bile acids, tryptophan metabolites, saccharin

From Weisburger, J. H. and Williams, G. M., in *Toxicology: The Basic Science of Poisons*, Doull, J., Klaassen, C., and Amdur, M., Eds., Macmillan, New York, 1980, 84. With permission.

nology and industrial development. Certainly it is true that a number of food additives, pesticides, insecticides, and industrial chemicals introduced commercially in the last 40 years have exhibited carcinogenic properties in animal models.[14,18,19] Historically, human cancers have also been associated with antecedent chemical exposure in an occupational environment, or with the use of specific drugs. However, current evidence documents that the main human cancers in the Western world do not stem from intentional or even inadvertent chemical contaminants in the environment. This is, in a sense, unfortunate, because if such an association could actually be demonstrated, prevention of the main human cancers through removal of the offending substances would be a readily achievable goal.

Cancer represents many different diseases and insight into the complex causes of cancer requires a detailed analysis of the factors inherent in the occurrence of each specific type of cancer. From world-wide statistics on the incidence of diverse cancers as well as the altered risk for migrants from areas of high to low incidence over several generations, and the corresponding analysis of data obtained under controlled conditions in animal models, a picture emerges permitting delineation of the multiple causative and modulating factors involved in some of the main human cancers (Table 2).

The introduction of a carcinogen to the general public, i.e., cigarette smoking (mostly in U.S. males during and after World War I and in females during and after World War II) has resulted, years later, in an impressive rise in the incidence of lung

Table 2
HUMAN CANCERS AND LIFESTYLE FACTORS

Organs affected	Carcinogens	Enhancing factors	Inhibiting factors
Oral cavity, esophagus	Tobacco smoke	Alcohol	Vitamins (?)
Stomach	Nitrite + specific foods (fish, beans, not meat)	Salt	Vitamin C, E
Colon, breast, prostate, pancreas	Fried foods	Fat	Fiber, vitamins (?)

cancer[20] as well as in an increase in other specific cancers such as those of the renal excretory pathway. On the other hand, the introduction of insecticides, pesticides, and food additives in the last 40 years appears to have had no detectable effect on the incidence of the main human cancers in the U.S., such as those of the respiratory tract, breast, prostate, and colon.[20-22] Breast and colon cancer have exhibited a slightly increased incidence, almost paralleling the increased daily intake of dietary fat in the U.S.[23-25] The incidence of pancreatic cancer also appears to have risen in the last few years, although one major risk factor is smoking and another is dietary fat.[26] Coffee drinking may be another modulating element.[27] On the other hand, stomach cancer has decreased appreciably, as has primary cancer of the liver. Occupational exposures, although clearly responsible for some cancers,[28] are a factor in only a small portion of all cancers.[29]

In this paper are presented summary statements of the voluntary lifestyle patterns of nutrition that contribute to the etiology of the main human cancers in the U.S., specifically, cancers of the breast, prostate, stomach, and colon. The relationship of smoking to health, specifically with regard to major killing and disabling diseases, e.g., myocardial infarction and cancer in the respiratory tract, kidneys, and bladder is reviewed elsewhere.[30-32]

A. Cancers Associated With Diet

Several lines of evidence suggest that cancers of the digestive tract, and in particular, cancers of the stomach, colon, and pancreas, are associated with dietary components.[33,34] Furthermore, cancers of the endocrine-related organs, such as breast, prostate, ovary, and endometrium, are also linked to dietary factors.[34,35] With the exception of lung cancer, these are the human cancers with the highest incidence in the U.S.

As discussed above in general terms, background information comes from observations of cancer incidence in various parts of the world. Cancer of the stomach is prevalent in Japan, the western part of South America, parts of Central America, and northern and eastern Europe, but it has a low incidence in the U.S. In each area the lower socioeconomic groups have a higher incidence of stomach cancer than more affluent groups. On the other hand, the pattern of cancer of the colon, breast, prostate, pancreas, ovary, and endometrium exhibits almost the opposite picture, being high in the U.S., Western Europe, Australia, New Zealand, and Argentina. This geographic distribution stems from the environment rather than genetic differences between populations, as revealed by studies of the changes in incidence found in first- and second-generation migrant groups. Thus, Japanese migrants have a decreasing risk for gastric cancer and an increasing risk for colon and breast cancer. In light of the high degree of industrialization in Japan, it is probably true that industrial activity is not responsible for the etiology of the main human cancers.

Individuals from northern or eastern Europe or Japan, who would normally have a high risk for gastric cancer, do not have this risk in successive generations in the U.S. but instead develop cancer in the colon, breast, or prostate. It is sometimes stated, with fatalistic resignation, that humans are destined to develop cancer of one type or another. Research documents that the etiologic factors for gastric cancer are of a totally different nature than those for colon or breast cancer and thus can be controlled independently to lower the risk for all these cancers. With developing knowledge in this field, it is hoped and expected that the risk for gastric cancer can be reduced without simultaneously increasing the risk for the other cancers.

The main factors considered to be responsible for the cancers noted above are dietary, namely, specific micronutrients and components (e.g., gastric cancer and perhaps also liver and esophageal cancer in certain populations) or macronutrients (e.g., colon, breast, prostate, ovary, endometrium, and perhaps pancreas cancer). Specifically, the Western diet supplies about 40% of its calories in fat, with more saturated fats, whereas a typical Japanese diet of 20 years ago supplied about 15% of its calories in fat, with more unsaturated oils.[25] Further research efforts are required to pinpoint the underlying mechanisms whereby nutrition translates to disease risk, and to secure data on interactions between macro- and micronutrients that might modulate or even control the overall effect.

B. Colon Cancer

Large bowel or colorectal cancer needs to be considered on the basis of several lines of evidence including international and intranational incidence patterns, sex ratio, and age distribution; and as two distinct diseases — colon cancer and rectal cancer. Moreover, right-sided colon cancer may be different from left-sided colon cancer. Not much is known about risk factors for rectal cancer.

On the basis of variations in incidence for different regions of the world and in view of the altered risk of migrant populations, it has been accepted that diet is a major etiologic factor in colon cancer.[33,34]

The specific dietary elements that have been documented to be relevant through studies in human and animal models are the amounts of dietary fat and fiber. In fact, one of the best arguments for these concepts is the changing incidence of colon cancer in Japan in recent years, as the Japanese nutritional intake progressively became Westernized. In addition, in many areas of the world, an association exists between colon cancer and coronary heart disease. Also, the amount of dietary fat and cholesterol have been shown to relate to risk for heart disease. A key exception to this rule is in Finland, where the risk for heart disease is high and that for colon cancer low. Several groups[34-37] have obtained some evidence that the lower risk of the Finnish people for colon cancer, despite a high fat intake, is attributable to the consumption of foods high in fiber, especially cereal bran fiber.

Laboratory research by a number of groups (see review, Reference 33), particularly by Reddy et al.,[34] has provided an explanation of the role played by fat in promoting colon cancer risk and by fiber in inhibiting colon carcinogenesis. The main effect of dietary fat appears to reside in a direct association between endogenous cholesterol biosynthesis that in turn leads to increased bile acid biosynthesis. Bile acids have been shown to be effective promoters in colon cancer. The effect of dietary fiber is to increase intestinal and stool bulk, thereby reducing the concentration of promoters.

C. Breast Cancer

Epidemiologic studies show that premenopausal American and Japanese women have comparable rates of breast cancer as a function of age, although Japanese women exhibit a somewhat lower incidence compared to American women. Furthermore, the incidence in Japanese women reaches a plateau at menopause and then decreases. In contrast, the incidence in Americans increases sharply during and following menopause, the curve showing a characteristic "hook". This biphasic incidence of breast cancer in pre- and postmenopausal women possibly reflects two independent disease factors.[38]

The key difference in environmental factors between Japan and Western countries such as the U.S. (particularly in Hawaii, where a large number of Americans of Japanese descent are concentrated) is also dietary, with the main variable being the quantity of dietary fat. High-risk populations consume a higher proportion of dietary fat, and the percentage of consumption has been increasing over the past 50 years.[23] This change may be related to the slight but definite increase in incidence of breast cancer in Western women over the same time span. De Waard[39] reported that obesity was a risk factor in Dutch women, but this apparently is not seen in epidemiologic studies of North American Caucasian women.[40]

In contrast to the findings from migrant studies that the first-generation migrants from a low-risk region have an appreciable increase in risk of colon cancer when transferred to a high-incidence area such as the U.S., first-generation migrants show only a slight increase in breast cancer. It is not until the second generation that a risk similar to that of the long-term population in the high-incidence region is attained. This aspect deserves further documentation since it suggests that, for breast cancer, residence in a high-risk area at the time of puberty and breast development is critical.[41] Female Sprague-Dawley rats are more sensitive to a single dose of hydrocarbon carcinogen at or around puberty than are older or younger animals and this susceptibility seems to be correlated with the rate of cell division and serum prolactin levels.[42]

Useful animal models have been those based on the induction of breast cancer in rats by specific chemicals, such as 7,12-dimethylbenz[*a*]anthracene (DMBA) and *N*-nitrosomethylurea (NMU). Using the DMBA model, Carroll[43] demonstrated that diets with a fat content between 10 and 20% bw significantly increased the frequency of breast cancer, as compared to diets containing 0.5 to 5% fat. The effect was found to occur primarily in the promotional phase of breast cancer development. Diets rich in polyunsaturated fats were more effective tumor promoters than were diets rich in saturated fats. Hill et al.[44] have extended Carroll's findings in the NMU model that possesses features that mimic human breast cancer more closely than does the DMBA tumor. Their work has shown that fat can increase carcinogenesis in ovariectomized rats, indicating a lack of primary involvement of ovarian hormones and offering a model for postmenopausal human breast cancer.

Evidence from experimental and epidemiologic investigations in breast cancer has reached sufficient proportions to warrant more systematic explorations into mechanisms. Postulated mechanisms by which dietary fat may influence breast cancer fall into two basic categories: those involving direct effects on host metabolism and those involving secondary effects of obesity.

In the Naylor Dana Institute for Disease Prevention, the authors and co-workers have focused on the possibility that dietary fat elicits its tumor-enhancing effect by altering host endocrine metabolism, in particular, the component of the endocrine system that regulates prolactin secretion. The Institute's interest in a fat-prolactin breast cancer relationship was prompted by two facts: prolactin is a classic known promoter in a number of murine breast tumor systems, and prolactin is a liporegulatory hormone in birds and lower mammals, rodents, and possibly in man. In addition, there is no question that other hormones are also involved in growth control of the mammary gland through general action and specific receptor systems. Thus, the problem to be solved is to define the nutritionally linked hormonal balances characteristic of high- or low-risk situations.

D. Prostate Cancer

Epidemiology has provided data on the incidence of prostate cancer as a function of a number of factors such as race, age, area of residence, and diet.[45-47] Thus, Caucasians in the Western world have a higher incidence of prostatic cancer than the Japanese. First-generation Japanese migrants to the U.S. retain their low risk, but later generations exhibit a higher risk. However, native Japanese have *in situ* lesions,[48] and it has been suggested that the difference in proliferative types of prostatic carcinoma as seen in Western men is due to promotion by environmental factors, among which diet deserves primary consideration.

For unknown reasons the risk for prostate cancer is higher in American blacks than in American whites. Nonetheless it is clear that, as a rule, population groups that have a high risk of breast, colon, and endometrial cancer also have a high risk of prostate cancer.[49] Since an association with diet has been documented for the former types of cancer, not only in epidemiologic studies, but also in metabolic and physiologic approaches, it is probable that dietary fat plays a role in the etiology of prostate cancer.[50] For breast and colon cancer, the dietary fat hypothesis is further supported by detailed studies in animal models, as described above. To date there has not been a reliable and realistic animal model available involving specific chemical carcinogens for prostate cancer amenable to mechanistic studies. However, the authors have developed one such model involving a systemically acting chemical carcinogen, 3,2′dimethyl-4-aminobiphenyl, and the effect of dietary modulators, such as the level of fat, is currently being explored.

E. Stomach Cancer

Cancer of the glandular stomach is an important neoplasm in Japan, Iceland, mountainous interior regions of central and western Latin America, and some Eastern European countries. In contrast, Western Europe, many Anglo-Saxon countries (except Wales in the U.K.), and the U.S. have a low incidence. In fact, over the last 50 years the gastric cancer rate in the U.S. has declined appreciably, from 30/100,000 for men and 22/100,000 for women, to 7.5 and 3.7/100,000 for men and women, respectively.

Dietary risk factors that show positive correlations in high-risk populations include the high consumption of dried salted fish, pickled vegetables, smoked fish, fewer vegetables, and a low vitamin C intake, particularly on a seasonal basis.[25,51,52] Another correlation was reported between elevated levels of nitrate in foods and drinking water due to high levels in the soil and water supply.[53] Crude salt or saltpeter, used for preserving certain foods such as fish, also contains nitrate. On the other hand, Hawaiian Japanese, with a higher intake of uncooked vegetables such as celery, lettuce, tomato, and fresh fruit juices, had a low gastric cancer risk as compared to indigenous Japanese or to the first-generation Japanese migrants to Hawaii.

Glandular stomach cancer has been induced in animal models with alkylnitrosoureido compounds such as *N*-methyl-*N*′-nitro-*N*-nitrosoguanidine (MNNG).[54] Kinetics of the reaction of the nitrosation of alkylamides such as methylurea has been studied by Mirvish et al.,[55] who also made the important discovery that ascorbic acid had an inhibitory effect on the nitrosation of methylurea and alkylamines.

It has been demonstrated that treating Sanma mackerel, a fish commonly eaten in Japan, with nitrite at pH 3, mimicking gastric conditions, yields high levels of mutagenic activity for *Salmonella typhimurium* TA1535.[56] Formation of the mutagen was inhibited by the simultaneous presence of vitamin C. When this mutagenic activity from the reaction of nitrite and Sanma was administered to Wistar rats by gavage, it proved carcinogenic, inducing adenocarcinomas of the glandular stomach.[57] The nature of the mutagen(s) obtained from the reaction is as yet unknown, but its mode of formation or inhibition and the mutagenicity/carcinogenicity results observed suggest it may be an alkylnitrosoureido compound.

Nitrite is required for the formation of gastric carcinogens. Nitrate present in foods is converted to nitrite during storage at room temperature for 24 hr, a reaction inhibited by refrigeration. Another source of nitrite is reduction of nitrate by the bacterial flora in the oral cavity.[58]

Thus, various lines of evidence lead to the conclusion that gastric cancer may result from the consumption of salted, pickled foods, that the active agent or agents has the properties of alkylnitrosoureido compounds formed from nitrite and undefined substrates, and most importantly, that the formation of the carcinogen may be blocked by vitamin C or by vitamin E.

Intake of foods containing vitamin C with each meal throughout the year, from childhood onward, may inhibit the formation of gastric carcinogens, a factor accounting for the sharp decline in the incidence of gastric cancer in the U.S.

Lower salt use would have similar benefits and would also reduce the risk of hypertension. In the MNNG-induced gastric cancer model, Tatematsu et al.[59] found that salt (sodium chloride) had a promoting effect. This observation may mimic the human environment. Joossens et al.[60] have documented a parallelism in international trends for gastric cancer and for hypertension.

III. MUTAGENS AND MODE OF COOKING

The genotoxic carcinogens responsible for nonoccupational human cancer in the general public in Western countries are not known, except for those found in tobacco smoke.[14,61] This is especially so for nutritionally linked cancers such as colon, breast, or prostate.

An important clue to the nature of such carcinogens came from the demonstration that charcoal broiling of meat or fish yielded mutagenic activity for *S. typhimurium* TA 98.[62,63] Since mutagenic activity is often an indicator of carcinogenic activity, the development of mutagenic activity as a function of mode and temperature of cooking was studied under typical, realistic, cooking conditions.[17] Frying meat together with soy protein flour yielded less mutagenic activity.[67]

One of the products from the pyrolysis of tryptophan was a γ-carboline derivative, an *o*-methylarylamine type of compound that has been shown to induce liver tumors and s.c. tumors in rodents.[62]

The principal mutagenic components of fried beef were resolved into two major fractions, distinct from any of the known mutagens from amino acid or protein pyrolysates. Mutagens from fried fish could be resolved into more fractions, and included small amounts of compounds such as the amino acid pyrolysates. One of the main components in fried sardines is identical to one of the mutagens in beef.[64] Its structure, 2-amino-3-methyl-imidazo-[4,5-*d*]quinoline, is similar sterically and structurally to known homocyclic carcinogenic arylamines, such as 3,2′-dimethyl-4-aminobiphenyl, which are colon and mammary gland carcinogens in rodents.[17] The structure of the mutagens in fried meat or fish most likely do not stem from the pyrolysis of amino acids or peptides, but rather from the formation of heterocyclic compounds from carbohydrate components and amino acids. Indeed, a model system for browning, developed to test this hypothesis, showed that reacting sugars with ammonium ion under reflux conditions results in the formation of strong mutagenic activity, with similar

Table 3
CAUSES OF HUMAN CANCERS

Type	% of total
Occupational cancers	
Various organs	1—5
Cryptogenic cancers	
Lymphomas, leukemias, sarcomas - cervic? Virus?	10—15
Lifestyle cancers	
Tobacco-related	21
Lung, pancreas, bladder, kidney	
Diet-related	
Nitrate-nitrite, low vitamin C, mycotoxin — stomach, liver	5
High fat, low fiber, broiled or fried foods(?) — large bowel, pancreas, breast, prostate, ovary, endometrium	45
Multifactorial	
Tobacco and alcohol — oral cavity, esophagus	5
Tobacco and asbestos, tobacco and mining, tobacco and uranium and radium — lung, respiratory tract	5
Iatrogenic cancers	
Radiation, drugs — diverse organs	1

From Weisburger, J. H. and Williams, G. M., *Cancer Medicine,* 2nd ed., Holland, J. F. and Frei, E., III, Eds., Lea & Febiger, Philadelphia, 1982, 42. With permission.

mobilities on high pressure chromatography systems as activities seen in fried meat.[17] These reactions, which produce many pyrazine derivatives, are base-catalyzed and can be inhibited by the antioxidant, propyl gallate.[65] Thus, the reactions leading to mutagens during frying may be akin to those that take place during the browning reactions.

Whether this mutagenic activity is relevant to human disease is the subject of current research. The fact that antioxidants such as propyl gallate or butylated hydroxyanisole (BHA) are effective inhibitors of mutagen formation, as was soy protein during the frying of meat, may provide practical ways of reducing the mutagen formation that may be associated with disease risk.

IV. CONCLUSION

The concepts outlined in this paper, based on various lines of evidence and multidisciplinary approaches, suggest that nutrition and specific nutritional components, as well as dietary habits in various parts of the world, may play an important role in the causation and development of a number of important cancer types. Weisburger and Williams[66] have calculated that nutrition may impact directly on the occurrence of 30 to 40% of cancers in men and 50 to 60% of cancers in women in the U.S. (Table 3) and other Western countries. Because of the rapid Westernization of nutritional customs in Japan, alterations in the incidence of specific cancers have taken place parallel to this change. These facts are excellent evidence for the rationale presented in this paper, namely that the mode of cooking and the level of dietary fat and fiber, are associated with the occurrence of cancer of the colon, breast, and perhaps of the prostate and pancreas. It is hoped, as further evidence accumulates from human studies through metabolic epidemiology and experimental systems, that a convincing case can be made for making relatively minor alterations in dietary habits, mainly involving a somewhat lower fat intake and a higher fiber consumption. If current research does further document that the mode of cooking, especially frying and broiling, yields carcinogens leading to these kinds of cancers, a means of preventing formation of such carcinogens may be found. This may eventually yield a lower risk. In addition research on optimal levels of vitamins, minerals, antioxidants, and other micronutrients in the current diet would provide a broad basis for chemoprevention. Over the last several years, research has provided new perspectives on the causes and modifiers of the main premature killing diseases. Data in the current monograph specifically record experimentation designed to yield understanding of underlying mechanisms as a sound reliable basis for the prevention of many important kinds of human cancer and for the long-term goal of disease prevention generally.

REFERENCES

1. **Griffin, A. C. and Shaw, C. R.**, Eds., *Carcinogens: Identification and Mechanisms of Action*, Raven Press, New York, 1979.
2. **Emmelot, P. and Kriek, E.**, Eds., *Environmental Carcinogenesis: Occurrence, Risk Evaluation and Mechanisms*, Elsevier/North-Holland, Amsterdam, 1979.
3. **Grover, P. L.**, Ed., *Chemical Carcinogens and DNA*, Vol. I and II, CRC Press, Boca Raton, Fla., 1979.
4. **Weisburger, J. H. and Williams, G. M.**, Chemical carcinogens, in *Toxicology: The Basic Science of Poisons*, 2nd ed., Doull, J., Klaassen, C., and Amdur, M., Eds., Macmillan, New York, 1980, 84.
5. **Tkeshelashvii, L. K., Shearman, C. W., Zakour, R. A., Koplitz, R. M., and Loeb, L. A.**, Effects of arsenic, selenium, and chromium on the fidelity of DNA synthesis, *Cancer Res.*, 40, 2455, 1980.
6. **Roberts, J. J.**, Carcinogen-induced DNA damage and its repair, *Br. Med. Bull.*, 36, 25, 1980.
7. **Setlow, R. B.**, Different basic mechanisms in DNA repair, *Arch. Toxicol.*, 3 (Suppl.), 217, 1980.
8. **Cleaver, J. E.**, DNA damage, repair systems, and human hypersensitive diseases, *J. Environ. Pathol. Toxicol.*, 3, 53, 1980.
9. **Slaga, T. J., Sivak, A., and Boutwell, R. K.**, Eds., *Mechanisms of Tumor Promotion and Cocarcinogens*, Raven Press, New York, 1978.
10. **Weinstein, I. B., Lee, L. S., Fisher, P. B., Mufson, A., and Yamasaki, H.**, The mechanism of action of tumor promoters and a molecular model of two stage carcinogenesis, in *Environmental Carcinogenesis: Occurrence, Risk Evaluation and Mechanisms*, Emmelot, P. and Kriek, E., Eds., Elsevier/North-Holland, New York, 1979, 265.
11. **Pitot, H. and Sirica, A.**, The stages of initiation and promotion in hepatocarcinogenesis, *Biochem. Biophys. Acta*, 605, 191, 1980.
12. **Williams, G. M. and Weisburger, J. H.**, Systematic carcinogen testing through the decision point approach, *Ann. Rev. Pharmacol. Toxicol.*, 21, 393, 1981.
13. **Wynder, E. L. and Gori, G. B.**, Contribution of the environment to cancer incidence: an epidemiologic exercise, *J. Natl. Cancer Inst.*, 58, 825, 1977.
14. **Hiatt, H. H., Watson, J. D., and Winsten, J. A.**, Eds., *Origins of Human Cancer*, Cold Spring Harbor Laboratory, Cold Spring Harbor, N.Y., 1977.
15. **Higginson, J.**, Environmental carcinogenesis: a global perspective, in *Environmental Carcinogenesis: Occurrence, Risk, Evaluation and Mechanisms*, Emmelot, P. and Kriek, E., Eds., Elsevier/North-Holland, Biomedical Press, Amsterdam, 1979, 9.
16. **Doll, R. and Peto, R.**, The causes of cancer: quantitative estimates of avoidable risks of cancer in the U.S. today, *J. Natl. Cancer Inst.*, 66, 1191, 1981.
17. **Weisburger, J. H., Reddy, B. S., Hill, P., Cohen, L. A., Wynder, E. L., and Spingarn, N. E.**,Nutrition and cancer — on the mechanisms bearing on causes of cancer of the colon, breast, prostate, and stomach, *Bull. N.Y. Acad. Med.*, 56, 673, 1980.
18. **Searle, C. E.**, Ed., *Chemical Carcinogenesis*, American Chemical Society, Washington, D.C., 1976.
19. International Agency for Research on Cancer, Chemicals and Industrial Processes Associated with Cancer in Humans, IARC Monogr. Vol. 1-20, IARC Monogr. Suppl. 1, Lyon, 1979.
20. **Silverberg, E.**, Cancer statistics, 1981, *Ca-A: Cancer J. Clin.*, 31, 20, 1981.
21. **Fraumeni, J. F., Jr.**, Ed., *Persons at High Risk of Cancer*, Academic Press, New York, 1975.
22. **Demopoulos, M. D. and Mehlman, M. A.**, Eds., Cancer and the environment, *J. Environ. Pathol. Toxicol.*, 3, 1, 1980.

23. **Gortner, W. A.**, Nutrition in the U.S., *Cancer Res.*, 35, 1900, 1975.
24. **Stamler, J.**, Public health aspects of optimal serum lipid-lipoprotein levels, *Prev. Med.*, 8, 733, 1979.
25. **Hirayama, T.**, Diet and cancer, *Nutr. Cancer*, 1, 67, 1979.
26. **Wynder, E. L.**, An epidemiological evaluation of the causes of cancer of the pancreas, *Cancer Res.*, 35, 2228, 1975.
27. **Macmahon, B., Yen, S., Trichopoulos, D., Warren, J. L., and Nardi, G.**, Coffee and cancer of pancreas, *N. Engl. J. Med.*, 304, 630, 1981.
28. **Saffiotti, U., and Wagoner, J. K.**, Eds., Occupational carcinogenesis, *Ann. N.Y. Acad. Sci.*, 271, 1, 1976.
29. **Higginson, J.**, Proportion of cancers due to occupation, *Prev. Med.*, 9, 180, 1980.
30. **Hammond, E. C. and Seidman, H.**, Smoking and cancer in the U.S., *Prev. Med.*, 9, 169, 1980.
31. **Dawber, T. R.**, *The Framingham Study*, Harvard University Press, Cambridge, 1980, 1.
32. **Greenwald, P.**, Assessment of risk factors for cancer, *Prev. Med.*, 9, 260, 1980.
33. **Wynder, E. L., Reddy, B. S., McCoy, G. D., Weisburger, J. H., and Williams, G. M.**, Diet and gastrointestinal cancer, *Clin. Gastroenterol.*, 5, 463, 1976.
34. **Reddy, B. S., Cohen, L. A., McCoy, G. D., Hill, P., Weisburger, J. H., and Wynder, E. L.**, Nutrition and its relationship to cancer, *Adv. Cancer Res.*, 32, 237, 1980.
35. **Williams, G. M. and Wynder, E. L.**, Nutrition and cancer, in *Nutrition and Metabolism in Cancer*, Kluthe, R. and Lohr, G. W., Eds., Georg Thieme Verlag, Stuttgart, 1981.
36. **Jensen, O. M. and McLennan, R.**, Dietary factors and colorectal cancer in Scandinavia, *Isr. J. Med. Sci.*, 15, 329, 1979.
37. International Agency for Research on Cancer Intestinal Microecology Group, Dietary fiber, transit time, fecal bacteria, and steroids in two Scandinavian populations, *Lancet*, 2, 207, 1977.
38. **De Waard, F.**, Premenopausal and postmenopausal breast cancer: one disease or two?, *J. Natl. Cancer Inst.*, 63, 549, 1979.
39. **De Waard, F.**, Breast cancer incidence and nutritional status with particular reference to body weight and height, *Cancer Res.*, 35, 3351, 1975.
40. **Wynder, E. L., McCoy, G. D., Reddy, B. S., Cohen, L., Hill, P., Spingarn, N. E., and Weisburger, J. H.**, Nutrition and metabolic epidemiology of cancers of the oral cavity, esophagus, colon, breast, prostate, and stomach, in *Nutrition and Cancer:* Etiology and Treatment, Newell, G. R. and Ellison, N. M., Eds., Raven Press, New York, 1981.
41. **Weisburger, J. H.**, Mechanism of action of diet as a carcinogen, *Cancer (Philadelphia)*, 43, 1987, 1979.
42. **Yanai, R. and Nagasawa, H.**, Effects of pituitary graft and 2-bromo-α-ergocryptine on mammary DNA synthesis in mice in relation to mammary tumorigenesis, *J. Natl. Cancer Inst.*, 56, 1055, 1976.
43. **Carroll, K. K.**, Experimental evidence of dietary factors and hormone-dependent cancers, *Cancer Res.*, 35, 3374, 1975.
44. **Hill, P., Chan, P. C., Cohen, L., Wynder, E. L., and Kuno, K.**, Diet and endocrine-related cancer, *Cancer (Philadelphia)*, 39, 1890, 1977.
45. **Wynder, E. L., Mabuchi, K., and Whitmore, W. F.**, Epidemiology of cancer of the prostate, *Cancer (Philadelphia)*, 28, 344, 1971.
46. **Guileyardo, J. M., Johnson, W. D., Welsh, R. A., Akazaki, K., and Correa, P.**, Prevalence of latent prostate carcinoma in two U.S. populations, *J. Natl. Cancer Inst.*, 65, 311, 1980.
47. **Hirayama, T.**, Epidemiology of prostate cancer with special reference to the role of diet, *Natl. Cancer Inst. Monogr.*, 53, 149, 1979.
48. **Akazaki, K. and Stemmermann, G. N.**, Comparative study of latent carcinoma of the prostate among Japanese in Japan and Hawaii, *J. Natl. Cancer Inst.*, 50, 1137, 1973.
49. **Armstrong, B. and Doll, R.**, Environmental factors and cancer incidence and mortality in different countries, with special reference to dietary factors, *Int. J. Cancer*, 15, 617, 1975.
50. **Hill, P., Wynder, E. L., Garnes, H., and Walker, A. R. P.**, Environmental factors, hormone status, and prostatic cancer, *Prev. Med.*, 9, 657, 1980.
51. **Bjelke, E.**, Epidemiologic studies of cancer of the stomach, colon and rectum with special emphasis on the role of diet, *Scand. J. Gastroenterol.*, 9(32), 235, 1974.
52. **Stemmermann, G. N.**, Gastric cancer in the Hawaii Japanese, *Gann*, 68, 525, 1977.
53. **Zaldivar, R. and Wetterstrand, W. H.**, Nitrate nitrogen levels in drinking water of urban areas with high and low risk populations from stomach cancer: an environmental epidemiology study, *Z. Krebsforsch.*, 92, 227, 1978.
54. **Sugimura, T. and Kawachi, T.**, Experimental stomach carcinogenesis, in *Gastrointestinal Tract Cancer*, Lipkin, M. and Good, R., Eds., Plenum Press, New York, 1978, 327.
55. **Mirvish, S., Karlowski, K., Birt, D., and Sams, J. P.**, Dietary and other factors affecting nitrosomethylurea (NMU) formation in the rat stomach, in *N-Nitroso Compounds: Analysis, Formation, and Occurrence*, Walker, E. A. et al., Eds., IARC Sci. Publ. 31, International Agency for Research on Cancer, Lyon, 1980, 271.

56. **Weisburger, J. H., Marquardt, H., Mower, H. F., Hirota, N., Mori, H., and Williams, G.,** Inhibition of carcinogenesis: vitamin C and the prevention of gastric cancer, *Prev. Med.,* 9, 352, 1980.
57. **Weisburger, J. H., Marquardt, H., Hirota, N., Mori, H., and Williams, G. M.,** Induction of glandular stomach cancer in rats with an extract of nitrite-treated fish, *J. Natl. Cancer Inst.,* 64, 163, 1980.
58. **Tannenbaum, S. R., Weisman, N., and Fett, D.,** The effect of nitrate intake on nitrite formation in human saliva, *Food Cosmet. Toxicol.,* 14, 459, 1976.
59. **Tatematsu, M., Takashashi, M., Fukishima, S., Hananouchi, M., and Shirai, T.,** Effects in rats of sodium chloride on experimental gastric cancers induced by *N*-methyl-*N'*-nitro-*N*-nitrosoguanidine or 4-nitroquinoline-1-oxide, *J. Natl. Cancer Inst.,* 55, 101, 1975.
60. **Joossens, J. V., Kesteloot, H., and Amery, A.,** Salt intake and mortality from stroke, *N. Engl. J. Med.,* 300, 1396, 1979.
61. **Wynder, E. L. and Hoffmann, D.,** Tobacco and health. A societal challenge, *N. Engl. J. Med.,* 300, 894, 1979.
62. **Sugimura, T., Kawachi, T., Nagao, M., and Yahagi, T.,** Mutagens in food as causes of cancer, in *Nutrition and Cancer: Etiology and Treatment,* Newell, G. R. and Ellison, N. M., Eds., Raven Press, New York, 1981, 59.
63. **Shelby, M. D. and Matsushima, T.,** Mutagens and carcinogens in the diet and digestive tract, *Mutat. Res.,* 85, 177, 1981.
64. **Kasai, H., Yamaizumi, Z., Wakabayashi, K., Nagao, M., Sugimura, T., Yokoyama, S., Miyazawa, T., Spingarn, N. E., Weisburger, J. H., and Nishimura, S.,** Potent novel mutagens produced by boiling fish under normal conditions, *Proc. Jpn. Acad.,* 56, 278, 1980.
65. **Spingarn, N. E. and Garvie, C. T.,** Formation of mutagens in sugar-ammonia model systems, *J. Agric. Food Chem.,* 27, 1319, 1979.
66. **Weisburger, J. H. and Williams, G. M.,** Chemical carcinogenesis, in *Cancer Medicine,* 2nd ed., Holland, J. F. and Frei, E., III, Eds., Lea & Febiger, Philadelphia, 1982, 42.
67. Naylor Dana Institute for Disease Prevention, American Health Foundation, Unpublished data.

Part II

Carcinogens and Mutagens in Food

Chapter 6

SOURCES OF CARCINOGENS AND MUTAGENS IN FOOD*

G. A. H. Elton

TABLE OF CONTENTS

I. INTRODUCTION

There is no doubt that substances known to be capable of causing cancer in experimental animals, or of producing mutations in cultures of microorganisms, can sometimes occur in particular food items.[1] There are even epidemiological links with human cancer in a small number of cases, one of the best examples being the relationship between the amount of aflatoxin B_1 ingested in certain countries and the incidence of primary liver cancer.[2] We all know that while the science of toxicology is advancing rapidly, the science of analytical chemistry is advancing still more rapidly. It is a cruel paradox that the greater sensitivity and precision of analysis that we now have has increased the perplexity of interpretation of the results. We tend to worry, perhaps rightly, perhaps not, about the presence of some substances in food at levels of a few ppb. It has been pointed out before that the enormity of a billion (10^9) as a number is difficult to appreciate: a billion seconds ago the Second World War had only recently ended, a billion minutes ago St. Paul was writing his epistles, and a billion hours ago civilization had not yet been born. As for trillions, a trillion seconds takes us back in one bound to about 30,000 B.C. When we are talking of contamination of food, therefore, at levels of 1 ppb or 50 ppt, we are dealing with almost unimaginably small concentrations of substances, and dose/response relationships down to such levels are very difficult, if not impossible, to establish.

II. QUANTIFYING RISKS: RADIONUCLIDES

The best-investigated area of dose/response relationship and risk assessment for carcinogenicity is probably that of radiation risks because of early human experience with ionizing radiation following the discovery of X-rays and radium, and more recently with medical uses of radiation and exposure from nuclear weapon tests and the peaceful uses of atomic energy. Furthermore, we all experience natural background radiation from cosmic rays, natural radioactivity in the environment (particularly soil), food, buildings, and the natural radioactivity of the human body. In the U.K. the typical dose equivalent for humans from natural background radiation is 1 millisievert (mSv; 1 sievert = 100 rem) corresponding to about 200 ionizing events per year for every cell in the body.[3] The International Commission on Radiological Protection (ICRP) recommends the assumption of a linear dose/response relationship with no threshold, with a mortality risk factor (averaged over both sexes and all ages) of 10^{-2}/Sv. On this basis, the ICRP considers that an occupational dose limit of 50 mSv/year will lead to occupational risks comparable with those in industries considered safe. In the U.K., the Central Electricity Generating Board (CEGB) nuclear plant is designed to limit doses to workers to a maximum of 1/5 of this, namely, 10 mSv/year. This ensures that the increment of risk of induction of cancer in radiation workers never exceeds a few tenths of 1% of the risk due to natural causes.

If we regard radionuclides in food as potential carcinogens, or at least as capable of promoting the development of potentially carcinogenic changes, then the risks from ingesting food containing them can be calculated provided that the types of nuclide present, their concentration, retention time in the body, and other relevant factors are known. Table 1 shows, for example, the radiation dose equivalent (as a percentage of ICRP recommended dose limit[4] for members of the public) received by consumers of fish taken from the Irish Sea,[5] that contain radionuclides discharged from nuclear installations. Typical members of the fish-eating public were receiving not more than 1.2% of the recommended ICRP dose limit in 1978, while heavy eaters of seafood in the local fishing community, who consumed fish, crustaceans, and mollusks were es-

Table 1
INDIVIDUAL RADIATION EXPOSURES DUE TO CONSUMPTION OF IRISH SEA FISH AND SHELLFISH 1978

Exposed population	Consumption rate used in assessment	Radiation dose equivalent (as % of ICRP-recommended dose limit for members of the public) Effective (ICRP-26)		
Consumers in local fishing community	170 g/d^{-1} fish	26	^{90}Sr	0.6
	15 g/d^{-1} crustaceans		^{106}Ru	1.5
	6 g/d^{-1} mollusks		^{134}Cs	2.0
			^{137}Cs	16.8
			^{239}Pu+^{240}Pu	0.9
			^{241}Am	3.7
Consumers associated with commercial fisheries (Whitehaven, Fleetwood, Morecambe Bay)	360 g/d^{-1} fish	15	^{90}Sr	0.9
	70 g/d^{-1} crustaceans		^{106}Ru	0.4
	50 g/d^{-1} mollusks		^{134}Cs	1.4
			^{137}Cs	12.1
			^{239}Pu+^{240}P	0.04
			^{241}Am	0.2
Typical member of the fish-eating public consuming fish landed at Whitehaven/ Fleetwood	40 g/d^{-1} fish	1.2	^{134}Cs	0.1
			^{137}Cs	1.0

Table 2
SOME MEDICAL RISKS OF DEATH[7]

Liver biopsy	2×10^{-4}
Vaccination	1×10^{-6}/vaccination
Medicaments	1×10^{-6}/prescription
Fluoroscopy	1×10^{-4}/examination
Pelvic X-rays	2×10^{-4}/fetus
Maternity	1×10^{-4}
Legal abortion	4×10^{-5}

timated to receive 26% of the recommended dose limit. The ICRP concludes that its system of dose limitation is likely to ensure that the average annual dose equivalent to members of the population from all sources, excluding natural and medical irradiation, will not exceed 0.1 mSv, and the National Radiological Protection Board[6] considers that the contribution from all U.K. waste management practices is unlikely to exceed half of this, i.e., 0.05 mSv.

Turning to other types of risk, the risk of death from various medical treatments, etc.,[7] is shown in Table 2. Pochin[7] has said, for example, that the break-even age for mass radiography is about 30. Below 30 the risk of inducing cancer by irradiation is greater than the benefit of early detection of tumors.

For some potential hazards then, including radionuclides in food, we can be reasonably precise about assessing risks. However, when it comes to chemical carcinogens in food, the quantitative data on which to base judgment are rarely available. Mechanisms may differ widely from one agent to another, and dose/risk relationships may be of different forms. With experimental animals, reasonably good quantitative information is available in a limited number of cases. For example,[8] DMNA and DENA

Table 3
SOME DATA ON SURVIVAL PERIODS AND TUMOR PRODUCTION AT DIFFERENT DOSE LEVELS OF DMNA AND DENA

Compound	Sex	Median survival time (days)	Lowest level increased tumor incidence observed
Controls	M	920	
	F	837	
DENA	M	> 900 at levels up to 1060 ppb	132 ppb
	F	> 800 at levels up to 264 ppb	264 ppb
DMNA	M	> 900 at levels up to 132 ppb	132 ppb
	F	> 800 at levels up to 264 ppb	264 ppb

From Crampton, R. F., *Oncology*, 37, 251, 1980. With permission.

have been administered in drinking water to rats (Table 3) at 15 dose levels from 33 to 1690 ppb, corresponding to 2 to 1080 μg/kg bw/day for male rats and 3 to 1470 μg/kg bw/day for females. Lowest levels at which carcinogenic responses were observed (using 60 animals in each test group) were 132 ppb for DMNA and DENA in male rats after median survival times in excess of 900 days. Hyperplastic nodules were observed with DMNA at the lowest dose level of 33 ppb or 2 μg/kg/day. However, when we turn to risks to humans from chemical carcinogens in food, we find ourselves descending fairly rapidly from precise mathematical relationships to qualitative, or at best semiquantitative, descriptions of possible risks.

III. NUTRITIONAL FACTORS

The only nutritional factor (other than aflatoxin) that has been established as contributing to the causation of cancer in man is alcohol [2](the author has heard alcohol described as the most dangerous mycotoxin of all). However, it may not be true to describe alcohol as a carcinogen in its own right; it is possible that alcohol acts simply by bringing the carcinogens from other sources more effectively in contact with the stem cells of the relevant organism. Certainly alcohol can act synergistically with cigarette smoke, multiplying the effect of the latter. Furthermore, Schoental[9] believes that mycotoxins, especially those produced by *Fusarium* molds, are responsible for some pathological conditions associated with alcoholic drinks, including fetal alcohol syndrome, although it seems unlikely that mycotoxins would carry over into distilled drinks.

Apart from aflatoxin and alcohol, firm evidence of food components promoting cancer in humans is sparse. Carroll[16] has discussed the well-known correlation between intake of dietary fat in different countries and the incidence of cancer of the endometrium, breast, and colon that may arise from a causal relationship or may be a secondary effect due to the general association of high fat consumption with high income per capita. It is obviously particularly important to the future well-being of mankind to establish whether high intakes of fat, of themselves, predispose to high levels of cancer. In recent years it has been shown experimentally that vitamin A and its analogues can diminish the risk of experimentally induced cancer in laboratory animals, and there are now several reports[10] that suggest a significant negative association between vitamin A intake in humans and lung cancer.

As already discussed, it is not sufficient to say simply that such-and-such a factor may increase risk. What we need to have is the best possible quantification of risk assessment. This is one of the most difficult problems,[11,12] generally involving difficult extrapolations, such as from high doses in animals to low doses in animals; from animals to man; or from microorganisms to man. Furthermore, the undoubted fact that calorie restriction inhibits formation of various types of tumors in mice and rats[13] implies the possibility that risks from a given level of a particular carcinogen in the diet may vary from one human population to another. At least as far as rats are concerned, aflatoxins seem to be more dangerous in an ample diet than in a frugal one. We also need to take into account the possibility that some substances that are apparently more or less harmless by themselves (at least in small amounts) may be converted into mutagens or carcinogens in the GI tract. For example, it was recently shown[14] that certain glycosides from tea, herbal tea, grape juice, and red wine are inactive in the Ames test but potently mutagenic in the presence of the enzyme, fecalase, isolated from human feces.

IV. SOME PROBLEMS FOR RESEARCH

The discussion, so far, leads us on to a whole array of questions that we need to answer. Some of these questions will be considered now, very briefly.

When trying to assess the carcinogenicity of dietary components in test animals, what should the base line be? For a given group of animals it might be the diet that produced the lowest incidence of cancers, all other things being equal. The quantity of food consumed, as well as its composition, will probably be important. It is not necessarily true that the standard laboratory rat diet is the best "control diet" to use.

When comparing different human populations epidemiologically how should we attempt to distinguish variations in cancer incidence arising from these factors?

1. Genetic differences
2. Lifestyle characteristics
3. Quantity of food consumed
4. Composition of the diet
5. Environmental characteristics
6. The presence, at varying levels, of the individual carcinogens in the diet

Because of the multivariate nature of the relationships, only fairly major effects are likely to be detectable.

In attempting to use experimental animals to assess hazards to man, how do we overcome the problem of interspecies variation in sensitivity? Do qualitative differences exist, i.e., are some compounds carcinogens in some species and not in others, or are the differences only quantitative? Obviously, if a compound is metabolized differently in man than in the test animal, the validity of any extrapolation becomes much more difficult to establish.

In assessing hazards to man, should we assume that the dose/response parameters for man are the mean of the dose/response parameters for the animals studied, or should we be more conservative and assume that man is as sensitive as the most sensitive of the animal species tested, and then incorporate a substantial safety factor in setting regulatory limits? So far as the author is aware, the validity of extrapolation of dose/response relationships from animals to man is not currently testable because of the lack of suitable human epidemiological data.

How do we assess the safety of food components that may form a substantial part of the human diet? Safety factors may sometimes be established for trace components.

For example, if an additive is to be used at 1 ppm in the food supply, then it may be acceptable to demonstrate that it gives no indication of carcinogenesis when fed at 1000 ppm (0.1%) or 10,000 ppm (1%) using, say, 100 test animals in each case. Subject to certain assumptions about the shape of the dose/response curve, we may be able to predict that at the projected level of use in the human food supply, the expected risk to rats is less than 10^{-6}, for example. However, for a component used at a higher level in the diet, say, 100 ppm, the safety factor is likely to be less, and we may only be able to guarantee that the expected risk is less than, say, 10^{-5}, though the real risk may in fact be zero. For substances likely to be major components of the diet (i.e., used at 1% or more) the safety factor obtainable in a rat-feeding experiment is going to be small, even if very large numbers of animals are used.

V. SUMMARY AND CONCLUSIONS

The taking of risk/benefit decisions in relation to food components, additives, and contaminants is an important duty of the regulatory agencies. The major problem is that it is often difficult, if not impossible, to measure either the risks or the benefits (the latter is, of course, a separate question altogether). Only very rough approximations are usually available, and these are sometimes the basis for statements such as "the risks from dietary levels of aflatoxin and saccharin can be compared with riding a bicycle. Eating grilled steak with any PAH contained therein exposes man to the same kind of risk as being struck by lightning."[12] Crude though these comparisons may seem, it is difficult to do much better in the present state of knowledge on risk assessment for foods. Of course, one can rightly say that one has a choice on whether to ride a bicycle or walk about in a thunderstorm, whereas consumers do not always know when they are being exposed to carcinogens in their food supply, or understand whether the risks are great or small. Furthermore, it may be that the important thing to consider is not the true risk, but the average man's perception of the risk, which may be higher or lower. Slovic et al.[15] analyzed this question in detail and pointed out that perceived risk (i.e., what people think the risk is) is usually much easier to measure than the true risk. Obviously, as we become better able to assess true risks we can improve the education and information of the public, but we are never likely to be able to make decisions about the safety of food constituents entirely on a cold-blooded quantitative basis and rightly so. The author believes regulatory agencies and the food and agriculture industries need to:

1. Ensure that food is as safe as possible in the light of existing scientific knowledge
2. Improve methods of identifying carcinogens, co-carcinogens, and mutagens in food
3. Quantify the concentrations of such compounds in foods
4. Establish dose/response relationships
5. Study their mode of action in animal species of various types
6. Study the ways in which carcinogens, co-carcinogens, and mutagens are formed in foods, and methods for reducing their concentrations
7. Do all we can to quantify risks and if these are appreciable in relation to the other risks of life, take all necessary steps to protect the consumer from these substances in the food supply

REFERENCES

1. **Doll, R.**, Strategy for detection of cancer hazards to man, *Nature (London)*, 265, 589, 1977.
2. **Doll, R.**, Nutrition and cancer: a review, *Nutr. Cancer*, 1, 35, 1979.
3. **Mayneord, W. V. and Wheatley, B. M.**, Nuclear power: putting the risks into perspective, *CEGB Res.*, 11, 31, 1981.
4. **Anon.**, *Ann. Int. Comm. Radiol. Prot.*, 1, 3(26), 1977.
5. **Hunt, G. J.**, MAFF Aquatic Environment Monitoring Report, Ministry of Agriculture, Fisheries and Food, Her Majesty's Stationery Office, London, 4, 1980.
6. **Anon.**, Report of the National Radiological Protection Board, Her Majesty's Stationery Office, (NRPB-ASP2), London, 1978.
7. **Pochin, E.**, The assessment and perception of risk, *A Royal Society Discussion*, The Royal Society, London, 1981, 87.
8. **Crampton, R. F.**, Carcinogenic dose-related response to nitrosamines, *Oncology*, 37, 251, 1980.
9. **Schoental, R.**, Relationships of *Fusarium* mycotoxins to disorders and tumors associated with alcoholic drinks, *Nutr. Cancer*, 2, 88, 1980.
10. **Gregor, A., Lee, P. N., Roe, F. J. C., Wilson, M. J., and Melton, A.**, Comparison of dietary histories in lung cancer cases and controls with special reference to vitamin A, *Nutr. Cancer*, 2, 93, 1980.
11. **Roe, F. J. C.**, Food and cancer, *J. Human Nutr.*, 33, 405, 1979.
12. **Turner, A.**, Food and public safety, *Proc. Inst. Food Sci. Technol., (U.K.)*, 13, 235, 1980.
13. **Carroll, K. K.**, Experimental evidence of dietary factors and hormone-dependent cancers, *Cancer Res.*, 35, 3374, 1975.
14. **Tamura, G., Gold, C., Ferro-Luzzi, A., and Ames, B. N.**, Fecalase: a model for activation of dietary glycosides to mutagens by intestinal flora, *Proc. Natl. Acad. Sci. U.S.A.*, 77, 4961, 1980.
15. **Slovic, P., Fischhoff, B., and Lichtenstein, S.**, The assessment and perception of risk, *A Royal Society Discussion*, The Royal Society, London, 1981, 17.
16. **Carroll, K. K.**, Dietary fat and its relationship to human cancer, in *Carcinogens and Mutagens in the Environment*, Volume I: *Food Products*, Stich, H. F., Ed., CRC Press, Boca Raton, Fla., 1982.

Chapter 7

SOURCES OF CARCINOGENS AND MUTAGENS IN EDIBLE PLANTS: PRODUCTION OF URINARY BLADDER AND INTESTINAL TUMORS BY BRACKEN FERN *(PTERIDIUM AQUILINUM)**

George T. Bryan and A. Mahir Pamukcu

TABLE OF CONTENTS

* Supported in part by Grants CA14520 and CA14523 through the National Bladder Cancer Project, and by CA20432.

I. INTRODUCTION

A number of edible plants consumed by domestic animals or humans have demonstrated experimental carcinogenic activity for several species and a variety of tissues.[1] Some of these plants and the implicated chemicals are cycad nut — cycasin; mushroom — hydrazines; *Petasites japonicum, Tussilago farfara* L., and *Symphytum officinale* L. — pyrrolizidine alkaloids; and Senecio — monocrotaline. It has been known for many years that plants contain carcinogens. However, it is only within the past decade that experimental oncologists have increased their attention in this interesting, challenging, and important area of inquiry. In some instances plant carcinogens structurally resemble synthetic chemical carcinogens and in other cases unique structures have been characterized. In many circumstances responsible plant carcinogens remain to be isolated and identified.[1] In this brief review it is not possible to adequately cover all of the interesting studies concerning mutagenic and carcinogenic substances of plant origin. Instead, the authors will summarize several recent developments concerning the plant carcinogen, bracken fern *(Pteridium aquilinum)* (BF). Periodic reviews of work conducted on BF carcinogenicity in the authors' laboratories have been published.[2-5]

II. SOURCES OF BF

BF and other botanical relatives grow well in moist, shaded areas, or hilly pasture lands in temperate and midtropic zones north and south of the equator.[2] Because these ferns are among the first to develop in the spring, they are frequently ingested by grazing domestic animals. In the fall, farmers in less well-developed countries harvest BF for use as animal bedding where it may also be eaten by animals. BF and related plants are produced commercially in Japan, Canada, and northeastern parts of the U.S. for human diets as greens or salads.[5] Milk and dairy products obtained from cows grazing on BF are also consumed by humans. Thus, some humans are directly exposed to BF in their diets and indirectly exposed to BF through products obtained from animals.

III. CARCINOGENICITY OF BF

In several areas of the world an association of BF ingestion has been made with the appearance of urinary bladder tumors in cows and water buffalo,[2,3] esophageal and intestinal tumors in cows,[6] and bladder and intestinal tumors in sheep.[5,7,8] The incidence of bladder tumors may vary from one area to another, but may be as high as 90%.[5] The epidemiologic and histologic features of naturally occurring bovine bladder tumors related to BF ingestion were described.[2,3,5]

Evidence that ingested BF is a primary determinant of bladder cancer in cows stems from several studies.[5] Deliberate, prolonged low-level (2 g/kg bw) feeding of BF produced bladder cancers in 20 of 30 cows.[3] A mean period of 550 days from the start of BF feeding to tumor appearance was determined.[3] Invasive bladder carcinomas appeared within 2.5 years.[3] The induced vesical neoplasms were histologically indistinguishable from the naturally occurring bladder tumors.[3]

Other species also exhibit a variety of neoplastic lesions when fed BF (1 part BF to 2 parts diet).[5] Rats developed vesical carcinoma and ileal adenoma, adenocarcinoma, and sarcoma;[9-13] mice demonstrated urinary bladder and intestinal carcinoma, pulmonary adenoma and adenocarcinoma, and leukemia;[14,15] guinea pigs had bladder and intestinal tumors;[5] and Japanese quail displayed intestinal adenocarcinoma[16] following oral administration of BF. Several strains of rats including a Norwegian-derived strain studied in Turkey,[17] Fischer, Sprague-Dawley,[18] and ACI[19] exhibit susceptibility to BF carcinogenesis although tumor incidence rates may vary. Male and female rats seem equally sensitive.[17,18]

A recent epidemiologic study in Japan suggested that daily intake of BF by humans significantly increased risk of development of esophageal cancer.[20] When combined with a hot tea gruel (chagayu) drink, the relative esophageal cancer risk was increased in males to 5.2 and in females to 4.7.[20] Addition of cigarette smoking to daily intake of BF and chagayu further increased the relative esophageal cancer risk in men to 9.2 and in women to 6.4.[20] Combined use of BF, chagayu, and cigarettes resulted in a 34-fold higher relative esophageal cancer risk for both sexes.[20] Combined attributable risk of these factors to esophageal cancer development was calculated as 80.2%.[20] It was suggested that BF acted as an initiator and the hot tea gruel as a promoter or localizer of these neoplasms in the human population studied.[20] Daily intake of meats and fruits tended to lower esophageal cancer risk.[20]

BF may not be the only plant possessing carcinogenic activity for the bovine bladder.[5] Some other ferns or other plants may also possess such capability. For example, it was reported recently that in the absence of BF, chronic ingestion of mugla fern or rock fern *(Cheilanthes sieberi)* was a major cause of bladder neoplasms in cows residing in Queensland.[21] Hirono and associates[22] reported the development of hepatic and vesical tumors in inbred ACI strain rats fed Russian comfrey (*Symphytum officinal* L.), used by humans as a green vegetable or tonic.

IV. FACTORS AFFECTING BF CARCINOGENICITY

A number of factors have been identified that alter carcinogenic risk of susceptible tissues to BF administered to experimental animals. BF carcinogenicity varies with the part of the plant studied.[5] Curled tops of young fronds have stronger activity than young fronds.[23] Fronds of mature BF, when compared to curled tops of immature BF, induced a significantly higher incidence of vesical, but not intestinal neoplasms.[24] BF carcinogenicity for the ileum of inbred ACI rats of both sexes was diminished by storage at 4°C or room temperature for up to two years.[25] Though vesical hyperplasia was found in some rats, no bladder tumors were detected.[25]

Young rats were more susceptible to BF carcinogenesis than were older ones.[16] A brief duration of BF feeding to rats produced only intestinal tumors,[13] whereas long-term feeding resulted in the presence of coexisting intestinal and vesical tumors.[10,17,18] Lifetime BF feeding was necessary to induce bladder and intestinal tumors in mice.[14,15]

Concomitant administration of various chemicals with BF to rats influenced the incidences of bladder or intestinal tumors.[5] Thus thiamin, nitrate, or nitrite supplementation increased urinary bladder neoplasm frequency.[5] Dietary phenothiazine,[26] BHA, disulfiram, calcium chloride,[27] or nicotinamide[28] significantly ($p < 0.01$) decreased the incidence of intestinal tumors, while phenothiazine,[26] calcium chloride, polyvinylpyrrolidone,[27] and nicotinamide[28] significantly ($p < 0.01$) decreased the incidence of vesical neoplasms in rats fed BF. Suggestions for possible inhibitory mechanisms of these chemicals on bladder or intestinal carcinogenesis have been advanced.[5,26-28] However, since the chemical carcinogens in BF responsible for bladder and intestinal carcinogenesis are incompletely known, and since the interactions of these biologically active substances present in BF are unknown, these suggested mechanisms await further study.

V. BF CARCINOGENS IN URINE AND MILK

One or more chemicals derived from ingested BF are excreted in the urine or milk of cows, and in the urine of rats fed this plant.[5] Fractions of urine obtained from cows fed BF demonstrated carcinogenic activity when implanted intravesically in mice.[5] Rats fed BF excreted two different biologically active compounds in urine.[29] One chemical was mutagenic for *Salmonella typhimurium* TA98 and TA100 and was identified as quercetin.[29] The other isolate was mutagenic for TA100 but not TA98, and its identification is in progress.[29] Administration of rutin or quercetin to rats enhanced the excretion of urinary quercetin, but had no effect on the mutagenic activity of the isolate active for TA100.[29]

Milk of cows fed BF contains mutagenic, toxic, and carcinogenic chemical(s) not present in milk of cows fed hay and grain.[30] A chloroform-methanol extract of milk from BF-fed cows was mutagenic for *S. typhimurium* TA100, but not for TA98.[30] This mutagen appeared identical to that present in urine of rats fed BF and active for TA100.[5] Evans and associates[31] reported the presence of a BF-derived toxin in cow's milk. Two solvent extracts of milk from BF-fed cows demonstrated acute toxicity for mice and one of these fractions displayed significant bladder carcinogenicity when assayed by the intravesical pellet implantation method.[30] Whole and powdered milk obtained from cows fed BF, compared with that from control cows, and incorporated into the diets of rats produced significant incidences of intestinal and urothelial carcinomas.[30] The mutagenic, toxic, and carcinogenic chemicals derived from milk of BF-fed cows await identification.[5,30]

VI. CHEMICALS ISOLATED FROM BF AND THEIR BIOLOGICAL ACTIVITY

The identification and characterization of the chemicals responsible for mutagenic and carcinogenic activities and present in BF has become a significant scientific problem for several research groups.[5,16,19,32] For many years progress was delayed due to lack of a suitable short-term biological assay that was sensitive and reflected carcinogenic potential of isolates. With the introduction of *S. typhimurium* mutagenesis assays by Ames and associates[33] and their application[34] to the isolation of active compounds in BF, or in samples obtained from animals fed BF, accelerated progress has occurred. Nevertheless, though several chemicals have been characterized in BF possessing mutagenic or carcinogenic activities, a single entity has yet to be identified that, when fed to a susceptible species, reproduces in all respects the carcinogenic effects displayed by intact BF.

Among the variety of chemicals identified in BF are isoquercitrin and rutin (glycosides of quercetin), astragalin and tiliroside (glycosides of kaempferol),[17,35,36] fumaric and succinic acids,[36] shikimic acid,[37] catecholamines and pteraquilin,[38] pterolactam,[39] pterosin and indanone,[32] and tannin.[40] No unconjugated quercetin or kaempferol were detected in BF.[17,29] Quantitative methods of analysis have been employed by us to obtain estimates of concentrations of some of these compounds in BF measured as g/kg dried weight: quercetin, 0.57 to 0.86 and kaempferol, 1.1 to 2.55 as liberated aglycones,[17,29] and tannin, 2.45.[40]

The authors first reported mutagenicity of quercetin for *S. typhimurium* TA98 and TA100.[34] This report has been amply confirmed by others.[41-45] Quercetin mutagenicity was enhanced by an Aroclor 1254-induced rat-liver metabolism system (S-9 mix), and kaempferol displayed mutagenicity with the incorporation of the S-9 mix.[42-44] Quercetin and kaempferol, the latter with metabolic activation, displayed mutagenic activity for V79 Chinese hamster cells.[45] Shikimic acid,[34] tannin,[34] rutin,[34,43,44] and known quercetin mammalian urinary metabolites, *m*-hydroxyphenylacetic acid, homovanillic acid, and 3,4-dihydroxyphenylacetic acid[29,41] were not mutagenic.

Shikimic acid was claimed to be a mutagen and a murine carcinogen.[37] Following i.p. injection of shikimic acid, tumors produced in 9/14 mice were mainly of the glandular stomach and lymphoreticular systems.[37] Shikimic acid fed to rats was not carcinogenic.[46] Shikimic acid is widespread in plants, serves as a metabolic precursor of aromatic ring structures, and does not structurally resemble known experimental carcinogens.[37] It is not considered to be the principal BF carcinogen.[34,46] Pterosin and pterosides were not carcinogenic when fed to rats.[47]

A tannin that was isolated from BF was highly toxic (LD_{50} 160 mg/kg) to mice following i.p. injection, and exhibited bladder carcinogenesis when tested by intravesical implantation.[40] Repeated s.c. injection of BF tannin produced an 80% incidence of fibrous histiocytoma in rats.[18] Tannin isolated from BF was not carcinogenic when fed to rats.[18] The role, if any, of tannin in the production of bladder or intestinal tumors is unknown.[18]

Solvent extracts of BF, not containing tannin, have been prepared and they display significant mutagenic,[48] or vesical or intestinal carcinogenic[18,19] activities when fed to rats. A tannin-free aqueous fraction as well as a chloroform fraction of BF produced a high incidence of intestinal tumors when fed to rats.[18] Quercetin and kaempferol, present as hydrolyzable glycosides, were the major sources of mutagens in this fraction.[29] Work is in progress on the isolation and identification of chemicals in the chloroform fraction displaying mutagenic activity.[49] Small quantities of quercetin and kaempferol glycosides are present in this fraction, perhaps trapped in the aqueous drops incompletely separated.[49] Their quantity is less than 5% of the total quercetin and kaempferol glycosides present in BF.[49] This quantity of these compounds in this fraction may not account for the carcinogenic activity displayed.[18] In addition, the chloroform fraction contains two or more substances that display mutagenic activity for *S. typhimurium* TA100, but not TA98.[49] Mutagenic activity is markedly reduced by incubation with the S-9 mix.[49] These substances may be similar to those isolated by van der Hoeven and van Leeuwen,[48] and may be precursors of compounds in the urine and milk of BF-fed animals that display similar mutagenic behavior. These substances have not been chemically characterized, however, their known physical and chemical properties do not resemble those exhibited by flavonoids.

The authors recently reported the intestinal and bladder carcinogenicity of quercetin fed at a level of 0.1% (1000 ppm) for 14 months to male and female albino noninbred weanling rats.[17] Incidences of tumors were intestinal adenoma and adenocarcinoma: 20/25 rats (80%), and vesical transitional cell carcinoma: 5/25 rats (20%).[17] The incidence of intestinal tumors was comparable to that usually obtained by the authors in a Norwegian-derived strain studied in Turkey,[17] and Fischer or Sprague-Dawley[18] rats fed BF in Wisconsin. However, the incidence of bladder tumors was much lower than that usually observed in BF-fed rats.[17,18,26-28] Quercetin was not carcinogenic for the bladder when tested by the intravesical method.[40] Quercetin fed to strain ddY mice at a dietary level of 2% was not judged as carcinogenic.[50] Male and female strain A mice (A/JJms) fed quercetin at a level of 5% for up to 23 weeks did not display an augmented lung tumor response.[51] BF is carcinogenic for mice but requires prolonged administration.[14,15] It is not known if ddY or strain A mice are susceptible to BF carcinogenic effects. Recently, it was reported[52] that inbred ACI strain rats fed a diet containing 1% or 5% quercetin, or 5% rutin for 540 days, or 10% quercetin or 10% rutin for 850 days, did not display a significant incidence of any tumor type. It is not yet known why these dichotomous results have occurred.[50,52] It is possible that the doses of quercetin fed to ACI rats may have been too high.[52] Substantial opportunity yet exists for workers interested in isolating and characterizing the mutagens and carcinogens in BF.

VII. CONCLUSIONS

BF, a primitive plant, contains chemicals that are carcinogenic for a variety of epithelial tissues of several species. Humans ingest this plant. It has been suggested that BF may contribute to the development of esophageal neoplasms in people.[20] Several chemicals in BF have been characterized. One of these, quercetin, present in BF as glycosides, exhibited mutagenicity,[34] and rat carcinogenicity in one study.[17] It has been estimated that humans in the U.S. consume about 1 gm per day of quercetin.[53] If this estimate is correct, it would suggest that people, during the course of their lifetime, might ingest in excess of 25 kg per person of quercetin. Other as yet unidentified mutagens are also present in BF. As these chemicals are identified and quantitative chemical methods of analysis are developed for them, it will be possible to investigate other plant foodstuffs eaten by humans, for their presence.

ACKNOWLEDGMENT

We thank Ms. S. Pertzborn for manuscript preparation and editorial assistance.

REFERENCES

1. **Weisburger, J. H. and Williams, G. M.,** Chemical carcinogens, in *Toxicology: The Basic Science of Poisons,* 2nd ed., Doull, J., Klaassen, C. D., and Amdur, M. O., Eds., Macmillan, New York, 1980, chap. 6.
2. **Pamukcu, A. M.,** Epidemiologic studies on urinary bladder tumors in Turkish cattle, *Ann. N.Y. Acad. Sci.,* 108, 938, 1963.
3. **Pamukcu, A. M., Price, J. M., and Bryan, G. T.,** Naturally occurring and bracken-fern-induced bovine urinary bladder tumors: clinical and morphological characteristics, *Vet. Pathol.,* 13, 110, 1976.
4. **Bryan, G. T. and Pamukcu, A. M.,** Bracken fern (BF), a natural urinary bladder carcinogen, in *Proc. 10th Int. Am. Conf. Toxicol. Occup. Med.,* Deichmann, W. B., Ed., Elsevier/North-Holland, New York, 1979, 229.
5. **Pamukcu, A. M. and Bryan, G. T.,** Bracken fern, a natural urinary bladder and intestinal carcinogen, in *Naturally Occurring Carcinogens-Mutagens and Modulators of Carcinogenesis,* Miller, E. C., Miller, J. A., Hirono, I., Sugimura, T., and Takayama, S., Eds., University Park Press, Baltimore, 1979, 89.
6. **Jarrett, W. F. H., McNeil, P. E., Grimshaw, W. T. R., Selman, I. E., and McIntyre, W. I. M.,** High incidence area of cattle cancer with a possible interaction between an environmental carcinogen and a papilloma virus, *Nature (London),* 274, 215, 1978.
7. **Dodd, D. C.,** Adenocarcinoma of the small intestine of sheep, *N.Z. Vet. J.,* 8, 109, 1960.
8. **Harburtt, R. P. and Leaver, D. D.,** Carcinoma of the bladder of sheep, *Aust. Vet. J.,* 45, 473, 1969.
9. **Evans, I. A. and Mason, J.,** Carcinogenic activity of bracken, *Nature (London),* 208, 913, 1965.
10. **Pamukcu, A. M. and Price, J. M.,** Induction of intestinal and urinary bladder cancer in rats by feeding bracken fern *(Pteris aquilina), J. Natl. Cancer Inst.,* 43, 275, 1969.
11. **Yunoki, K., Hayashi, T., and Morita, N.,** Induction of intestinal tumors in rats with feeding of bracken fern, *Acta Med. Univ. Kagoshima.,* 14, 249, 1973.
12. **Hirono, I., Fushimi, K., Mori, H., Miwa, T., and Haga, M.,** Comparative study of carcinogenic activity in each part of bracken, *J. Natl. Cancer Inst.,* 50, 1367, 1973.
13. **Pamukcu, A. M., Ertürk, E., Yalçiner, Ş., and Bryan, G. T.,** Histogenesis of urinary bladder cancer induced in rats by bracken fern, *Invest. Urol.,* 14, 213, 1976.
14. **Pamukcu, A. M., Ertürk, E., Price, J. M., and Bryan, G. T.,** Lymphatic leukemia and pulmonary tumors in female Swiss mice fed bracken fern *(Pteris aquilina), Cancer Res.,* 32, 1442, 1972.

15. **Pamukcu, A. M., Ertürk, E., and Bryan, G. T.,** Induction of urinary bladder and intestinal tumors in mice by feeding bracken fern, *Ankara Univ. Vet. Fak. Derg.*, 24, 28, 1977.
16. **Evans, I. A.,** The radiomimetic nature of bracken toxin, *Cancer Res.*, 28, 2252, 1968.
17. **Pamukcu, A. M., Yalçiner, Ş., Hatcher, J. F., and Bryan, G. T.,** Quercetin, a rat intestinal and bladder carcinogen present in bracken fern *(Pteridium aquilinum), Cancer Res.*, 40, 3468, 1980.
18. **Pamukcu, A. M., Wang, C. Y., Hatcher, J., and Bryan, G. T.,** Carcinogenicity of tannin and tannin-free extracts of bracken fern *(Pteridium aquilinum)* in rats, *J. Natl. Cancer Inst.*, 65, 131, 1980.
19. **Hirono, I., Ushimaru, Y., Kato, K., Mori, H., and Sasaoka, I.,** Carcinogenicity of boiling water extract of bracken, *Pteridium aquilinum, Gann*, 69, 383, 1978.
20. **Hirayama, T.,** Epidemiological evaluation of the role of naturally occurring carcinogens and modulators of carcinogenesis, in *Naturally Occurring Carcinogens-Mutagens and Modulators of Carcinogenesis*, Miller, E. C., Miller, J. A., Hirono, I., Sugimura, T., and Takayama, S., Eds., University Park Press, Baltimore, 1979, 359.
21. **McKenzie, R. A.,** Bovine enzootic haematuria in Queensland, *Aust. Vet. J.*, 54, 61, 1978.
22. **Hirono, I., Mori, H., Haga, M., Fujii, M., Yamada, K., Hirata, Y., Takanashi, H., Uchida, E., Hosaka, S., Ueno, I., Matsushima, T., Umezawa, K., and Shirai, A.,** Edible plants containing carcinogenic pyrrolizidine alkaloids in Japan, in *Naturally Occurring Carcinogens-Mutagens and Modulators of Carcinogenesis*, Miller, E. C., Miller, J. A., Hirono, I., Sugimura, T., and Takayama, S., Eds., University Park Press, Baltimore, 1979, 79.
23. **Hirono, I., Fushimi, K., Mori, H., Miwa, T., and Haga, M.,** Comparative study of carcinogenic activity in each part of bracken, *J. Natl. Cancer Inst.*, 50, 1367, 1973.
24. **Pamukcu, A. M., Ertürk, E., Milli, U., and Bryan, G. T.,** Variation in carcinogenic activity of bracken fern as a bladder carcinogen, *Firat Univ. Vet. Fak. Derg.*, 4, 1, 1977.
25. **Kawai, T., Takanashi, H., Nakayama, M., Mori, H., and Hirono, I.,** Effect of storage on carcinogenic activity of bracken fern, *Cancer Lett.*, 12, 29, 1981.
26. **Pamukcu, A. M., Wattenberg, L. W., Price, J. M., and Bryan, G. T.,** Phenothiazine inhibition of intestinal and urinary bladder tumors induced in rats by bracken fern, *J. Natl. Cancer Inst.*, 47, 155, 1971.
27. **Pamukcu, A. M., Yalçiner, Ş., and Bryan, G. T.,** Inhibition of carcinogenic effect of bracken fern *(Pteridium aquilinum)* by various chemicals, *Cancer (Philadelphia)*, 40, 2450, 1977.
28. **Pamukcu, A. M., Milli, U., and Bryan, G. T.,** Protective effect of nicotinamide on bracken fern induced carcinogenicity in rats, *Nutr. Cancer Int. J.*, 3, 86, 1981.
29. **Hatcher, J. F., Pamukcu, A. M., and Bryan, G. T.,** Quercetin (Q) and kaempferol (K) content of bracken fern (BF) and mutagenic activity in urine of rats ingesting Q, rutin (R), or BF, *Proc. Am. Assoc. Cancer Res.*, 22, 114, 1981.
30. **Pamukcu, A. M., Ertürk, E., Yalçiner, Ş., Milli, U., and Bryan, G. T.,** Carcinogenic and mutagenic activities of milk from cows fed bracken fern *(Pteridium aquilinum), Cancer Res.*, 38, 1556, 1978.
31. **Evans, I. A., Jones, R. S., and Mainwaring-Burton, R.,** Passage of bracken fern toxicity into milk, *Nature (London)*, 237, 107, 1972.
32. **Yoshihira, K., Fukuoka, M., Kuroyanagi, M., and Natori, S.,** Further characterization of 1-indanone derivatives from bracken, *Pteridium aquilinum* var. *latiusculum, Chem. Pharm. Bull.*, 20, 426, 1972.
33. **Ames, B. N., McCann, J., and Yamasaki, E.,** Methods for detecting carcinogens and mutagens with the *Salmonella*/mammalian-microsome mutagenicity test, *Mutation Res.*, 31, 347, 1975.
34. **Wang, C. Y., Pamukcu, A. M., and Bryan, G. T.,** Bracken fern, a naturally occurring carcinogen, in *Médecine Biologie Environnement Intern. J. Bulletin European Institute Ecology Cancer (INEC)*, Vol. 4, Stock, C. C., Santamaria, L., Mariani, P., and Gorini, S., Eds., TECNOS, Cremona, Italy, 1976, 565.
35. **Nakabayashi, T.,** Isolation of astragalin and isoquercitrin from bracken, *Pteridium aquilinum, Bull. Agric. Chem. Soc. Jpn.*, 19, 104, 1955.
36. **Wang, C. Y., Pamukcu, A. M., and Bryan, G. T.,** Isolation of fumaric acid, succinic acid, astragalin, isoquercitrin and tiliroside from *Pteridium aquilinum, Phytochemistry*, 12, 2298, 1973.
37. **Evans, I. A. and Osman, M. A.,** Carcinogenicity of bracken and shikimic acid, *Nature (London)*, 250, 348, 1974.
38. **Kwasniewski, V.,** Die Inhaltstoffe der Rhizome des Adlerfarns, *Pteridium aquilinum* (L.), *Kuhn. Arch. Pharmakol. (Weinheim)*, 288, 307, 1955.
39. **Takatori, K., Nakano, S., Nagata, S., Okumura, K., Hirono, I., and Shimizu, M.,** Pterolactam, a new compound isolated from bracken, *Chem. Pharm. Bull.*, 20, 1087, 1972.
40. **Wang, C. Y., Chiu, C. W., Pamukcu, A. M., and Bryan, G. T.,** Identification of carcinogenic tannin isolated from bracken fern *(Pteridium aquilinum), J. Natl. Cancer Inst.*, 56, 33, 1976.
41. **Bjeldanes, L. F. and Chang, G. W.,** Mutagenic activity of quercetin and related compounds, *Science*, 197, 577, 1977.

42. **Sugimura, T., Nagao, M., Matsushima, T., Yahagi, T., Seino, Y., Shirai, A., Sawamura, M., Natori, S., Yoshihira, K., Fukuoka, M., and Kuroyanagi, M.**, Mutagenicity of flavone derivatives, *Proc. Jpn. Acad., Ser. B.*, 53, 194, 1977.
43. **MacGregor, J. T. and Jurd, L.**, Mutagenicity of plant flavonoids: structural requirements for mutagenic activity in *Salmonella typhimurium, Mutat. Res.*, 54, 297, 1978.
44. **Brown, J. P. and Dietrich, P. S.**, Mutagenicity of plant flavonols in the *Salmonella*/mammalian microsome test. Activation of flavonol glycosides by mixed glycosidases from rat cecal bacteria and other sources, *Mutat. Res.*, 66, 223, 1979.
45. **Maruta, A., Enaka, K., and Umeda, M.**, Mutagenicity of quercetin and kaempferol on cultured mammalian cells, *Gann*, 70, 273, 1979.
46. **Hirono, I., Fushimi, K., and Matsubara, N.**, Carcinogenicity test of shikimic acid in rats, *Toxicology*, 1, 9, 1977.
47. **Saito, M., Umeda, M., Enomoto, M., Hatanaka, Y., Natori, S., Yoshihira, K., Fukuoka, M., and Kuroyanagi, M.**, Cytotoxicity and carcinogenicity of pterosins and pterosides, 1-indanone derivatives from bracken *(Pteridium aquilinum), Experientia*, 31, 829, 1975.
48. **van der Hoeven, J. C. M. and van Leeuwen, F. E.**, Isolation of a mutagenic fraction from bracken *(Pteridium aquilinum), Mutat. Res.*, 79, 377, 1980.
49. **Hatcher, J. F., Kitagawa, T., Pamukcu, A. M., and Bryan, G. T.**, Unpublished data, 1981.
50. **Sugimura, T.**, Naturally occurring genotoxic carcinogens, in *Naturally Occurring Carcinogens-Mutagens and Modulators of Carcinogenesis*, Miller, E. C., Miller, J. A., Hirono, I., Sugimura, T., and Takayama, S., Eds., University Park Press, Baltimore, 1979, 241.
51. **Hosaka, S. and Hirono, I.**, Carcinogenicity test of quercetin by pulmonary-adenoma bioassay in strain A mice, *Gann*, 72, 327, 1981.
52. **Hirono, I., Ueno, I., Hosaka, S., Takanashi, H., Matsushima, T., Sugimura, T., and Natori, S.**, Carcinogenicity examination of quercetin and rutin in ACI rats, *Cancer Lett.*, 13, 15, 1981.
53. **MacGregor, J.**, Testimony at November 6, 1980, Natl. Acad. Sci. public hearing, *The Blue Sheet*, p. 24, November 12, 1980.

Chapter 8

THE METABOLIC ACTIVATION AND CARCINOGENICITY OF ALKENYLBENZENES THAT OCCUR NATURALLY IN MANY SPICES

James A. Miller, Elizabeth C. Miller, and David H. Phillips

TABLE OF CONTENTS

I. INTRODUCTION

The use of plant parts to improve the taste and odor of bland and spoiled human food is an age-old practice.[1] In modern times spice flavors have also been available in the form of oleoresins, extracts, and especially essential oils (volatile oils obtained by steam distillation, solvent extraction, or expression).[2-5] During the past century the active flavor components of many spices have been isolated and chemically characterized. The acute and subacute pharmacologic and toxicologic properties of many of these agents have also been better defined. However, it is only in the past two decades that some spice components have been examined for their mutagenicity and long-term toxicity, including carcinogenicity.

Many modern cuisines, domestic and commercial, depend extensively on the use of crude (leaf, fruit, seeds, bark and root pieces, and powders), refined, and synthetic spice flavors. Alkenylbenzenes are prominent as active components of many of these spice flavor additives, and they also occur naturally in some nonspice foods. Most of these compounds are relatively simple allyl- or propenylbenzenes with methoxy and methylenedioxy ring substituents.

The carcinogenic alkenylbenzenes constitute a relatively new class of chemical carcinogens. As is the case for the other dozen or more classes of chemical carcinogens, no common structural features are shared by the carcinogenic alkenylbenzenes with any of the other classes. However, all classes of chemical carcinogens share a common type of reactive electronic structure that is essential to their covalent interactions in vivo with cellular molecules.[6] The reactive structures are electrophilic reactants, most of which are generated metabolically. These electrophilic reactants initiate chemical carcinogenesis through some of their covalent interactions with critical cellular informational macromolecules. This primary event of initiation of carcinogenesis is followed by the more prolonged stage of promotion, when apparently irreversible alterations induced during initiation are expressed phenotypically by the development of gross tumors. These overall steps in chemical carcinogenesis are shown in general terms in Figure 1.[6]

The carcinogenic alkenylbenzenes are active as complete carcinogens, as exemplified by the production of hepatic tumors in the long-term feeding of these compounds to adult rats and mice.[7-10] In small doses administered only during the preweanling period, these compounds appear to act primarily as initiators of hepatocarcinogenesis in mice.[7-12] Under these conditions, growth of the liver in a normal male hormonal environment apparently provides the promotion that leads to the appearance approximately a year later, of gross liver tumors, without further treatment.

II. THE KNOWN CARCINOGENIC ALKENYLBENZENES

Four carcinogenic alkenylbenzenes have been described previously. One that has recently been uncovered, methyl eugenol, is noted in this review. Their structures and some major natural sources are shown in Figure 2.

Safrole (1-allyl-3,4-methylenedioxybenzene) was the first alkenylbenzene discovered to have carcinogenic properties. This fragrant oil is a major component ($\geqslant 80\%$) of the essential oil from the root and bark of the sassafras tree *(Sassafras albidum)*,[7] common in the eastern U.S. Its root was among the first of the money crops sent by the North American colonies to England, where in the 16th and 17th centuries it was prized as a source of teas, tonics, and cure-all drugs. Natural and synthetic safrole was used to flavor root beers and other soft drinks in the U.S. until 1960, when safrole, sassafras, and sassafras oil were prohibited from commercial use in foods.[2] The drink-

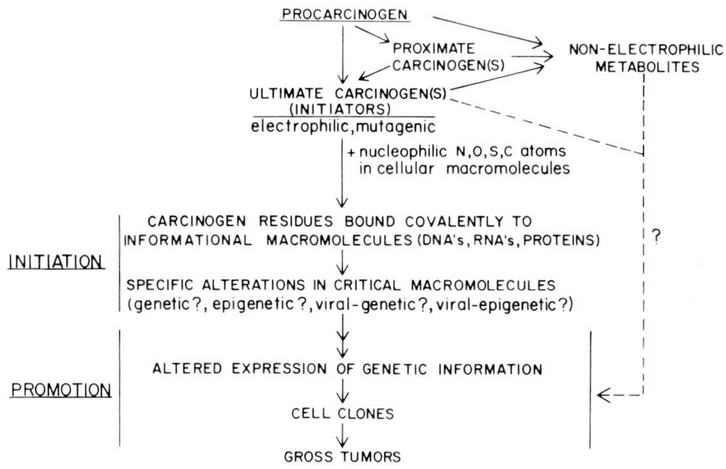

FIGURE 1. A current view of the overall mechanism of chemical carcinogenesis. A majority of chemical carcinogens are procarcinogens.[6]

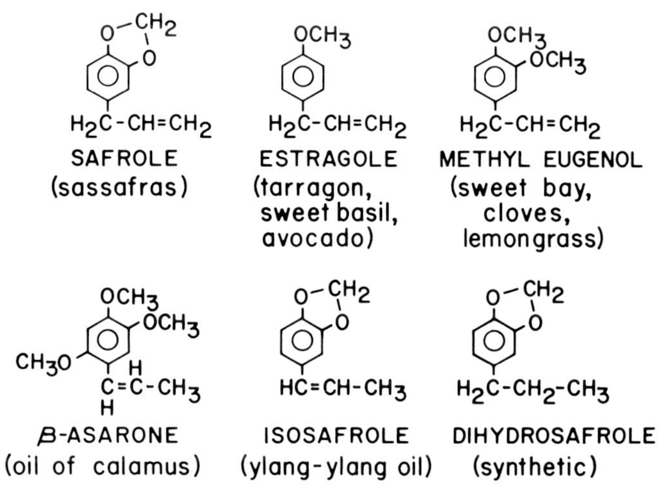

FIGURE 2. The known carcinogenic alkenylbenzene spice flavors.

ing of sassafras tea apparently still occurs to a limited extent in the U.S. Safrole is also a major component of oils from the *Heterotropa* genus, a minor component of some other essential oils, and a very minor component (about 0.1%) of the oils of nutmeg, star anise, mace, and cinnamon (see Leung[2] and Borchert et al.[13]).

The accidental and deliberate dosing of animals and humans early demonstrated that large doses of safrole were hepatotoxic and the chronic administration of safrole to rats and mice was shown to eventually produce liver tumors. Under these conditions safrole is a weak carcinogen. Dietary concentrations of 0.5 to 1% for a year or more are required to induce appreciable incidences of liver tumors in adult rats or mice. However, total doses of only about 0.5 to 1.5 mg injected i.p. or administered by stomach tube to infant male mice, can induce high incidences of liver tumor-bearing mice within a year.[9,11,14] Some carcinogenic activity is retained after isomerization or saturation of the allyl group in safrole. Isosafrole (1-propenyl-3,4-methylenedioxyben-

Table 1
LIVER TUMOR DEVELOPMENT IN MICE TREATED PRIOR TO WEANING WITH ALKENYLBENZENES

Compound[a]	No. mice examined	% with hepatic tumors	Average no. of tumors/liver
Estragole	23	56	1.2
1′-Hydroxyestragole	21	100	4.8
Methyl eugenol	23	70	1.5
Elemicin	19	21	0.3
Myristicin	20	10	0.2
Dill apiol	18	6	0.1
Parsley apiol	21	0	0
Anethole	14	7	0.1
3′-Hydroxyanethole[b]	23	13	0.3
(Trioctanoin only)	20	5	0.1
(Uninjected)	26	12	0.2

Note: (C57B1/6 × C3H/He)F_1 male mice were injected i.p. on days 1, 8, 15, and 22 after birth with 0.25, 0.50, 1.0, and 3.0 μmol, respectively, of the indicated compounds dissolved in trioctanoin. The total doses were 4.75 μmol/mouse. The data were obtained by examination of the mice by laparotomy at 13 months; the experiment is still in progress.[14] Representative tumors have been examined histologically and were diagnosed as hepatomas type A, type B, and mixed type A-type B.[12]

[a] See Figures 2, 3, and 4 for the structures of these compounds.
[b] A metabolite of anethole.[26,31]

zene), which occurs rarely in essential oils,[2] was weakly hepatocarcinogenic upon feeding to male mice.[8] Dihydrosafrole (1-propyl-3,4-methylenedioxybenzene), a synthetic derivative, was also less hepatocarcinogenic than safrole when fed to male and female mice.[8] Surprisingly, administration of high levels of dihydrosafrole in the diet produced esophageal tumors, but not liver tumors, in rats.[15]

Estragole (1-allyl-4-methoxybenzene) occurs in a wide variety of essential oils, occasionally as a major component and frequently as a minor constituent.[2] This alkenylbenzene is essentially equivalent to safrole in its hepatocarcinogenic activity in mice;[10,12] its carcinogenicity for rats has not been studied.

β-Asarone (*cis*-1-propenyl-2,4,5-trimethoxybenzene) is a major component of oil of calamus, a bitters flavor. The administration of high levels of this compound in the diet of rats induced mesenchymal tumors of the small intestine.[16] The commercial use of oil of calamus as a flavor is no longer permitted in the U.S.[2]

Recently preliminary results in this laboratory indicate that methyl eugenol (1-allyl-3,4-dimethoxybenzene) is about equal to safrole and estragole in the induction of liver tumors by administration to infant male mice (Table 1).[14] Methyl eugenol occurs widely in many essential oils,[2] and it has been employed agriculturally as a lure in the control of the male oriental fruit fly.[17]

III. OTHER ALKENYLBENZENE COMPONENTS OF ESSENTIAL OILS

Although the majority of the other two dozen or more known alkenylbenzenes, some of which are widely used as spice flavors,[2-5] have not yet been tested for carcinogenic

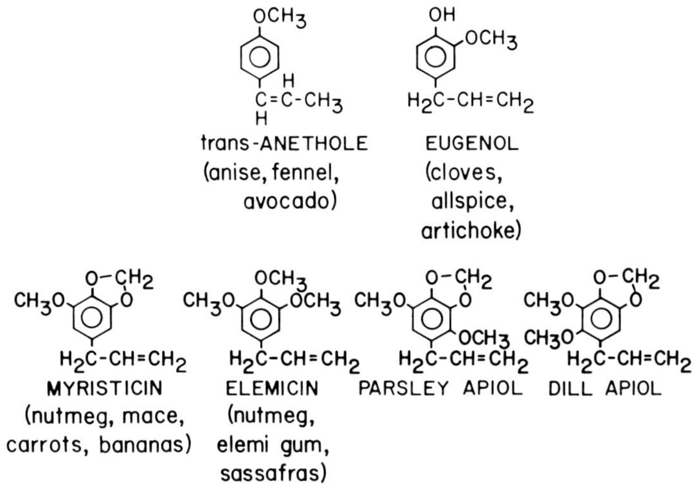

FIGURE 3. Some other naturally occurring alkenylbenzenes.

activity, several of these spice components are presently being studied in the authors' laboratory.[14,18] The structures of these compounds are shown in Figure 3. Anethole (*trans*-1-propenyl-4-methoxybenzene, an isomer of estragole) and eugenol (1-allyl-3-methoxy-4-hydroxybenzene) are major spice flavors in the U.S. at the present time. In laboratory studies neither of these compounds showed significant carcinogenic activity, either when fed for a year in the diet of female mice or when injected i.p. in infant male mice.[18] A preliminary report from the National Toxicology Program indicated that they had equivocal evidence for an increased liver tumor incidence on feeding 0.3% or 0.6% of eugenol for up to two years to (C57Bl/6 × C3H)F_1 mice; similar studies on F344 rats were negative.[19] Furthermore, in tests still in progress in this laboratory, myristicin (1-allyl-3-methoxy-4,5-methylenedioxybenzene), elemicin (1-allyl-3,4,5-trimethoxybenzene), parsley apiol (1-allyl-2,5-dimethoxy-3,4-methylenedioxybenzene), and dill apiol (1-allyl-2,3-dimethoxy-4,5-methylenedioxybenzene), all of which are found in common foods or spice flavors, appear to have little or no significant hepatocarcinogenic activity when administered to infant male mice in doses at which safrole and estragole are active (Table 1).[14]

IV. THE METABOLIC ACTIVATION OF SAFROLE AND ESTRAGOLE IN VIVO AND IN VITRO

A. 1′-Hydroxy Metabolites as Proximate Carcinogens

The generalization that the ultimate carcinogenic forms of most, if not all, chemical carcinogens are strong electrophilic reactants[20-22] formed the basis of the initial studies in this laboratory on safrole and estragole. The lack of electrophilic groups in these compounds and their carcinogenicity in a tissue distant from the site of administration suggested that these agents must be metabolically activated. Attention centered on the 1′-carbon atoms in the allyl groups of these carcinogens since they might be the site of formation of metabolic allylic and benzylic esters. Esters of this type may exhibit electrophilic activity, and the pyrrole metabolites of the hepatocarcinogenic pyrrolizidine alkaloids provided examples of strongly electrophilic allylic esters that were presumed to be ultimate carcinogens.[23-25] The prediction that safrole and estragole were metabolized to 1′-hydroxy derivatives (Figure 4), both stronger carcinogens than the parent

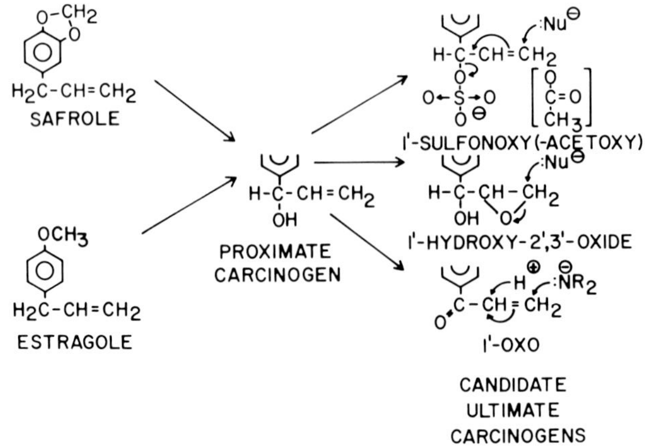

FIGURE 4. Possible pathways of metabolic activation of safrole and estragole. A fourth pathway of direct oxidation to the 2′,3′-oxides is discussed in the text.

compounds, proved to be correct. Urinary excretion of glucuronides of these 1′-hydroxy metabolites occurred when safrole or estragole was administered to rats or mice.[12,13,26] Most importantly, both 1′-hydroxy metabolites were stronger carcinogens in the liver than the parent allylbenzenes.[9,10,12] Safrole, estragole, and their 1′-hydroxy metabolites also induced interscapular angiosarcomas when fed to mice.[9,10,14]

B. Electrophilic Metabolites as Candidate Ultimate Carcinogens

The greater carcinogenicity of the 1′-hydroxy metabolites of safrole and estragole as compared to the parent carcinogens further suggested that these nonelecrophilic metabolites might yield one or more electrophilic metabolites in vivo that could initiate carcinogenesis (Figure 4). Under favorable circumstances, such electrophilic metabolites might exhibit still greater carcinogenicity than the 1′-hydroxy precursors.

This approach was first tested with the synthetic 1′-acetoxy derivatives of safrole and estragole.[12,13] These model esters were electrophilic in nonenzymatic reactions with tissue nucleophiles. 1′-Acetoxysafrole was also more carcinogenic than 1′-hydroxysafrole at local sites of administration, e.g., the forestomach upon feeding, and the s.c. tissue upon injection.[9] Two other electrophilic metabolites were suggested by studies in other laboratories. The possible oxidation of the 1′-hydroxy metabolites to the 1′-oxo derivatives was indicated by the observation of Oswald et al.[27] that small amounts of addition products (Mannich bases) of 1′-oxosafrole, and certain secondary amines, occurred in the urine of rats and guinea pigs given safrole. Evidence for urinary 1′-oxoestragole was presented by Solheim and Scheline.[26] Evidence for two other electrophilic metabolites, safrole-2′,3′-oxide and 1′-hydroxysafrole-2′,3′-oxide, was provided by Stillwell et al.[28] who found 2′,3′-dihydro-2′,3′-dihydroxysafrole and 2′,3′-dihydro-1′,2′,3′-trihydroxysafrole in the urine of rats and guinea pigs injected with safrole. Delaforge et al.[29] have noted the formation of safrole-2′,3′-oxide from safrole in rat liver cells in culture. So far, however, tests in the authors' laboratory of 1′-oxosafrole in the rat and mouse[10] and of safrole-2′,3′-oxide in the mouse[18] have not revealed any activity of these compounds as hepatocarcinogens. Safrole-2′,3′-oxide and 1′-hydroxysafrole-2′,3′-oxide have weak initiating activity for mouse skin, and the latter compound has also initiated hepatic tumors when administered to infant mice.[10,18]

Chapter 9

MYCOTOXINS AS POTENTIAL ENVIRONMENTAL CARCINOGENS

Leonard Stoloff

TABLE OF CONTENTS

I. INTRODUCTION

The concept developed by Higginson and Oettlé, and quantified in 1960 by Higginson[1] — that the cause of a large proportion of the cancers in many areas of the world probably has an environmental component, although frequently misquoted and misconstrued,[2] has been the basis for a considerable search for "environmental" carcinogens. An even older concept, that molds are capable of producing wonderfully complex organic molecules, had been the subject of a systematic investigation by Raistrick and colleagues at the London School of Hygiene and Tropical Medicine since 1931.[3] It was generally known that a number of these mold-produced products, such as the ergot alkaloids, were biologically active, with the potential for causing harm when ingested by man. It was also known that a history of numerous toxicoses from consumption of moldy food or feed attested to a relation between some molds and the presence of a toxin. But it was not until the properties of penicillin were fully described by Florey and colleagues[4] in 1941, and the ensuing era of antibiotics, that mold metabolites were systematically examined for biological activity in relation to both the target microorganisms and the host animal to be protected. Both lines of investigation, the systematic search for novel organic compounds and the search for antibiotics, have produced a considerable body of information about the structures and the acute biological activity of mold metabolites. Even though several proposed antibiotics were shown to have carcinogenic potential, the idea that some mold metabolites might be part of the environmental component of cancer causation was not generally adopted until the demonstration that the potent, mold-produced carcinogens, the aflatoxins, could be found as natural contaminants in a number of widely used dietary staples.[5] Perhaps the difference in concern lies in the difference between a compound that is the product of a laboratory culture and one found occurring in a product ready for consumption.

The search has progressed in two directions: a determination of the level and incidence of occurrence of those known mold metabolites for which there was evidence of natural presence in a foodstuff and some evidence of potential oncogenicity, and the isolation and characterization of candidate metabolites from molds implicated in some human or animal toxicosis. Ultimately, because exposure to mold metabolites has been a recurrent phenomenon, epidemiological support has been sought for the hypothesis that mold metabolites may be environmental carcinogens for man.[6] This presentation, however, will be restricted to an enumeration of candidate mold metabolites and information on their natural occurrence.

A committee organized by the IARC was assigned the task to develop a manual[7] of selected analytical methods for mycotoxins that might be of interest for study as possible environmental carcinogens. The list of mold metabolites compiled by that committee (Table 1, Figure 1) was based on published evidence of the ability of each compound to produce some type of cellular aberration. As an aid to the selection of mold metabolites most likely to be found as contaminants of food or feed, the list was supplemented by a compilation of those molds that had been reported as sources of each listed compound (Table 2). The normal substrate(s), abundance, and vigor of each mold provide clues as to where and whether natural occurrence of the compound is possible. In addition to the aflatoxins, the following had already been detected as natural contaminants: sterigmatocystin, T-2 toxin, ochratoxin A, citrinin, alternariol methyl ether, mycophenolic acid, patulin, penicillic acid, penitrem A, PR toxin, zearalenone, and zearalenol. Compounds selected for the manual were restricted initially to those shown capable of producing a cancer in some mammalian species after oral administration, and ultimately to those for which suitable methods were available.

Table 1
MOLD METABOLITES THAT HAVE BEEN REPORTED TO PRODUCE VARIOUS TYPES OF CELLULAR ABERRATIONS

	Mammalian cancer		Terato-genicity		Muta-genicity			
Compound	Oral route	Other route	Oral route	Other route	Ames test	Other test	Ref.	Structure[a]
Difurano Compounds								
Aflatoxin B_1	+	+		+	+	+	8—16	XXII
Aflatoxin B_2	+	+		±			10—12, 17	XXIII
Aflatoxin G_1	+	+			+		10—12, 16, 17	XXV
Aflatoxin G_2					+		12	XXVI
Aflatoxin M_1	+				+		12, 17, 18	XXII
Sterigmatocystin	+	+			+	+	11, 13—17, 19, 20	XVIII
5-Methoxysterigmatocystin					+		19	XVIII
Sterigmatin					+		19	XXI
Austocystins A & D					+		15	XXI
Versicolorins A & B					+		15, 19, 20	XXIX,XX
Trichothecenes								
T-2 toxin	+			+	—		16, 21, 22	XII
Fusarenon	+	+			±	—	12—14, 19, 23	XII
Crotocin					+		16	XIII
Penicillium islandicum Toxins								
Cyclochlorotine	+				—		16, 24, 25	XVII
Emodin					±		25—27	VII
Chrysophanol					±		25, 28	VII
Rugulosin	+				±	+	16, 25, 29—31	XXVII
Luteoskyrin	+				−	±	14—16, 19, 23—25, 29	XXVII
Rubroskyrin					+		25	XXIV
Simatoxin					+		25	NS[b]
Penicillium viridicatum Toxins								
Ochratoxin A	+		+	+	—	—	15, 17, 19—21, 32—35	XVI
Viridicatumtoxin					+		15	XXIX
Citrinin	+				±	+	15—17, 19, 30, 36, 37	IV
Alternaria Toxins								
Altertoxin I					+		38	NS
Alternariol methyl ether					+		38	VIII
Penicillium roquefortii Toxins								
PR toxin					±	+	15, 16, 19, 30, 39	XI
Mycophenolic acid					—	+	13—15, 19, 39	VI
Miscellaneous Compounds								
Austdiol					+		15	V
Chaetoglobosin A					+		19	XIV
Cytochalasin D			+	+			40	XV
Erythroglaucin					+		41	VII
Griseofulvin	+	+	+		—	—	15, 16, 19, 42—46	IX
Kojic acid					+		15	I
Mollicellins C & E					+		47	XXX
Patulin	—	+		+	—	+	13—17, 19, 30, 48—50	II
Penicillic acid		+			—	+	13—17, 19, 30, 47	III
Penitrem A				+	—		51	NS
Physcion					+		41	VII
Physcionanthrone					+		41	VII
Rubratoxin B	—		—	+	—	+	17, 19, 52—55	XXVIII
Zearalenone	—		±		—	+	16, 30, 56—58	X
Zearalenol						+	30	X

Note: + : positive observation, −: negative observation; negative observations are reported only in opposition to positive observations for the same compound.

[a] Roman numerals are keyed to Figure 1.

[b] NS: No Structure published.

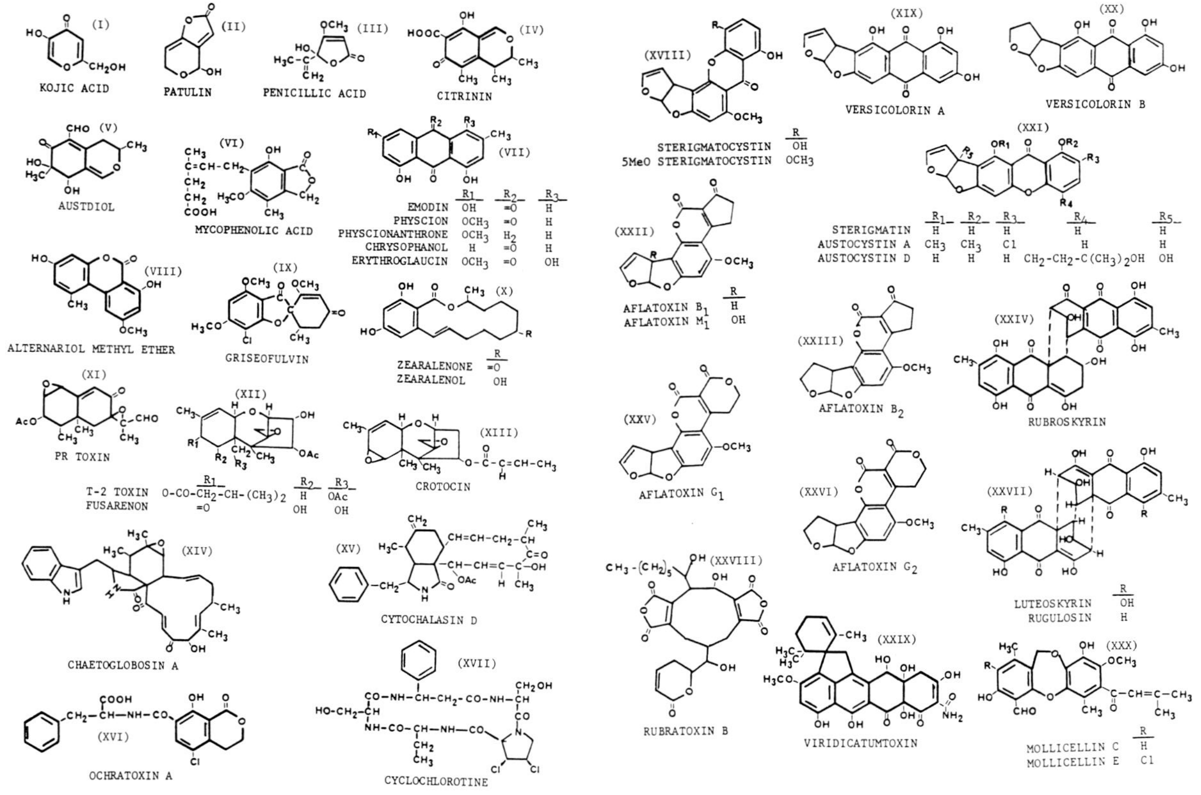

FIGURE 1. Chemical structures of mold metabolites that have been reported to produce various types of cellular aberrations. Roman numerals are keyed to Table 1.

Table 2
REPORTED MOLD SOURCES OF MOLD METABOLITES THAT HAVE BEEN REPORTED TO PRODUCE VARIOUS TYPES OF CELLULAR ABERRATIONS

Compound	Producing Mold(s)
Difurano Compounds	
Aflatoxins B_1, B_2, G_1, G_2, M_1	*Aspergillus flavus, A. parasiticus*
Sterigmatocystin	*Aspergillus versicolor, A. flavus, A. parasiticus, A. nidulans, Penicillium luteum, Bipolaris* sp.
5-Ethoxysterigmatocystin	*A. flavus*
Sterigmatin	*A. versicolor, A. parasiticus*
Austocystins A & D	*Aspergillus ustus*
Versicolorins A & B	*A. versicolor, A. parasiticus*
Trichothecenes	
T-2 toxin	*Fusarium sporotrichioides, F. poae*
Fusarenon	*Fusarium nivale*
Crotocin	*Trichotecium roseum*
Penicillium islandicum Toxins	
Cyclochlorotine	*Penicillium islandicum*
Emodin	*P. islandicum, P. rugulosum, Aspergillus wentii, A. ochraceus, Cladosporium fulvum*
Chrysophanol	*P. islandicum*
Rugulosin	*P. islandicum, P. rugulosum, P. brunneum, P. tardum*
Rubroskyrin	*P. islandicum, P. rugulosum, P. brunneum*
Luteoskyrin	*P. islandicum*
Simatoxin	*P. islandicum*
Penicillium viridicatum Toxins	
Ochratoxin A	*Aspergillus ochraceus, A. melleus, A. sulphureus, A. alliaceus A. ostianus, A. sclerotiorum, A. petrakii, Penicillium viridicatum, P. palitans, P. commune, P. variabile, P. purpurescens, P. cyclopium*
Viridicatumtoxin	*P. viridicatum*
Citrinin	*Aspergillus terreus, A. candidus, A. niveus, Penicillium viridicatum, P. fellutanum, P. lividum, P. implicatum, P. jenseni, P. steckii, P. citreo-viride, P. notatum, P. expansum, P. velutinum, P. canescens, P. palitans, P. claviforme, Clavariopsis aquatica, Blennoria* sp.
Alternaria Toxins	
Altertoxin I	*Alternaria alternata*
Alternariol methyl ether	*A. alternata*
Penicillium roquefortii Toxins	
PR toxin	*Penicillium roquefortii*
Mycophenolic acid	*P. roquefortii, P. viridicatum, P. brevi-compactum*
Miscellaneous Compounds	
Austdiol	*A. ustus*
Chaetoglobosin A	*Chaetomium globosum*
Cytochalasin D	*Metarrhizium anisopliae*
Erythroglaucin	*Aspergillus chevalieri*
Griseofulvin	*Penicillium urticae, P. albidum, P. raciborskii, P. melinii, P. raistrickii, P. nigricans, P. brefeldianum*
Kojic acid	*A. flavus, A. parasiticus, A. oryzae, A. tamarii, A. glaucus*
Mollicellins C & E	*Chaetomium mollicellum*
Patulin	*Aspergillus clavatus, A. giganteus, A. terreus, Penicillium claviforme P. patulum, P. expansum, P. melenii, P. equinum, P. novae-zeelandiae, P. divergens, P. urticae, P. leucopus, P. cyclopium, P. lapidosum, Byssochlamys nivea, Paecilomyces* sp.
Penicillic acid	*A. ochraceus, A. quercinus, A. sulphureus, A. melleus, A. ostianus, A. alliaceus, A. sclerotiorum, Penicillium olivino-viride, P. puberulum, P. cyclopium, P. thomii, P. suavolens, P. martensii, P. baarnense, P. urticae, P. madriti, P. simplicissimum, P. fennelliae, P. palitans, P. aurantio-virens, P. viridicatum, P. expansum, P. janthinellum, P. commune*

Table 2 (continued)
REPORTED MOLD SOURCES OF MOLD METABOLITES THAT HAVE BEEN REPORTED TO PRODUCE VARIOUS TYPES OF CELLULAR ABERRATIONS

Compound	Producing Mold(s)
Penitrem A	*Penicillium crustosum, P. spinulosum, P. puberulum, P. cyclopium*
Physcion	*A. chevalieri, Penicillium charlesii*
Physcionanthrone	*A. chevalieri*
Rubratoxin B	*Penicillium rubrum, P. purpurogenum*
Zearalenone	*Fusarium roseum*
Zearalenol	*F. roseum*

The extent to which existing methods have been applied determines the amount of information available on potential human exposure to each mycotoxin. By far the greatest volume of data and the most comprehensive surveys are concerned with the aflatoxins. These data are sufficiently extensive that some estimates of human exposure have been attempted. At the other end of the spectrum we have only simple notes that penitrem A and alternariol methyl ether have been identified as present in naturally molded commodities.

II. AFLATOXINS

The aflatoxin-producing molds, *Aspergillus flavus* and *A. parasiticus*, are ubiquitous and vigorous competitors on stored products, frequently encountered as outgrowths when the temperature is relatively high (30 to 40°C) and the substrate moisture relatively low (a_w = 0.85 to 0.90), such as occurs in grains and other seeds held under conditions prevailing in many tropical areas. These molds may also be found growing on prepared foods stored without refrigeration. Much to the surprise of many mycologists, these so-called storage molds were found to be facultative parasites to a much greater extent than anyone had imagined. In a temperate, developed area, such as the U.S., most contamination with aflatoxin is restricted to those crops that are invaded by the producing molds before harvest. The levels and incidence of aflatoxin found in the freshly harvested commodity are sufficiently high to account for a major portion of the contamination encountered. In the U.S. aflatoxin is a frequent contaminant of peanuts, maize, and cottonseed, with a strong regional bias to the occurrence. In undeveloped tropical areas, the little information available shows frequent contamination of these same commodities overlaid by random contamination of various other commodities,[59] such as rice, caused by improper household storage of the prepared foods. Significant aflatoxin occurrence has been seen in copra, Brazil nuts, and pistachio nuts imported into the U.S., and also in the pistachio nuts from California groves that have recently come into bearing. There is evidence from the FDA and USDA surveys of the potential for aflatoxin contamination of almonds, pecans, walnuts, and sunflower seed produced in the U.S. and for aflatoxin as a contaminant in an occasional sample of a small grain such as rye, sorghum, or wheat, but the incidence and level of aflatoxin found in these commodities is inconsequential compared to peanuts, maize, and cottonseed. Aflatoxin has also been found at relatively low levels in filberts, pumpkin seed, melon seed, ginger root, and nutmeg imported into the U.S.

Human exposure to aflatoxin from most of these sources is direct, but the potential also exists for exposure to aflatoxin, or one of its still-active metabolites, as a residue in the edible product of animals that may have consumed aflatoxin-contaminated feed. Studies of aflatoxin metabolism in swine, cattle, and poultry led to the conclusion that levels of aflatoxin below those that could have an adverse effect on the animal would result in no detectable residue of aflatoxin in eggs, meat, or organ tissue.[60] The conclusion from this academic exercise in reasoning was confirmed by two field studies carried out by the USDA Food Safety and Quality Service (FSQS).[61] In 1977, and again in 1980, the maize crops in the southeastern U.S. suffered unusually severe aflatoxin contamination. Much of this crop is used for local feed purposes, therefore the probability is very good that in this region, after harvest of each of these crops, many animals going to slaughter and many layers had consumed maize with unusually high aflatoxin levels. For 12 months, starting in October 1977, swine, beef, and chicken livers from slaughterhouses, and eggs from cracking plants in the Southeast were routinely sent to a central laboratory for aflatoxin analysis by a method demonstrated capable of detecting either aflatoxin B_1 or M_1 at a level as low as 0.05 ng/g. There was no detectable aflatoxin in any of 1453 samples that were almost equally divided among the 4 sample types. Early in this same period the FDA had analyzed 101 samples of shell eggs from retail outlets in the Southeast without detecting any aflatoxin. The FSQS survey was repeated for three months, beginning in October 1980, with a modified sampling plan: only samples of swine and turkey livers were taken. Based on experimental metabolism data, the greatest likelihood of finding aflatoxin in liver was in swine, and turkeys had not been sampled previously. No detectable aflatoxin was found in any of 251 swine livers or 114 turkey livers sampled. The only finding of aflatoxin in a commercial sample of liver or eggs was in an FDA survey, early in 1977, of eggs from FSQS-inspected cracking plants located in the southern U.S. Aflatoxin B_1 was found in 1 of 112 samples at a level of 0.06 ng/g. Transmission of an active aflatoxin metabolite, aflatoxin M_1, to milk is a different story that will be recounted separately.

A. Aflatoxins in Peanuts

Because of the unique marketing situation for peanuts in the U.S., it was possible to establish mandatory sampling and analysis of peanuts for aflatoxin once a regulatory decision had been made. The data from this certification process have provided a comprehensive picture of the aflatoxin distribution pattern for essentially the entire crop each year since the system was put into effect in 1968. The distribution of lots tested, by contamination category, for the period 1973 to 1979 is shown in Table 3 for total production and by region. This period is used because after constant improvement up to 1973 reporting detail has since remained constant, facilitating the averaging process. There were yearly fluctuations in contamination distribution, attributed to weather and insects, but the regional pattern clearly persisted each year. Unfortunately, the areas that produce the most peanuts also produce the most highly contaminated peanuts.

There is little systematic survey information about aflatoxin contamination of peanuts from other major producing areas of the world. However, the short 1980 crop in the U.S. resulted in an unprecedented volume of peanut imports, all of which were subject to the same testing for aflatoxin as the domestic product. The aflatoxin contamination distribution by country of origin for these imports is also shown in Table 3. The same distribution of aflatoxin contamination by climate seen within the U.S. is seen in other countries.

Since peanuts are seldom eaten raw, it is more interesting to determine the distribution of aflatoxin contamination in the product as prepared for consumption. The normal roasting process for peanuts results in the destruction of 40 to 80% of the aflatoxin that may have been present. Manufacturers normally had removed defective kernels to ensure flavor; however, after defective kernels were found more likely to contain aflatoxin than sound-appearing kernels, this operation was intensified. A significant

Table 3
OCCURRENCE OF AFLATOXINS IN RAW SHELLED PEANUTS IMPORTED INTO THE U.S. IN 1981 BY COUNTRY OF ORIGIN; IN THE U. S. CROPS HARVESTED 1973 TO 1979, BY YEAR AND THE AVERAGE OF THE U.S. 1973 TO 1979 CROPS, BY REGION[a]

			Total aflatoxins (ng/g) (% of lots)					
Area of origin	Crop year	No. of lots	ND[b]	>4	>15	>26	>50	>100
PR China		2,585	85	15	5	2.5	—	0.5
India		1,453	8	92	73	58	—	10
Sudan		932	6	94	85	78	—	60
Argentina		446	60	40	11.5	4.5	—	0.5
South Africa		112	59	41	26	21	—	8
Malawi		80	40	60	21	10	—	0
Australia		52	90	10	4	4	—	0
Brazil		44	0	100	95	95	—	77
Egypt		41	86	14	2	2	—	0
Taiwan		37	73	27	3	0	—	0
U.S.	1973	24,862	58	42	13	5	—	—
U.S.	1974	18,247	70	30	6	2	0.3	—
U.S.	1975	20,559	69	31	10	4	1.1	—
U.S.	1976	21,239	79	21	4	1	0.1	—
U.S.	1977	27,222	77	23	8	4	1.4	—
U.S.	1978	30,320	65	35	13	5	1.1	—
U.S.	1979	30,220	82	18	5	1	0.3	—
U.S.	1973—79	24,667	71	29	8	3	0.7	—
Virginia-Carolina	1973—79	3,213	87	11	4	1.4	0.5	—
Southeast	1973—79	16,276	69	31	9	3	0.5	—
Southwest	1973—79	4,807	66	34	11	5	1.6	—

[a] Peanut Administrative Committee data.
[b] ND: Not Detectable.

reduction in aflatoxin occurrence in the consumer peanut product compared to the raw shelled peanut should be evident and this is seen in Table 4 from both FDA and Canadian surveillance data. The FDA data for 1964 to 1965 are for a period before there were any controls for aflatoxin and should be compared to the potential use of the total crop. The data since 1973 are for a period when controls were well-established and should be compared to that portion of the crop containing total aflatoxin levels less than 25 ng/g, the maximum permitted level for raw shelled peanuts destined for further processing.

From disappearance data, per capita consumption of peanuts in the U.S. is currently around 0.1 lb/week.[62] Household food consumption data[63] indicate that this consumption is relatively uniform across all survey parameters.

Peanut oil is not likely to be a source of aflatoxin in the occidental diet. It is not widely used in Europe and America, and what little is used is refined; a process that effectively removes any aflatoxin that may be present.[64] The opposite situation prevails in the Orient, where the flavor of crude peanut oil is preferred. In India and Taiwan more aflatoxin may be consumed through cooking oil than by direct consumption of peanuts.[65,66]

B. Aflatoxins in Maize

Maize is by far the most susceptible to aflatoxin contamination of all grains grown

Table 4
OCCURRENCE OF AFLATOXINS IN CONSUMER PEANUT PRODUCTS[a] IN THE U.S. FOR PRECONTROL (1964 TO 1965) AND POSTCONTROL (1973 TO 1980) PERIODS, AND IN CANADA FOR A POSTCONTROL (1974 TO 1981) PERIOD[b]

			Total aflatoxins (ng/g)						
Country	Period	No. of samples	ND[c]	> T[d]	> 5	> 10	> 15	> 20	> 50
			(% of samples)						
U.S.	1964—65	159	69	31	25	20	18	16	6
U.S.	1973—80	1,692	87	13	9	6	4	2.5	0.5
Canada	1974—81	2,788			7	3.5	1.2	0.6	—

[a] U.S. samples include peanut butter and salted roast peanuts. Canadian samples are peanut butter only.

[b] U.S. data from FDA surveillance, and Canadian data from Health and Welfare Canada surveillance.

[c] ND: Not Detectable.

[d] T: Trace.

in the U.S. This susceptibility is markedly related to region, the contamination being most prevalent in the southeastern states (Table 5). The degree of contamination and the size of the area involved varies from year to year, apparently related to environmental factors such as drought and temperature that stress the plant, affect insect activity, and influence inoculum density. There had been some speculation that aflatoxin contamination of maize might partly be a result of advanced agricultural practices, but that theory is largely negated by the results of a study[74] in which currently used hybrids were grown in parallel with the open-pollinated varieties previously planted. Preharvest aflatoxin contamination was the same in hybrids and varieties. The widespread occurrence of aflatoxin in maize grown in less advanced regions[59] would also counter that contention. Spot checks around the world indicate that aflatoxin may be found at some level in maize wherever it is grown, but nowhere do we have as comprehensive a sampling and analysis system as we have for peanuts in the U.S. The closest we can come to a comprehensive picture in the U.S. is by pooling survey, surveillance, service, and compliance data from all possible federal and state sources (Table 5). These data, influenced by the sampling period and the definition of the area included in the Southeast, show a range of aflatoxin incidence in Southeastern maize from 42 to 93%, an average level in contaminated samples frequently greater than 100 ng/g, and maximum levels in the micrograms per gram range. In contrast, maize from the Midwest, where more than 90% of the crop marketed in the U.S. is produced, can be said to be aflatoxin-free.

Typically, about 80% of the maize produced is used for animal feed. Unfortunately, in the Southeast a disproportionate portion of the crop is raised for local consumption as meal and grits, which in the rural parts of this region can be considered an important dietary item, at a 1965 per capita rate of 0.7 lb/week.[63] Some selection by grade for milling into food can be anticipated and some reduction in aflatoxin level is expected when germ and bran are separated from the food millstream. Even after this process, approximately one-third of the maize products from the Southeast that were analyzed had detectable aflatoxins at an average level of about 30 ng/g, with one extreme lot containing total aflatoxins at 1870 ng/g (Table 5). Some further reduction in the aflatoxin level can be expected when the maize meal or grits is prepared for the table.[75] This loss amounts to about one-fifth of the aflatoxin present when the meal is used in

Table 5
OCCURRENCE OF AFLATOXIN IN SHELLED MAIZE AND DRY-MILLED MAIZE PRODUCTS IN THE U.S. BY REGION AND PERIOD, FROM VARIOUS DATA SOURCES

Area	Period	No. of samples	Aflatoxins (ng/g) ND[a]	> 20	> 100	Detectable aflatoxins av.	max.	Ref.
			(% of samples)			(ng/g)		
			Shelled Maize					
Southeast[b]	1969—75	1,927	58	28	9	103	3,300	67
Southeast	Aug 77—Oct 77	—	—	56	26	—	—	68
Southeast	Oct 77—Mar 78	202	26	66	41	350	2,150	69
Southeast	Apr 78—Sep 78	37	24	51	38	170	725	69
Southeast	Oct 78—Sep 79	70	20	31	9	55	701	69
Southeast	Oct 79—Sep 80	64	45	34	17	60	405	69
Southeast	Oct 80—Jun 81	108	37	33	12	57	321	69
Southeast	Dec 80—Apr 81	908	35	46	16	92	2,003	70
N.C.	Fall 1978	228	54	13	1	17	141	71
N.C.	Fall 1979	205	32	33	4	37	475	71
N.C.	Fall 1980	233	8	73	17	64	558	71
Ga.	Fall 1977	30	7	90	78	662	4,780	72
Ga.	Fall 1978	45	7	40	18	55	620	72
Ga.	Fall 1979	45	20	28	11	60	3,029	72
Ga.	Fall 1980	45	7	74	45	241	1,620	72
Southeast	Fall 1978	243	54	34	19	—	—	73
Ga.	Fall 1979	57	63	18	5	98	1,100	73
Ga.	Fall 1980	67	7	75	33	—	—	73
Southeast	Fall 1979	168	51	27	15	327	4,630	74
Southeast	Fall 1980	168	10	76	37	215	5,200	74
Midwest[c]	1964—73	1,763	98	≪ 1	0	8	37	67
Midwest	Oct 77—Sep 80	68	94	1	0	16	28	69
Midwest	Fall 1978	634	98	< 1	0	—	—	73
			Milled Maize Products					
Southeast	Oct 77—Mar 78	40	52	28	—	46	115	69
Southeast	Apr 78—Sep 78	22	86	9	—	18	27	69
Southeast	Oct 78—Sep 79	92	72	12		24	125	69
Southeast	Oct 79—Sep 80	92	74	4	—	15	73	69
Southeast	Oct 80—Jun 81	120	40	16	—	28	242	69
Southeast	Dec 80—Apr 81	1,038	69	14	—	41	1,870	70
North and West[d]	Oct 77—Sep 80	278	98	0	—	6	11	69

[a] ND: Not Detectable.

[b] Southeast is here broadly defined as the region east of the Mississippi river and south of Washington, D.C.

[c] Midwest is here broadly defined as the band of states from Ohio and Michigan west to Kansas, Nebraska, and the Dakotas.

[d] North and West is all of the continental U.S. outside of the Southeast.

baked goods or the grits are boiled, and about one-half of the aflatoxin when the boiled grits are fried.

C. Aflatoxins in Cottonseed

Except for a small-volume specialty snack item, cottonseed is not consumed directly

Table 6
OCCURRENCE OF AFLATOXINS IN COTTONSEED AND COTTONSEED MEAL PRODUCED IN THE U.S., BY REGION FOR VARIOUS PERIODS

Region	Period	No. of samples	Aflatoxins (ng/g) ND[a]	> 20	> 100	Detectable aflatoxins Av.	Max.	Ref.
			(% of samples)			(ng/g)		
Cottonseed								
N.M., Ariz., Calif.	1964—66	803	80	13	6	—	—	67
Ariz., Imperial Valley Calif.	1971	88	15	83	—	450	2,000	67
Ariz., Imperial Valley Calif.	1974—75	79	70	20	9	145	1,060	67
Ariz., Imperial Valley Calif.	1976	2,013	18	43	11	—	—	76
Ariz., Imperial Valley Calif.	1977	2,389	0	82	55	—	—	76
Ariz., Imperial Valley Calif.	1978	1,671	0	67	40	—	—	76
U.S. other than above	1964—66	2,504	96	1.5	0.5	—	—	67
U.S. other than above	1974—75	52	96	0	0	7	7	67
Cottonseed Meal								
N.M., Ariz., Calif.	1964—66	728	59	30	14	—	—	67
U.S. other than above	1964—66	2,606	90	4	1	—	—	67

[a] ND: Not Detectable.

as food, and except for a specialty colorant for baked goods, neither is cottonseed meal, the residual cake from cottonseed oil mills. Human exposure to aflatoxin from cottonseed is, therefore, primarily through transmission to animal tissue from use of the seed and meal in animal feed. As previously noted, refining processes effectively remove any aflatoxin that may be present in the crude oil. As also previously noted, the only animal source product in which measurable levels of an aflatoxin are likely to be found in market samples is bovine milk. In the U.S. the whole seed is preferred for dairy rations. In feed importing countries, such as those in western Europe, for logistic and economic reasons the meal will be used. For the same reasons, seed is most likely to be used close to the areas in which it is produced.

After cottonseed meal had been implicated as the ration ingredient associated with hepatoma in rainbow trout and aflatoxin had been found as a contaminant of the meal, the USDA, with the cooperation of the National Cottonseed Products Association, carried out a comprehensive three-year survey for aflatoxin in cottonseed and cottonseed meal in all producing areas of the country. Aflatoxin was found in seed and meal from all areas with the expected year-to-year fluctuation, but in each year seed and meal from the Far West (New Mexico, Arizona, and California) had by far the highest incidence and levels (Table 6). This regional pattern showed up in limited FDA surveys in subsequent years. Because of this pattern, when a 1973 FDA survey for aflatoxin in dairy products found contamination in all samples of cottage cheese from Arizona, cottonseed was the suggested source.

State and industry investigations have narrowed the major problem area to Arizona and southern California, where there has been systematic monitoring for occurrence and searching for causes and remedies. Comprehensive data generated from these studies (Table 6) show the extent of the local problem and the year-to-year variation in the extent of aflatoxin occurrence.

Spot checks for aflatoxin in cottonseed and cottonseed meal from other areas of the world indicate that the problem is not restricted to the U.S.[59]

D. Aflatoxin M_1 in Milk

Aflatoxin M_1, a hydroxylated metabolite of aflatoxin B_1, appears in the milk of mammals that have ingested aflatoxin B_1. From compiled transmission studies, an estimate[60] has been made that, on the average, an aflatoxin B_1 level of 30 ng/g in the feed of a cow will result in an aflatoxin M_1 level of 0.1 ng/mℓ in the milk from that cow. Surveys of milk and dairy products in a number of countries[77,78] have shown that aflatoxin M_1 can be found in marketed milk in relation to the potential for aflatoxin contamination of the dairy rations. The incidence and level obviously depend on the point in the collection and distribution system at which the sampling is done. In pooled milk the incidence of samples with detectable aflatoxin M_1 (>0.02 ng/mℓ) has been as high as 80%. The level of aflatoxin M_1 has seldom exceeded 0.5 ng/mℓ, although individual values as high as 8 ng/mℓ have been recorded.

Aflatoxin M_1 will be carried through into products made from contaminated fluid milk in proportion to the concentration effect, except for butter and cheese where solubility and adsorption to protein create a partition unrelated to product weight.[77] As expected, aflatoxin M_1 has been found in milk powder and cheeses produced in areas where aflatoxin M_1 has been found in milk.

Because the occurrence of aflatoxin in dairy rations can be controlled, the publicity and regulatory action attendant on these survey results should produce a reduction in this source of exposure to aflatoxin, as has already taken place in the U.S. An action level for aflatoxin M_1 in fluid milk had been set in 1977 at 0.5 ppb.[79] In a 1979 survey of dairy products (992 samples) manufactured in the U.S.,[80] one sample, a cottage cheese, was found to contain detectable aflatoxin M_1 at a level equivalent to 0.08 ng/mℓ in the milk from which the product was made. The sampling had been weighted to double the representation of establishments in the southern tier of states where contamination was most likely to occur.

E. Aflatoxins in Tree Nuts

The occurrence of aflatoxins in almonds, filberts, pecans, and walnuts is negligible compared to Brazil nuts and pistachio nuts, particularly after growers and shellers in the U.S. and Turkey, on being advised of the potential, determined causes and took remedial action. The initial surveillance for aflatoxin in Brazil nuts and pistachio nuts offered for import into the U.S. produced such high rates of rejection for excessive aflatoxin that the importers requested, and were permitted, to enter into voluntary certification agreements. Under these agreements, essentially all Brazil and pistachio nuts entering the U.S. have been sampled and tested for aflatoxins. No similar system exists in other world markets.

In recent years, pistachio trees planted in California have come into bearing. Aflatoxin has been found in California nuts harvested in the last two years, but it is still too early to know whether the source of contamination has been located and controlled.

As a source of aflatoxin in the human diet, tree nuts probably can be ignored. Per capita consumption in the U.S. of all these nuts together, at around 0.03 lb/week is small compared to peanuts or maize, and, except for Brazil nuts and pistachio nuts, the incidence of aflatoxin contamination is very low. This generality does not exclude regional peculiarities as was found with Aleppo pine nuts in Tunisia.[81] Of 50 samples of these nuts used in a popular local pudding, 26 contained aflatoxins, with levels in 8 of the samples in the micrograms per gram range.

III. STERIGMATOCYSTIN

No significant occurrence of sterigmatocystin has been reported. The first survey, in 1968, was in relation to an epidemiological study in Mozambique,[82] in which no sterigmatocystin could be detected in any of 173 samples of various foods subject to mold attack. Neither was sterigmatocystin detected in any of 574 samples of U.S. grains sampled by the FDA over a 2-year period in 1974 and 1975, nor in 142 maize, fruit, and vegetable samples taken in East Germany;[83] 30 of this last group of samples were taken because of obvious mold growth. There have been reports of finding sterigmatocystin in samples of moldy wheat,[84] moldy feed,[85] moldy green coffee,[86] and in-shell pecans.[87]

Based on the finding in green coffee, surveys[88] for sterigmatocystin were made of green coffee received in the U.S. (15 samples by industry and 201 by FDA) and in Italy (502 samples). Sample selection in all cases was biased to the lower grades. Sterigmatocystin was detected in only one heavily molded sample of beans taken in Italy. Roasting of the green beans can be expected to destroy about 70% of any sterigmatocystin present.[88]

The finding of sterigmatocystin most approaching significance was in 9 of 39 samples from hard cheeses ripening in Dutch warehouses.[89] The analytical method used had been modified to achieve a detection limit of 5 ng/g, and levels as high as 600 ng/g were found. All samples were from the outer 1 cm layer of cheese, and had been taken from cheeses with a surface outgrowth of *Aspergillus versicolor*, a known sterigmatocystin-producing mold. A follow-up study on sterigmatocystin penetration into the cheese, showed the contamination was essentially confined to the rind.[90] When informed of the relation of sterigmatocystin to the presence of *A. versicolor* on ripening cheeses, the FDA examined 100 samples of moldy cheese rind from a large commercial operation in the U.S. No *A. versicolor* was detected.

IV. T-2 TOXIN

T-2 toxin has been found as a natural contaminant, most frequently in maize[91-95] and also in barley,[96] safflower seed,[97] and sorghum.[98] In all instances a single sample was involved, and in all but one instance the sample had been examined because of mold damage and/or a veterinary problem. Since animals are much more likely to be exposed to T-2 toxin than are humans, the possibility has been explored that T-2 toxin might be transmitted to edible animal tissue.[99-102] These studies showed that T-2 toxin is rapidly cleared in the form of highly polar metabolites, that milk is the tissue in which a residue of T-2 toxin is most likely to be found, and that this residue in milk will probably be much less than 1% of the administered dose. Should any T-2 toxin-contaminated maize be accepted for wet milling, most of the toxin will follow the steepwater, germ, and bran. If the germ is used for production of maize oil, the refining process will remove any T-2 toxin that might be present.[103] Characteristic of similar findings for aflatoxin and zearalenone-contaminated maize, the mycotoxin tends to concentrate in the feed fractions of the milling process.

In an FDA survey of U.S. grains for the incidence of specific *Fusaria* and their trichothecene-producing capabilities, T-2 toxin production was observed in cultures only of isolates of *F. sporotrichioides* or *F. poae* (classified according to Wollenwerber and Reinking). Because analysis for T-2 toxin and identification of the *Fusaria* are equally difficult, association of T-2 toxin with any but the two *Fusarium* species above should bring into question both mold and toxin identification.

V. OCHRATOXIN A

Surveys for ochratoxin A show a worldwide occurrence (Table 7) with very large

Table 7
OCCURRENCE OF OCHRATOXIN A IN VARIOUS COMMODITIES, BY COUNTRY AND YEAR

Commodity	Year	Country	No. of samples	Contaminated (%)	Ochratoxin level (ng/g)	Ref.
Maize	1967	U.S.	283	0.4	130	104
Maize	1968—69	U.S.	293	1.0	83—166	105
Wheat	1970—73	U.S.	848	1.3	<15—115	106
Sorghum	1975—76	U.S.	197	0	—	107
Barley	1972	U.S.	159	14	<10—29	[a]
Malt barley	1973	U.S.	138	0	—	[a]
Green coffee	1972	U.S.	335	10	<20—360	108
Feedstuffs	1972—77	Canada	2,022	0.1	—	109
Feedstuffs	1973—74	Britain	131	1.5	Trace	110
Maize	1973—74	France	924	1.9	15—200	111
Barley & oats (feed)	1971	Denmark	116	36	9—27,500	112
Barley & oats	1972	Sweden	84	8	16—410	113
Barley & oats	1976	Sweden	189	4	<5—11	114
Barley (feed)	1966—77	Sweden	269	7	2—20	115
Maize (feed)	1975	Yugoslavia	191	26	45—5,100	116
Food grains	1972—76	Yugoslavia	768	13	14—140	117
Feed grains	1974	Poland	302	5	50—200	118
Mixed feeds	1976	Poland	234	5	10—50	119
Food grains	1977	India	180	6	30—2,000	120

[a] Unpublished data, U.S. Food and Drug Administration.

differences by country in incidence and level. Large differences are also found within countries. When obviously moldy samples have been taken, ochratoxin A has been found in 18/29 samples of heated grain, in 1/3 samples of mixed feeds, in 3/3 samples of white beans, and 1/2 samples of moldy peanuts.[84] The highest level, 27,000 ng/g was in one sample of grain. Similarly, ochratoxin A was found in 4/5 samples of moldy green coffee beans,[108] with a high level of 400 ng/g; and in 4/4 samples of moldy wheat,[121] with a high level of 100 ng/g.

In Denmark and in Sweden, where nephropathy in swine at slaughter has been related to the occurrence of ochratoxin in feed grains in the affected regions, ochratoxin A residues have been found in the diseased kidneys,[122-124] related to lower ochratoxin A levels in the muscle.[122] Transmission of ochratoxin A to chicken tissue and eggs has also been observed in experimental feeding.[125] As with swine, the organ in which the highest level was found was the kidney. In both swine and chickens only a small fraction of the ingested ochratoxin A can be found as a residue. There is no accumulation and within days after cessation of exposure the tissues are clear.

Because of the high incidence of occurrence in barley, the effect on ochratoxin A of conversion of barley into beer was investigated.[126] In the finished beer 2 to 7% of the initial amount of ochratoxin A in the barley could be detected. Similarly, because of the finding in green coffee beans, the effect on ochratoxin A of roasting was determined and about 80% of the added ochratoxin A was destroyed.[108]

VI. CITRININ

There is no reliable analytical method for citrinin, but because citrinin-producing mold species have a wide distribution and are generally vigorous, reports of citrinin occurrence are not to be discounted. Reports of levels, however, cannot be accepted as valid.

Table 8
OCCURRENCE OF ZEARALENONE IN VARIOUS COMMODITIES IN THE U.S., BY YEAR[a]

Commodity	Year	No. of samples	Zearalenone (μ/g) ND[b]	>0.1	>0.4	>1.0	>5.0
			(% of samples)				
Maize	1967	283	99	1	1	0	0
Maize	1968—69	293	98	2	2	0	0
Maize[c]	1972	223	83	17	8	4	0
Maize	1973	315	94	6	0	0	0
Maize meal	1975—76	119	100	0	0	0	0
Wheat[d]	1975	112	83	17	16	14	2
Soybeans	1975	180	0	0	0	0	0
Sorghum	1975—76	197	71	29	27	19	1

[a] Data from Ref. 130.
[b] ND: Not Detectable.
[c] Sampling concentrated in area where infection of maize with *Fusarium roseum* was unusually severe.
[d] Associated with "tombstone" kernels caused by *Fusarium roseum* scab.

The first report of the possible occurrence of citrinin in a foodstuff is by inference. One of the molds frequently associated with yellowed rice disease in Japan[127] is *Penicillium citrinum.* Isolates of this mold from rice produced large quantities of citrinin, and the yellow color of the rice could be associated in laboratory experiments with the production of this yellow compound. Most of the subsequent reports of finding citrinin[84,112,115,121] relate to the simultaneous occurrence of ochratoxin A in grains and its production by *P. viridicatum.* One report[128] identifies citrinin in yellow peanut kernels from damaged pods. A number of citrinin-producing mold strains have been isolated from fermented sausage, but citrinin could not be detected in sausage inoculated with these same isolates.[129]

VII. ZEARALENONE AND ZEARALENOL

A recent review[130] provides a good compilation of survey data on the occurrence of zearalenone in commodities and problem feeds. The data in Tables 8 and 9 are derived from that compilation. Since zearalenol was not originally expected as a direct metabolic product of the mold, convenient methodology for its determination had not been developed. There is now evidence that zearalenol can occur with zearalenone,[131] and, equally important, that zearalenol is a major metabolite of zearalenone in man.[132] Exposure data for either compound can therefore be considered as equivalent to exposure to the other. One report,[133] not included in the tabulation, is the unexpected finding of zearalenone in 3 of 60 samples of walnuts at levels of 50, 105, and 450 ng/g.

Occurrence of zearalenone in problem grains and feeds is relatively common. The detection of zearalenone in food grade grains by the analytical method used in most surveys is relatively rare. This method has a low limit of detection, around 100 ng/g. However, with recently developed methods using HPLC rather than TLC for the determinative step, the limit of detection was improved to 5 to 10 ng/g.[134,135] When these methods were applied to dry milled maize food products, zearalenone was found in 13 of 21 random samples analyzed at levels of 14 to 69 ng/g. These observations are

Table 9
OCCURRENCE OF ZEARALENONE IN FEED GRAINS AND FEEDS ASSOCIATED WITH VETERINARY PROBLEMS,[a] BY COMMODITY AND COUNTRY[b]

Problem	No. of reports	Commodity	Country	Zearalenone level (μg/g)
Porcine hyperestrogenism	5	Maize	U.S.	0.12—6.4
Porcine hyperestrogenism	1	Maize	Canada	0.2
Porcine hyperestrogenism	2	Maize	Yugoslavia	18—37
Porcine hyperestrogenism	1	Maize	France	2.3
Porcine feed refusal	1	Maize	U.S.	2.5
Porcine feed refusal	1	Maize	Yugoslavia	0.7—15
Porcine abortion	1	Maize	U.S.	32
Porcine abortion	1	Maize	England	306
Porcine poisoning	1	Maize	Yugoslavia	2.5—36
Bovine abortion	1	Sorghum	U.S.	12
Porcine hyperestrogenism	1	Sorghum	U.S.	2.0—5.6
Porcine hyperestrogenism	1	Sesame meal	U.S.	1.5
Swine reproduction disorders	1	Barley	Scotland	0.5—0.75
Porcine hyperestrogenism	4	Feed	U.S.	0.1—2,900
Porcine hyperestrogenism	5	Feed	Canada	0.07—1.0
Swine infertility, abortion	1	Feed	U.S.	0.01
Swine and cattle infertility	1	Feed	Finland	25
Bovine feed refusal	1	Feed	U.S.	1
Porcine hemorrhage	1	Feed	U.S.	0.1
Cattle toxicosis	1	Feed	Hungary	5—75

[a] The presence of zearalenone is not necessarily related to the observed problem.
[b] Data from Ref. 130.

surprising in light of maize dry milling experiments[136] that show the prime product to have a zearalenone level about 20% of that in the original maize, with most of the zearalenone appearing in the feed fractions. Maize wet milling experiments[137] showed no detectable zearalenone in the starch, which is the prime product of that operation.

Because zearalenone is more likely to be a contaminant of feed than food, the possibility of its transmission to edible tissue of swine,[138] cattle,[132,139] sheep,[139] and hens[140] has been studied. In general, zearalenone is rapidly metabolized, principally to zearalenol, and eliminated as water-soluble conjugates. However, sizeable transmission of both zearalenone and zearalenol to the milk of all species studied was observed. The level for the combined compounds in bovine milk was about 5% of the level of zearalenone in the feed. This 20/1 ratio of feed level to milk level is far different from the 300/1 ratio found for aflatoxin M_1 in bovine milk.

VIII. PATULIN

Considering the number of common storage molds that produce patulin and the number of times patulin was rediscovered as a novel compound, it is not surprising that, once a simple analytical method had been devised,[141] the subsequent surveys would produce positive evidence of natural contamination. Because information had already been developed that a high percentage of natural apple rots contained patulin,[142] and that patulin was relatively stable in apple juice,[143] the original and subsequent surveys concentrated on that product. The incidence of detectable contamination

Table 10
OCCURRENCE OF PATULIN IN APPLE JUICE IN VARIOUS COUNTRIES, USING POOLED DATA FROM VARIOUS SOURCES

Country	No. of samples	Contaminated (%)	Patulin level (ng/mℓ)	Ref.
U.S.	289	31	10—45,000	144—146
Canada	92	57	11—1,000	141, 147 [a]
Sweden	115	60	2—73	148, 149
Switzerland	42	45	5—50	150
East Germany	416	67	20—400	151—153
U.K.	20	70	1—38	78
France	21	42	100—300	154

[a] Department of Health and Welfare, Canada, unpublished data, 1981.

(Table 10) was high (31 to 70%) in all countries where apple products were analyzed. The magnitude of the incidence depended to some degree on the detection limit of the method employed and the maximum value depended on the size of the operation from which the sample was taken. The highest level (45,000 ng/g) was in juice obtained from a roadside stand in the U.S.

The occurrence of patulin is possible in any fruit subject to soft rot, provided that the fruit does not contain sulfhydryl compounds in an amount sufficient to bind all the patulin produced.[143] The same mold that causes soft rot of fruit is a common member of the mycoflora of fermented sausage.[155] However, after patulin-producing strains of the mold have been grown on sausage, patulin cannot be recovered, presumably because of interaction with sulfhydryl groups in the protein. Patulin bound in this manner is no longer biologically active.[156]

Fruit juices converted to wine or vinegar will contain no patulin, since the compound does not survive an alcoholic fermentation.[157,158]

IX. PENICILLIC ACID

Although the list of commonly found mold species that produce penicillic acid rivals that for patulin, the survey effort has been much more limited. One reported occurrence[159] was in 7 of 20 samples of maize, selected because of a high proportion of "blue eye" kernels, at levels of 5 to 230 ng/g. Some of the molds causing the "blue eye" condition are known to produce penicillic acid. Another reported finding was in 5 of 20 samples of commercial dried beans[159] at levels of 11 to 179 ng/g.

Penicillic acid reacts with sulfhydryl groups in the same manner as patulin. It is, therefore, not surprising that although penicillic acid can be detected immediately after a sample of "blue eye" corn has been ground, it can not be found by the same method 24 hr later. This reaction with patulin had been shown to be moisture-dependent.[160]

A number of molds found on fermented sausage are penicillic acid producers,[161] and as had been found in similar studies with patulin and citrinin, penicillic acid could not be detected in sausage on which producing molds were grown, nor was the reaction product of penicillic acid with cysteine or glutathione biologically active.[162]

X. MYCOPHENOLIC ACID

Because of the observation that a number of strains of *Penicillium roquefortii* isolated from cheeses produced mycophenolic acid, and because of the widespread use of cheeses ripened with this mold, 100 fully ripened cheeses from commercial sources were sampled and analyzed for this metabolite.[163] Mycophenolic acid was found in 38 of the samples by an analytical method having a low limit of detection around 0.02 μg/g. The highest level, found in 2 of the samples, was 10 μg/g.

XI. ALTERNARIOL METHYL ETHER

Analyses of a sample of sorghum[164] and a sample of pecans,[165] both invaded by *Alternaria* sp., for known toxic metabolites of these molds resulted in the detection and conclusive confirmation of identity of alternariol methyl ether among the other *Alternaria* metabolites found. Quantitation was not attempted in either instance and there is currently no adequate analytical methodology for this compound.

XII. PENITREM A

The only reported natural occurrence of penitrem A that has come to our attention was in a sample of badly molded cream cheese that had been involved in the intoxication of two dogs.[166] The identification of penitrem A in the sample extract was based on a comparison with an authentic sample of the compound on TLC and on the tremors typical of penitrem A intoxication induced in mice by injection of the extract.

XIII. PR TOXIN

Because formation of PR toxin by a number of *Penicillium roquefortii* strains used in the manufacture of cheeses had been demonstrated, an attempt was made to develop an analytical method for this compound.[167] In the course of this effort, evidence was accumulated that interaction of PR toxin with normal cheese components results in products with altered biological activity. The investigators properly concluded that any further experiments to assess the hazard from PR toxin will need to take into account the cheese as a whole.

XIV. POSTSCRIPT

The only serious attempt, so far, to link exposure to a specific mycotoxin with a specific cancer in man is in regard to aflatoxin and liver cancer. There has been speculation that exposure to one or more of the trichothecenes, not necessarily T-2 toxin, may be related to esophageal cancer, and that ochratoxin A may be a factor in the endemic nephritis observed in the Balkans with the concurrent increased risk of kidney tumors. In no case has causation yet been established.

REFERENCES

1. **Higginson, J.,** Population studies on cancer, *Acta Unio Int. Contra Cancrum,* 16, 1667, 1960.
2. **Maugh, T. H.,** Cancer and environment: Higginson speaks out (an interview), *Science,* 205, 1363, 1979.
3. **Raistrick, H., Birkinshaw, J. H., Charles, J. H. V., Clutterbuck, P. W., Coyne, F. P., Hetherson, A. C., Lilly, C. H., Rintoul, M. L., Rintoul, W., Robertson, R., Stoyle, J. A. R., Thom, C., and Young, W.,** Studies in the biochemistry of microorganisms, *Phil. Trans. R. Soc. London, Ser. B.,* 220, 1931.

4. Florey, H. W., Abraham, E. P., Chain, E., Fletcher, C. M., Gardner, A. D., Heatley, N. G., and Jennings, M. A., Further observations on penicillin, *Lancet*, 2, 177, 1941.
5. Stoloff, L., Aflatoxins — an overview, in *Mycotoxins in Human and Animal Health*, Rodricks, J. V., Hesseltine, C. W., and Mehlman, M. A., Eds., Pathotox Publishers, Park Forest South, Ill., 1977, 7.
6. Linsell, C. A., Environmental chemical carcinogens and liver cancer, *J. Toxicol. Environ. Health*, 5, 173, 1979.
7. International Agency for Research on Cancer, Environmental Carcinogens, Selected Methods of Analysis, Vol. 5, Analysis of Mycotoxins in Foods, IARC Sci. Publ. 44, IARC, Lyon, 1982.
8. Elis, J. and DiPaolo, J. A., Aflatoxin B_1 (induction of malformations), *Arch. Pathol.*, 83, 53, 1967.
9. DiPaolo, J. A., Elis, J., and Erwin, H., Teratogenic response by hamsters, rats and mice to aflatoxin B_1, *Nature (London)*, 215, 638, 1967.
10. Butler, W. H., Greenblatt, M., and Lijinsky, W., Carcinogenesis in rats by aflatoxins B_1, G_1, and B_2, *Cancer Res.*, 29, 2206, 1969.
11. International Agency for Research on Cancer, Monograph on the Evaluation of Carcinogenic Risk of Chemicals to Man, Vol. 10, IARC, Lyon, 1976.
12. Wong, J. J. and Hsieh, D. P. H., Mutagenicity of aflatoxins related to their metabolites and carcinogenic potential, *Proc. Natl. Acad. Sci. U.S.A.*, 73, 2241, 1976.
13. Umeda, M., Cytotoxicity of mycotoxins, in *Mycotoxins in Human and Animal Health*, Rodricks, J. V., Hesseltine, C. W., and Mehlman, M. A., Eds., Pathotox Publishers, Park Forest South, Ill., 1977, 713.
14. Umeda, M, Tsutsui, T., and Saito, M., Mutagenicity and inducibility of DNA single-strand breaks and chromosome aberrations by various mycotoxins, *Gann*, 68, 619, 1977.
15. Wehner, F. C., Thiel, P. G., van Rensburg, S. J., and Demasius, I. P. C., Mutagenicity to *Salmonella typhimurium* of some *Aspergillus* and *Penicillium* mycotoxins, *Mutat. Res.*, 58, 193, 1978.
16. Ueno, Y., Kubota, K., Ito, T., and Nakamura, Y., Mutagenicity of carcinogenic mycotoxins in *Salmonella typhimurium*, *Cancer Res.*, 38, 636, 1978.
17. Engel, G. and von Milczewski, K. E., Zum Nachweis von Mykotoxinen nach Aktwierung mit Rattenleberhomogenaten mittels Histidinmangelmutanten von *Salmonella typhimurium*, *Kiel. Milchwirtsch. Forschungsber.*, 28, 359, 1976.
18. Wogan, G. N. and Paglialunga, S., Carcinogenicity of synthetic aflatoxin M_1 in rats, *Food Cosmet. Toxicol.*, 12, 381, 1974.
19. Nagao, M., Honda, M., Hamasaki, T., Natori, S., Ueno, Y., Yamasaki, M., Seino, Y., Yahagi, T., and Sugimura, T., Mutagenicity of mycotoxins in *Salmonella*, *Proc. Jpn. Assoc. Mycotoxicol.*, 3/4, 41, 1976 (in Japanese).
20. Wong, J. J., Singh, R., and Hsieh, D. P. H., Mutagenicity of fungal metabolites related to aflatoxin biosynthesis, *Mutat. Res.*, 44, 447, 1977.
21. Hood, R. D., Kuczuk, M. H., and Szczech, G. M., Effects in mice of simultaneous prenatal exposure to ochratoxin A and T-2 toxin, *Teratology*, 17, 25, 1978.
22. Schoental, R., Joffe, A. Z., and Yagen, B., Cardiovascular lesions and various tumors found in rats given T-2 toxin, a trichothecene metabolite of *Fusarium*, *Cancer Res.*, 39, 2179, 1979.
23. Saito, M., Horiuchi, T., Ohtsubo, K., Hatanaka, Y., and Ueno, Y., Low tumor incidence in rats with long-term feeding of fusarenon-X, a cytotoxic trichothecene produced by *Fusarium nivale*, *Jpn. J. Exp. Med.*, 50, 293, 1980.
24. Uraguchi, K., Saito, M., Noguchi, Y., Takahashi, K., Enomoto, M., and Tatsuno, T., Chronic toxicity and carcinogenicity in mice of the purified mycotoxins, luteoskyrin and cyclochloritine, *Food Cosmet. Toxicol.*, 10, 193, 1972.
25. Stark, A. A., Townsend, J. M., Wogan, G. N., Demain, A. L., Mannade, A., and Ghosh, A. L., Mutagenicity and antibacterial activity of mycotoxins produced by *Penicillium islandicum* Sopp and *Penicillium rugulosum*, *J. Environ. Pathol. Toxicol.*, 2, 313, 1978.
26. Brown, J. P. and Brown, R. J., Mutagenesis by 9, 10-anthraquinone derivatives and related compounds in *Salmonella typhimurium*, *Mutat. Res.*, 40, 203, 1979.
27. Wehner, F. C., Thiel, P. G., and duRand, M., Mutagenicity of the mycotoxin emodin in the *Salmonella* microsome system, *Appl. Environ. Microbiol.*, 37, 658, 1979.
28. Liberman, D. F., Schaefer, F. L., Fink, R. C., Ramgopal, M., Ghosh, A. C., and Mulcahy, R., Mutagenicity of islandicin and chrysophanol in the *Salmonella*/microsome system, *Appl. Environ. Microbiol.*, 40, 476, 1980.
29. Ueno, Y., and Nakajima, M., Production of respiratory-deficient mutants of *Saccharomyces cerevisiae* by (−)luteoskyrin and (+)rugulosin, *Chem. Pharm. Bull.*, 22, 2258, 1974.
30. Ueno, Y. and Kubota, K., DNA-attacking ability of carcinogenic mycotoxins in recombination-deficient mutant cells of *Bacillus subtilis*, *Cancer Res.*, 36, 445, 1976.

31. **Ueno, Y., Sato, N., Ito, T., Ueno, I., Enomoto, M., and Tsunoda, H.,** Chronic toxicity and hepatocarcinogenicity of (+) rugulosin, an anthraquinoid mycotoxin from *Penicillium* species: preliminary surveys in mice, *J. Toxicol. Sci.*, 5, 295, 1980.
32. **Hayes, A. W., Hood, R. D., and Lee, H. L.,** Teratogenic effects of ochratoxin A in mice, *Teratology*, 9, 93, 1974.
33. **Brown, M. H., Szczech, G. M., and Purmalis, B. P.,** Teratogenic and toxic effects of ochratoxin A in rats, *Toxicol. Appl. Pharmacol.*, 37, 331, 1976.
34. **Kanisawa, M. and Suzuki, S.,** Induction of renal and hepatic tumors in mice by ochratoxin A, a mycotoxin, *Gann*, 69, 599, 1978.
35. **Hood, R. D., Naughton, M. J., and Hayes, A. W.,** Prenatal effects of ochratoxin A in hamsters, *Teratology*, 13, 11, 1978.
36. **Shinohara, Y., Arai, M., Hirao, K., Sugihara, S., Nakanishi, K., Tsunoda, H., and Ito, N.,** Combination effect of citrinin and other chemicals on rat kidney tumorigenesis, *Gann*, 67, 147, 1976.
37. **Thust, R. and Kneist, S.,** Activity of citrinin metabolized by rat and human microsome fractions in clastogenicity and SCE assays on Chinese hamster V79-E cells, *Mutat. Res.*, 67, 321, 1979.
38. **Scott, P. M. and Stoltz, D. R.,** Mutagens produced by *Alternaria alternata, Mutat. Res.*, 78, 33, 1980.
39. **Wei, R-D., Ong, T-M., Whone, W-Z., Frezza, D., Bronzetti, G., and Zeiger, E.,** Genetic effects of PR toxin in eukaryotic microorganisms, *Environ. Mutagenesis*, 1, 45, 1979.
40. **Shepard, T. H. and Greenaway, J. C.,** Teratogenicity of cytochalasin D in the mouse, *Teratology*, 16, 131, 1977.
41. **Bachmann, M., Luthy, J., and Schlatter, C.,** Toxicity and mutagenicity of molds of the *Aspergillus glaucus* group. Identification of physicion and three related anthraquinones as main toxic constituents of *Aspergillus chevalieri, J. Agric. Food Chem.*, 27, 1342, 1979.
42. **Hurst, E. W. and Paget, G. E.,** Protoporphyrin, cirrhosis, and hepatoma in the livers of mice given griseofulvin, *Br. J. Dermatol.*, 75, 105, 1963.
43. **De Matteis, F., Donnelly, A. J., and Runge, W. J.,** The effects of prolonged administration of griseofulvin in mice with reference to sex differences, *Cancer Res.*, 26, 721, 1966.
44. **Scott, F. W, de Lahunta, A., Schultz, R. D., Bistner, S. I., and Riis, R. C.,** Teratogenesis in cats associated with griseofulvin therapy, *Teratology*, 11, 79, 1975.
45. **Epstein, S. S., Andrea, J., Joshi, S., and Mantel, N.,** Hepatocarcinogenicity of griseofulvin following parenteral administration in infant mice, *Cancer Res.*, 27, 1900, 1967.
46. **Leonard, A., Poncelet, F., Grutman, G., Carbonelle, E., and Fabry, L.,** Mutagenicity tests with griseofulvin, *Mutat. Res.*, 68, 225, 1979.
47. **Stark, A. A., Kobbe, B., Matsuo, K., Büchi, G., Wogan, G. N., and Demain, A. L.,** Mollicellins: mutagenic and antibacterial mycotoxins, *Appl. Environ. Microbiol.*, 36, 412, 1978.
48. **Dickens, F. and Jones, H. E. H.,** Carcinogenic activity of a series of reactive lactones and related substances, *Br. J. Cancer*, 15, 85, 1961.
49. **Ciegler, A., Beckwith, A. C., and Jackson, L. K.,** Teratogenicity of patulin and patulin adducts formed with cysteine, *Appl. Environ. Microbiol.*, 31, 664, 1976.
50. **Becci, P. J., Cox, G. E., Dailey, R. E., and Parent, R. A.,** Long term carcinogenicity and toxicity studies of patulin in the rat, *J. Appl. Toxicol.*, 1, 256, 1981.
51. **Hayes, A. W. and Hood, R. D.,** Effects of prenatal administration of penicillic acid and penitrem A to mice, *Toxicon*, 16, 92, 1978.
52. **Wogan, G. N., Edwards, G. S., and Newberne, P. M.,** Acute and chronic toxicity of rubratoxin B, *Toxicol. Appl. Pharmacol.*, 19, 712, 1971.
53. **Hood, R. D., Innes, J. E., and Hayes, A. W.,** Effects of rubratoxin B on prenatal development in mice, *Bull. Environ. Contam. Toxicol.*, 10, 200, 1973.
54. **Koshakji, R. P., Wilson, B. J., and Harbison, R. D.,** Effect of rubratoxin B on prenatal growth and development in mice, *Res. Commun. Chem. Pathol. Pharmacol.*, 5, 584, 1973.
55. **Evans, M. A. and Harbison, R. D.,** Prenatal toxicity of rubratoxin B and its hydrogenated analog, *Toxicol. Appl. Pharmacol.*, 39, 13, 1977.
56. **Ruddick, J. A., Scott, P. M., and Harwig, J.,** Teratological evaluation of zearalenone administered orally to the rat, *Bull. Environ. Contam. Toxicol.*, 15, 678, 1976.
57. **Becci, P. J., Cox, G. E., Hess, F. G., Knickerbocker, M., Taylor, J. M., and Parent, R. A.,** Combined two generation reproduction-teratology study of zearalenone in the rat, *J. Appl. Toxicol.*, in press.
58. **Becci, P. J., Cox, G. E., Taylor, J. M., and Parent, R. A.,** Long term carcinogenicity and toxicity studies of zearalenone in the rat, *J. Appl. Toxicol.*, in press.
59. FAO/WHO/UNEP, Global perspective on mycotoxins, Doc. W/G5491, Agenda Item 4, Conf. on Mycotoxins, Nairobi, Kenya, 1977.

60. **Rodricks, J. V. and Stoloff, L.**, Aflatoxin residues from contaminated feed in edible tissues of food-producing animals, in *Mycotoxins in Human and Animal Health,* Rodricks, J. V., Hesseltine, C. W., and Mehlman, M. A., Eds., Pathotox Publishers, Park Forest South, Ill., 1977, 67.
61. **Trabosh, H. M.**, Personal communication, 1981.
62. **Anon.**, Economics, Statistics, and Cooperatives Service: Food Consumption, Prices, Expenditures, Suppl. for 1976, Agric. Econ. Rep. No. 138, U.S. Department of Agriculture, Washington, D.C., 1978.
63. **Anon.**, Agricultural Research Service Household Food Consump. Surv., 1965-6. Rep. No. 12: Food consumption of households in the United States — season and year 1965-6, U.S. Department of Agriculture, Washington, D.C., 1972.
64. **Parker, W. A. and Melnick, D.**, Absence of aflatoxin from refined vegetable oils, *J. Am. Oil Chem. Soc.,* 43, 635, 1966.
65. **Ling, K.-H., Tung, C.-M., Sheh, I.-F., Wang, J.-J., and Tung, T.-C.**, Aflatoxin B_1 in unrefined peanut oil and peanut products in Taiwan, *J. Formosan Med. Assoc.,* 67, 309, 1968.
66. **Peers, F. G. and Linsell, C. A.**, Aflatoxin contamination and its heat stability in Indian cooking oils, *Trop. Sci.,* 17, 229, 1975.
67. **Stoloff, L.**, Incidence, distribution, and disposition of products containing aflatoxins, *Proc. Am. Phytopathol. Soc.,* 3, 156, 1976.
68. **Olson, M. C. and Stoloff, L.** Analysis of Aflatoxin Problem in Corn in Southeast U.S., unpublished data, Bureau of Foods, U.S. Food and Drug Administration, Washington, D.C., 1977.
69. **Stoloff, L. and Wood, G. E.**, Evaluation of the Bureau of Foods Compliance Programs, Aflatoxins in Foods, FY 75 to FY 81, unpublished data, U.S. Food and Drug Administration, Washington, D.C., 1981.
70. **Stoloff, L. and Wood, G. E.**, Compilation of Aflatoxin Analyses from State Surveillance, Service, and Compliance Activity, Bureau of Foods, U.S. Food and Drug Administration, Washington, D.C., unpublished data, 1981.
71. **Blanton, L.**, Personal communication, 1981.
72. **Wilson, D.**, Personal communication, 1981.
73. **Shotwell, O. L., Kwolek, W. F., and Hesseltine, C. W.**, Aflatoxin in freshly harvested 1979 Georgia dent corn and formation after harvest, *J. Am. Oil Chem. Soc.,* 58, 980A, 1981.
74. **Stoloff, L. and Lillehoj, E. B.**, Effect of genotype (open pollinated vs. hybrid) and environment on preharvest aflatoxin contamination of maize grown in the southeastern United States, *J. Am. Oil Chem. Soc.,* 58, 976A, 1981.
75. **Stoloff, L. and Truckess, M. W.**, Effect of boiling, frying, and baking on recovery of aflatoxin from naturally contaminated corn grits or corn meal, *J. Assoc. Off. Anal. Chem.,* 64, 678, 1981.
76. **Russell, T. E.**, Personal communication, 1979.
77. **Stoloff, L.**, Aflatoxin M in perspective, *J. Food Prot.,* 43, 226, 1980.
78. **Anon.**, Survey of mycotoxins in the United Kingdom, Food Surveillance Paper No. 4, Ministry of Agriculture, Fisheries and Food, Her Majesty's Stationery Office, London, 1980.
79. **Anon.**, Aflatoxin Contamination of Milk: establishment of Action Level, Fed. Regist., 42, 61630, U.S. Food and Drug Administration, Washington, D.C., 1977.
80. **Stoloff, L., Wood, G., and Carter, L., Jr.**, Survey for aflatoxin M_1 in manufactured dairy products produced in the United States in 1979, *J. Dairy Sci.,* 64, 2426, 1981.
81. **Boutrif, E., Jemmali, M., Pohland, A. E., and Campbell, A. D.**, Aflatoxin in Tunisian Aleppo pine nuts, *J. Assoc. Off. Anal. Chem.,* 60, 747, 1977.
82. **Purchase, I. F. H. and Gonçalves, T.**, Preliminary results from food analyses in the Inhambane area, in *Mycotoxins in Human Health,* Purchase, I. F. H., Ed., Macmillan, London, 1971, 263.
83. **Thurm, V., Koch, Ch. E., and Paul, P.**, Zur hygienischen Bedeutung von Sterigmatocystin in pflanzlichen Lebensmitteln, *Nahrung,* 23, 121, 1979.
84. **Scott, P. M., van Walbeek, W., Kennedy, B., and Anyeti, D.**, Mycotoxins (ochratoxin A, citrinin, and sterigmatocystin) and toxigenic fungi in grains and other agricultural products, *J. Agric. Food Chem.,* 20, 1103, 1972.
85. **Shreeve, B. J., Patterson, D. S. P., and Roberts, B. A.**, Investigation of suspected cases of mycotoxicosis in farm animals in Britain, *Vet. Rec.,* 97, 275, 1975.
86. **Purchase, I. F. H. and Pretorius, M. E.**, Sterigmatocystin in coffee beans, *J. Assoc. Off. Anal. Chem.,* 56, 225, 1973.
87. **Schroeder, H. W. and Hein, H., Jr.**, Natural occurrence of sterigmatocystin in in-shell pecans, *Can. J. Microbiol.,* 23, 639, 1977.
88. **Levi, C.**, Mycotoxins in coffee, *J. Assoc. Off. Anal. Chem.,* 63, 1282, 1980.
89. **Northolt, M. D., van Egmond, H. P., Soentoro, P., and Deijll, E.**, Fungal growth and the presence of sterigmatocystin in hard cheese, *J. Assoc. Off. Anal. Chem.,* 63, 115, 1980.

90. **van Egmond, H. P. and Northolt, M. D.**, Distribution and stability of sterigmatocystin in hard cheese, *Proc. 4th Meet. Mycotoxins Anim. Dis.*, Weybridge, Her Majesty's Stationery Office, R. V. G. 10, 1979.
91. **Hsu, I.-C., Smalley, E. B., Strong, F. M., and Ribelin, W. E.**, Identification of T-2 toxin in moldy corn associated with a lethal toxicosis in dairy cattle, *Appl. Microbiol.*, 24, 684, 1972.
92. **Ghosal, S., Biswas, K., Srivastava, R. S., Chakrabarti, D. K., and Basu Chaudhary, K. C.**, Toxic substances produced by *Fusarium*, V: occurrence of zearalenone, diacetoxyscirpenol, and T-2 toxin in moldy corn infected with *Fusarium moniliforme* Sheld., *J. Pharm. Sci.*, 67, 1768, 1978.
93. **Jemmali, M., Ueno, Y., Ishii, K., Frayssinet, C., and Etienne, M.**, Natural occurrence of trichothecenes (nivalenol, deoxynivalenol, T2) and zearalenone in corn, *Experientia*, 34, 1333, 1978.
94. **Funnell, H. S.**, Mycotoxins in animal feedstuffs in Ontario 1972 to 1977, *Can. J. Comp. Med.*, 43, 243, 1979.
95. **Mirocha, C. J., Schauerhamer, B., Christensen, C. M., and Kommedahl, T.**, Zearalenone, deoxynivalenol, and T-2 toxin associated with stalk rot of corn, *Appl. Environ. Microbiol.*, 38, 557, 1979.
96. **Puls, R. and Greenway, J. A.**, Fusariotoxicosis from barley in British Columbia. II. Analysis and toxicity of suspected barley, *Can. J. Comp. Med.*, 40, 16, 1976.
97. **Ghosal, S., Chakrabarti, D. K., and Basu Chaudhary, K. C.**, The occurrence of 12,13-epoxy-trichothecenes in seeds of safflower infected with *Fusarium oxysporum* f. sp. carthami, *Experientia*, 33, 574, 1977.
98. **Rukmini, C. and Bhat, R. V.**, Occurrence of T-2 toxin in *Fusarium*-infected sorghum from India, *J. Agric. Food Chem.*, 26, 647, 1978.
99. **Chi, M. S., Robison, T. S., Mirocha, C. J., and Shimoda, W.**, Transmission of radioactivity into eggs from laying hens *(Gallus domesticus)* administered tritium labeled T-2 toxin, *Poultry Sci.*, 57, 1234, 1978.
100. **Robison, T. S., Mirocha, C. J., Kurtz, H. J., Behrens, J. C., Chi, M. S., Weaver, G. A., and Nystrom, S. D.**, Transmission of T-2 toxin into bovine and porcine milk, *J. Dairy Sci.*, 62, 637, 1979.
101. **Yoshizawa, T., Swanson, S. P., and Mirocha, C. J.**, T-2 metabolites in the excreta of broiler chickens administered ^{3}H-labeled T-2 toxin, *Appl. Environ. Microbiol.*, 39, 1172, 1980.
102. **Yoshizawa, T., Mirocha, C. J., Behrens, J. C., and Swanson, S. P.**, Metabolic fate of T-2 toxin in a lactating cow, *Food Cosmet. Toxicol.*, 19, 31, 1981.
103. **Collins, G. J.**, The Fate of T-2 Toxin in Corn Wet Milling, Ph.D. dissertation, Rutgers University, New Brunswick, N.J., 1979.
104. **Shotwell, O. L., Hesseltine, C. W., and Goulden, M. L.**, Ochratoxin A: occurrence as natural contaminant of a corn sample, *Appl. Microbiol.*, 17, 765, 1969.
105. **Shotwell, O. L., Hesseltine, C. W., Vandegraft, E. E., and Goulden, M. L.**, Survey of corn from different regions for aflatoxin, ochratoxin, and zearalenone, *Cereal Sci. Today*, 16, 266, 1971.
106. **Shotwell, O. L., Goulden, M. L., and Hesseltine, C. W.**, Survey of U.S. wheat for ochratoxin and aflatoxin, *J. Assoc. Off. Anal. Chem.*, 59, 122, 1976.
107. **Shotwell, O. L., Bennett, G. A., Goulden, M. L., Plattner, R. D., and Hesseltine, C. W.**, Survey for zearalenone, aflatoxin, and ochratoxin in U.S. grain sorghum from 1975 and 1976 crops, *J. Assoc. Off. Anal. Chem.*, 63, 922, 1980.
108. **Levi, C. P., Trenk, H. L., and Mohr, H. K.**, Study of the occurrence of ochratoxin A in green coffee beans, *J. Assoc. Off. Anal. Chem.*, 57, 866, 1974.
109. **Funnell, H. S.**, Mycotoxins in animal feedstuffs in Ontario, *Can. J. Comp. Med.*, 43, 243, 1979.
110. **Shreeve, B. J., Patterson, D. S. P., and Roberts, B. A.**, Investigation of suspected cases of mycotoxicosis in farm animals in Britain, *Vet. Rec.*, 97, 275, 1975.
111. **Galtier, P., Jemmali, M., and Larrieu, G.**, Enquete sur la présence éventuelle d'aflatoxine et d'ochratoxine A dans des mais recoltes en France en 1973 et 1974, *Ann. Nutr. Alim.*, 31, 381, 1977.
112. **Krogh, P., Hald, B., and Pedersen, E. J.**, Occurrence of ochratoxin A and citrinin in cereals associated with mycotoxic porcine nephropathy, *Acta Pathol. Microbiol. Scand.*, B81, 689, 1973.
113. **Krogh, P., Hald, B., England, P., Rutqvist, L., and Swahn, O.**, Contamination of Swedish cereals with ochratoxin A, *Acta Pathol. Microbiol., Scand.*, B82, 301, 1974.
114. **Josefsson, E.**, Ochratoxin in products of barley and oats, *Var Foeda*, 28, 206, 1976 (in Swedish).
115. **Hokby, E., Hult, K., Gatenbeck, S., and Rutqvist, L.**, Ochratoxin A and citrinin in 1976 crop of barley stored on farms in Sweden, *Acta Agric. Scand.*, 29, 174, 1979.
116. **Balzer, I., Bogdanić, C., and Mužic, S.**, Natural contamination of corn (*Zea mais*) with mycotoxins in Yugoslavia, *Ann. Nutr. Alim.*, 31, 425, 1977.
117. **Pavlovic, M., Plestina, R., and Krogh, P.**, Ochratoxin A contamination of foodstuffs in an area with Balkan (endemic) nephropathy, *Acta Pathol. Microbiol. Scand.*, B87, 243, 1979.
118. **Juszkiewicz, T. and Piskorska-Pliszczynska, J.**, Mycotoxins in grains for animal feeds, *Med. Weter.*, 32, 617, 1976 (in Polish).

119. **Juszkiewicz, T. and Piskorska-Pliszczynska, J.**, Occurrence of mycotoxins in mixed feeds and concentrates, *Med. Weter.*, 33, 193, 1977.
120. **Rao, E. R., Basappa, S. C., and Sreenivasa Murthy, V.**, Studies on the occurrence of ochratoxins in food grains, *J. Food Sci. Technol.*, 16, 113, 1979.
121. **Scott, P. M., van Walbeek, W., and Fennell, D. I.**, Occurrence of a mycotoxin, ochratoxin A, in wheat and isolation of ochratoxin A and citrinin-producing strains of *Penicillium viridicatum*, *Can. J. Plant Sci.*, 50, 583, 1970.
122. **Krogh, P.**, Ochratoxin A residues in tissues of slaughter pigs with nephropathy, *Nord. Vet. Med.*, 29, 402, 1977.
123. **Rutqvist, L., Bjorkland, N.-E., Hult, K., and Gatenbeck, S.**, Spontaneous occurrence of ochratoxin residues in kidneys of fattening pigs, *Zentralbl. Veterinsermed.*, 24, 402, 1977.
124. **Josefsson, E.**, Study of ochratoxin A in pig kidneys, *Var. Foeda*, 31, 415, 1979 (in Swedish).
125. **Frye, C. E. and Chu, F. S.**, Distribution of ochratoxin A in chicken tissues and eggs, *J. Food Safety*, 1, 147, 1977.
126. **Krogh, P., Hald, B., Gjertsen, P., and Myken, F.**, Fate of ochratoxin A and citrinin during malting and brewing experiments, *Appl. Microbiol.*, 28, 31, 1974.
127. **Saito, M., Enomoto, M., and Tatsuno, T.**, Yellowed rice toxins III. Citrinin, in *Microbial Toxins*, Vol. VI, Ciegler, A., Kadis, S., and Ajl, S. A., Eds., Academic Press, New York, 1971, 357.
128. **Subrahmanyam, P. and Rao, A. S.**, Occurrence of aflatoxins and citrinin in groundnut (*Arachis hypogaea* L.) at harvest in relation to pod condition and kernel moisture content, *Current Sci.*, 43, 707, 1974.
129. **Leistner, L., Mintzlaff, H.-J., and Maing, L.-Y.**, Mykotoxin-Bildung in Fleischwaren, in *Ber. Kolloquim v. Bioziden u. Umweltchemikalien in Fleisch (Bonn)*, 1974, 141.
130. **Bennett, G. A. and Shotwell, O. L.**, Zearalenone in cereal grains, *J. Am. Oil Chem. Soc.*, 56, 812, 1979.
131. **Mirocha, C. J., Schauerhamer, B., Christensen, C. M., Niku-Paavola, M. L., and Nummi, M.**, Incidence of zearalenone (*Fusarium* mycotoxin) in animal feed, *Appl. Environ. Microbiol.*, 38, 749, 1979.
132. **Mirocha, C. J., Pathre, S. V., and Robison, T. S.**, Comparative metabolism of zearalenone and transmission into bovine milk, *Food Cosmet. Toxicol.*, 19, 25, 1981.
133. **Jemmali, M., and Mazerand, C.**, Presence de zearalenone ou F_2 dans les noix du commerce, *Ann. Microbiol. (Inst. Pasteur)*, 131B, 319, 1980.
134. **Scott, P. M., Panalaks, T., Kanhere, S., and Miles, W. F.**, Determination of zearalenone in corn flakes and other corn-based foods by thin layer chromatography, high pressure liquid chromatography, and gas-liquid chromatography/high resolution mass spectrometry, *J. Assoc. Off. Anal. Chem.*, 61, 593, 1978.
135. **Ware, G. M. and Thorpe, C. W.**, Determination of zearalenone in corn by high pressure liquid chromatography and fluorescence detection, *J. Assoc. Off. Anal. Chem.*, 61, 1058, 1978.
136. **Bennett, G. A., Peplinski, A. J., Brekke, O. L., and Jackson, L. K.**, Zearalenone, distribution in dry-milled fractions of contaminated corn, *Cereal Chem.*, 53, 299, 1976.
137. **Bennett, G. A., Vandegraft, E. E., Shotwell, O. L., Watson, S. A., and Bocan, B. J.**, Zearalenone: distribution in wet milling fractions from contaminated corn, *Cereal Chem.*, 55, 455, 1978.
138. **Palyusik, M., Harrach, B., Mirocha, C. J., and Pathre, S. V.**, Transmission of zearalenone and zearalenol into porcine milk, *Acta Vet. Hung.*, 28, 217, 1980.
139. **Hagler, W. M., Danko, G., Horvath, L., Palyusik, M., and Mirocha, C. J.**, Transmission of zearalenone and its metabolite into ruminant milk, *Acta Vet. Acad. Sci. Hung.*, 28, 209, 1980.
140. **Dailey, R. E., Reese, R. E., and Brouwer, E. A.**, Metabolism of [^{14}C] zearalenone in laying hens, *J. Agric. Food Chem.*, 28, 286, 1980.
141. **Scott, P. M. and Kennedy, B. P. C.**, Improved method for the thin layer chromatographic determination of patulin in apple juice, *J. Assoc. Off. Anal. Chem.*, 56, 813, 1973.
142. **Harwig, J., Chen, Y.-K., Kennedy, B. P. C., and Scott, P. M.**, Occurrence of patulin-producing strains of *Penicillium expansum* in natural rots of apple in Canada, *Can. Inst. Food Sci. Technol. J.*, 6, 22, 1973.
143. **Scott, P. M. and Somers, E.**, Stability of patulin and penicillic acid in fruit juices and flour, *J. Agric. Food Chem.*, 16, 483, 1968.
144. **Stoloff, L.**, Patulin, a contaminant of apple juice, *N.Y. State Agric. Exp. St. Spec. Rep.*, 19, 51, 1975.
145. **Ware, G. M., Thorpe, C. W., and Pohland, A. E.**, Liquid chromatographic method for determination of patulin in apple juice, *J. Assoc. Off. Anal. Chem.*, 57, 1111, 1975.
146. **Brackett, R. E. and Marth, E. H.**, Patulin in apple juice from roadside stands in Wisconsin, *J. Food Prot.*, 42, 862, 1979.
147. **Scott, P. M., Miles, W. F., Toft, P., and Dube, J. G.**, Occurrence of patulin in apple juice, *J. Agric. Food Chem.*, 20, 450, 1972.

148. **Josefsson, E. and Andersson, A.**, Analysis of patulin in apple beverages sold in Sweden, *Var Foeda,* 28, 189, 1976 (in Swedish).
149. **Andersson, A. and Josefsson, E.**, Patulin in fruit, berries and juices. Methods for elimination of patulin, *Var Foeda,* 31, 365, 1979, (in Swedish).
150. **Tanner, H. and Zanier, C.**, Ueber eine neue Patulinbestimmung in fruchtsaften und Konzentraten, *Schweiz. Z. Obst. Weinbau,* 112, 656, 1976.
151. **Meyer, R. A.**, Zum Vorkommen des Mykotoxins Patulin in Obst und Obstprodukten, *Lebensmittelindustrie,* 25, 224, 1978.
152. **Thurm, V., Paul, P., and Koch, C. E.**, Zur hygienischen Bedeutung von Patulin in Lebënsmitteln 2. Mitt. Zum Vorkommen von Patulin in Obst und Gemüse, *Nahrung,* 23, 131, 1979.
153. **Fritz, W., Buthig, C., and Engst, R.**, Zur Analytik und lebensmittelhygienisch-toxikologischen Bedeutung von Patulin in Obst und Obstprodukten, *Nahrung,* 23, 159, 1979.
154. **Drilleau, J.-F. and Bohuon, G.**, La patuline dans les produits cidricoles, *CR Seances Acad. Agric. Fr.,* 59, 1031, 1973.
155. **Alperden, I., Mintzlaff, H.-J., Tauchmann, F., and Leistner, L.**, Untersuchungen über die Bildung des Mykotoxins Patulin in Rohwurst, *Fleischwirtschaft,* 53, 566, 1973.
156. **Hofmann, K., Mintzlaff, H.-J., Alperden, I., and Leistner, L.**, Untersuchung über die Inaktivierung des Mykotoxins Patulin durch Sulfhydrylgruppen, *Fleischwirtschaft,* 51, 1534, 1971.
157. **Stinson, E. E., Osman, S. F., Huhtanen, C. N., and Bills, D. D.**, Disappearance of patulin during alcoholic fermentation of apple juice, *Appl. Environ. Microbiol.,* 36, 620, 1978.
158. **Scott, P. M., Fuleki, T., and Harwig, J.**, Patulin content of juice and wine produced from moldy grapes, *J. Agric. Food Chem.,* 25, 434, 1977.
159. **Thorpe, C. W. and Johnson, R. L.**, Analysis of penicillic acid by gas-liquid chromatography, *J. Assoc. Off. Anal. Chem.,* 57, 861, 1974.
160. **Pohland, A. E. and Allen, R.**, Stability studies with patulin, *J. Assoc. Off. Anal. Chem.,* 53, 688, 1970.
161. **Ciegler, A., Mintzlaff, H.-J., Weisleder, D., and Leistner, L.**, Potential production and detoxification of penicillic acid in mold-fermented sausage (salami), *Appl. Microbiol.,* 24, 114, 1972.
162. **Ciegler, A., Mintzlaff, H.-J., Machnick, W., and Leistner, L.**, Untersuchungen über das Toxinbildungsvermögen von Rohwürsten isolierter Schimmelpilze der Gattung *Penicillium, Fleischwirtschaft,* 52, 1311, 1972.
163. **Lafont, P., Siriwardana, M. G., Combemale, I., and Lafont, J.**, Mycophenolic acid in marketed cheeses, *Food Cosmet. Toxicol.,* 17, 147, 1979.
164. **Seitz, L. M., Sauer, D. B., Mohr, H. E., Burroughs, R., and Paukstelis, J. V.**, Metabolites of *Alternaria* in grain sorghum, compounds which could be mistaken for zearalenone and aflatoxin, *J. Agric. Food Chem.,* 23, 1, 1975.
165. **Schroeder, H. W. and Cole, R. J.**, Natural occurrence of alternariols in discolored pecans, *J. Agric. Food Chem.,* 25, 204, 1977.
166. **Richard, J. L. and Arp, L. H.**, Natural occurrence of the mycotoxin penitrem A in moldy cream cheese, *Mycopathologia,* 67, 107, 1979.
167. **Scott, P. M. and Kanhere, S. R.**, Instability of PR toxin in blue cheese, *J. Assoc. Off. Anal. Chem.,* 62, 141, 1979.

Chapter 10

BROWNING REACTION SYSTEMS AS SOURCES OF MUTAGENS AND MODULATORS

W. D. Powrie, C. H. Wu, and H. F. Stich

TABLE OF CONTENTS

I. INTRODUCTION

Heat treatment of food has been in existence since the time when ancient man used hot coals for his cooking. Baking, roasting, broiling, frying, and boiling of food increases palatability through the creation of flavor compounds and the alteration in texture. Flavor compounds, including volatile aromatic substances and nonvolatile bitter material, are formed by degradation of sugars and amino acids. With some food products, brown pigments produced during heating are desirable visual attributes. The brown color of a product conveys flavor expectation. Textural characteristics, such as crispness, elasticity, and resistance to bite can be attained by heat denaturation and coagulation of protein and by gelatinization and retrogradation of starch granules.

The development of baked and fried flavors, together with brown pigment formation in food, is brought about by thermal degradation of carbohydrates and amino acids at sites where extensive evaporative water loss has occurred during heat treatment. Such heat treatment of foods includes exposure to hot oil (frying), heating by hot air (baking and roasting), and direct exposure to radiant energy sources such as a hot electrical element, gas flame, or glowing coals (broiling). Infrared radiation from red-hot elements and glowing coals is a major source of heat energy in the broiling process. The heating media temperatures involved in baking, roasting, deep-fat frying, and shallow-fat frying are presented in Table 1. The highest temperature used in such heat treatments is about 250°C. With metal pan and air heating, as well as infrared heating of food, a temperature gradient develops, since heat is transferred inwardly from the surface. Furthermore, mass transfer of water (as vapor and sometimes as drip) progresses during baking, broiling, and frying of food, and a moisture gradient is established with the highest moisture content in the center of the food product. Bengtsson et al.[1] and Godsalve et al.[2] have studied the unsteady heat and mass transfer in beef muscle during oven roasting. Temperature profiles, as well as fat and water losses in meat patties during double-sided frying at 140° and 180°C, have been presented by Dagerskog.[3] In that study, the temperature in the surface region of the meat patties rose above 100°C after a period of 10 sec (with 180°C frying temperature) and reached about 160°C in 5 min. On the other hand, the temperature at the centers of the patties increased to about 60°C in 5 min.

The rapid evaporation of water in the surface region of meat patties leads to the creation of a crust that has no free water but some residual oil that can readily absorb heat energy from the metal pan to bring the crust temperature above 100°C.[4] The combination of high temperature, absence of free water, and high concentration of reactants is responsible for the rapid browning of the crust through the Maillard reactions.[4] Maillard reactions do not occur normally in the interior of meat patties because of the relatively low temperature and high moisture content.[5] Dagerskog[6] found that the brown color intensity of the crust of fried meat patties was directly related to the pan temperature and frying time. He found that fat cover on the meat patties did not significantly influence the degree of crust color. The two-sided pan frier required less frying time than the single-sided frying pan to produce the same brown color intensity in the crust.

Pyrolysis is a term used to denote thermal degradation of constituents in food heated to temperatures much higher than 100°C in the absence of free water. During the roasting of coffee and cocoa beans, amino acids and sugars are decomposed at temperatures up to about 220°C. The crust of baked bread and fried steak become brown because of the pyrolysis process. Presumably, nonenzymic reaction pathways are involved in the pyrolysis treatment of food systems.

Table 1
HEAT TREATMENT OF FOOD

Type of heat treatment	Location of browning	Temperature of heat transfer medium (°C)	Final temperature of heated food (°C)
Baking			
Bread	Surface	230—240	91 (Interior)
Cakes	Surface	175—180	
Deep-Fat Frying			
Potatoes	Surface	160—190	
Shallow-Fat Frying			
Meat	Surface	140—200	
Roasting			
Coffee	Entire bean	240—250	220 (Interior)
Beef	Surface	163—177	
Cocoa	Entire bean	105—150	

II. NONENZYMIC BROWNING REACTIONS

The browning of food systems can be classified as nonenzymic and enzymic browning. Under the nonenzymic browning category, reactions responsible for browning are commonly designated as Maillard or carbonyl-amine reactions, caramelization, and ascorbic acid oxidation.

Maillard reactions in aqueous systems involve the degradation of carbonyl compounds, particularly reducing sugars, in the presence of amines such as amino acids in the free peptide and polypeptide forms. With respect to caramelization in aqueous systems, reducing sugars are decomposed in the absence of amines. However, it should be noted that caramelization reactions can occur in amino acid-containing sugar systems. Several reviews on the nonenzymic browning reactions have recently been published.[7-10]

Most of the initial chemical steps of the Maillard reaction sequence are the same as those for caramelization. However, the rates of the reactions (enolization and dehydration) of reducing sugars are increased considerably when amino acids are added. As the pH of the aqueous medium rises, the reaction rates in Maillard and caramelization events increase. The pH effect can be attributed, in part, to the more rapid conversion of cyclic hemiacetal forms of reducing sugars to acyclic (open-chain) carbonyls that are the initial reactive isomers in the Maillard and caramelization pathways.[11] Furthermore, enolization proceeds more rapidly at pH values above 4, with bases being more effective catalysts than acids. The amount of carbonyl form in a sugar solution is relatively small. For example, only 0.02% aldehydo form is present in a solution of D-glucose at about 25°C, and in a D-xylose solution about 0.17% of the aldehydo form exists. The highest percentage for an aldehydo form has been reported to be in a D-ribose solution. Pentoses bring about browning at a greater rate than hexoses because of the differences in aldehydo form concentrations and ring-opening rates.[12]

The pathway of caramelization involves enolization of open-chain reducing sugars, dehydration, fragmentation, and polymerization of reaction products to form brown pigments. In neutral and weakly alkaline media, enolization of sugars is rapid and dehydration is slow, whereas in acidic media, enolization is slow and dehydration is rapid. The speed of brown pigment formation is governed by the types of acidic and basic catalysts, the bases being the most effective for browning. Acidic products are formed during fragmentation and thus the pH drops during caramelization. High temperatures are required to attain within 1 hr, significant amounts of brown pigments in aqueous sugar systems or dry sugars (pyrolysis).

Regardless of pH, the first chemical event in caramelization involves the formation of 1,2-enediol by abstraction of the hydrogen atom on C-2 of a sugar molecule by a base to produce a hybrid monoanion. The monoanion may produce either a *cis-* or a *trans-*enediol depending on the substrate sugar. Incidentally, the enediols are intermediates in the Lobry de Bruyn-Alberda von Ekenstein transformation of sugars to a variety of isomers.[13] For example, D-glucose added to water can be converted, in part, to D-mannose and D-fructose, the extent being dependent on pH and temperature of solution. This transformation is important in nonenzymic browning reactions from the standpoint of the amounts of acyclic (open-ring) sugars.

The next step in the caramelization pathway stems from the dehydration of the 1,2-enediols by:

1. Elimination of H and OH groups from the α- and β-carbon atoms adjacent to each carbonyl group to form a conjugated, α,β-unsaturated carbonyl compound called 3-deoxyosone (Figure 1)
2. Elimination of H and OH groups on C-3 and C-4 to form 3-deoxyosone-1,4-ene (Figure 2)
3. Cyclization with a loss of water to form a furfural compound

In the three dehydration events, three molecules of water are lost. With 2-ketose such as D-fructose, the 2,3-enediol can lead to 2-hydroxyacetyl furan and isomaltol under acidic conditions. The methyl-α-dicarbonyl intermediate is not only the precursor of isomaltol but it can, under alkaline conditions, be involved in the formation of furans, acids, alcohols, and carbonyl compounds.

Acids formed in melted sugars or added as salts catalyze dehydration to furans and glycosans with little fragmentation, whereas alkaline salts catalyze fragmentation reactions at the expense of furan and glycosan formation.

Under alkaline conditions, volatile fragmentation compounds, such as acetol, acetoin, diacetyl, and γ-butyrolactone, are formed from sugars.

In the first step of the Maillard reactions, an amino acid condenses with a reducing sugar to form a Schiff's base (imine) that undergoes cyclization to N-substituted glycosylamine, as shown in Figure 3. Subsequently, the glycosylamine is coverted to 1,2-eneaminol that is transformed to N-substituted 1-amino-1-deoxy-2-ketose (Figure 4).[14,15] The transformation of the glycosylamine to the ketoseamine is known as the Amadori rearrangement. Amadori ketose compounds have been identified in several foods.[16]

A major pathway of the Maillard reaction leads from the 1,2-eneaminol to 3-deoxyhexosone intermediate. 5-Hydroxymethyl furfural and 5-methyl furfural are formed by the dehydration of 3-deoxyhexosone. Another pathway of the Maillard reaction is the conversion of the 2,3-enediol to methyl-α-dicarbonyl intermediate (1-deoxyhexosone), which undergoes transformation to aldehydes, ketones, furans, pyrroles, and thiophenes, Several cyclic flavor compounds, such as maltol and isomaltol, are formed from 2,3-enediol.

A third important pathway of the Maillard reaction is the Strecker degradation of amino acids (Figure 5). When α-dicarbonyl compounds are heated with amino acids, the amino group reacts with a keto group to form a nitrogen-containing carboxy aldehyde. The reaction products include aldehydes from amino acids, sugar fragments, carbon dioxide, and pyrazines from the transamination reaction. At elevated temperature conditions, 2,5-dimethylpyrazine is formed by the reaction of amino acids with pyrrolaldehyde, and tetramethyl pyrazine is formed when amines react with diacetyl.[14]

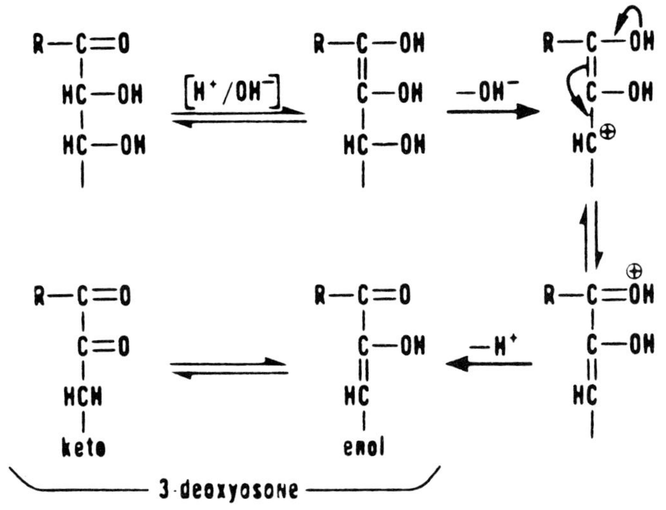

FIGURE 1. Formation of 3-deoxyosone. (From Hodge, J. E. and Osman, E. M., *Principles of Food Science*, Fennema, O. R., Ed., Marcel Dekker, New York, 1976, 76. With permission.)

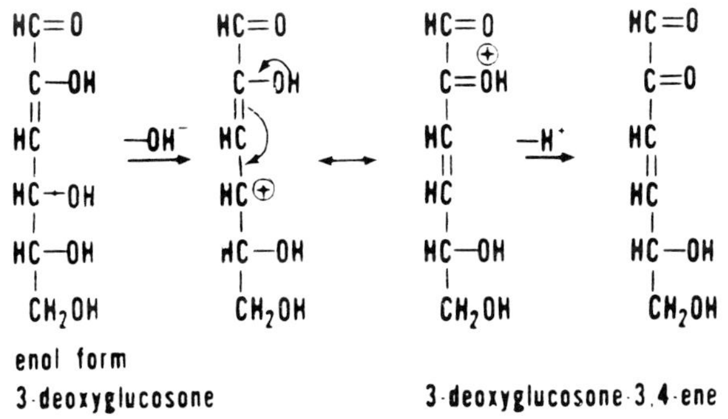

FIGURE 2. Sugar dehydration leading to 3-deoxyglucosone-3,4-ene. (From Hodge, J. E. and Osman, E. M., *Principles of Food Science*, Fennema, O. R., Ed., Marcel Dekker, New York, 1976, 77. With permission.)

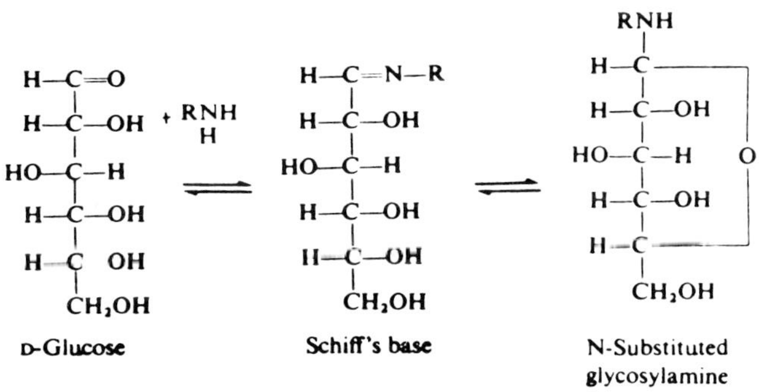

FIGURE 3. Formation of N-substituted glucosylamine.

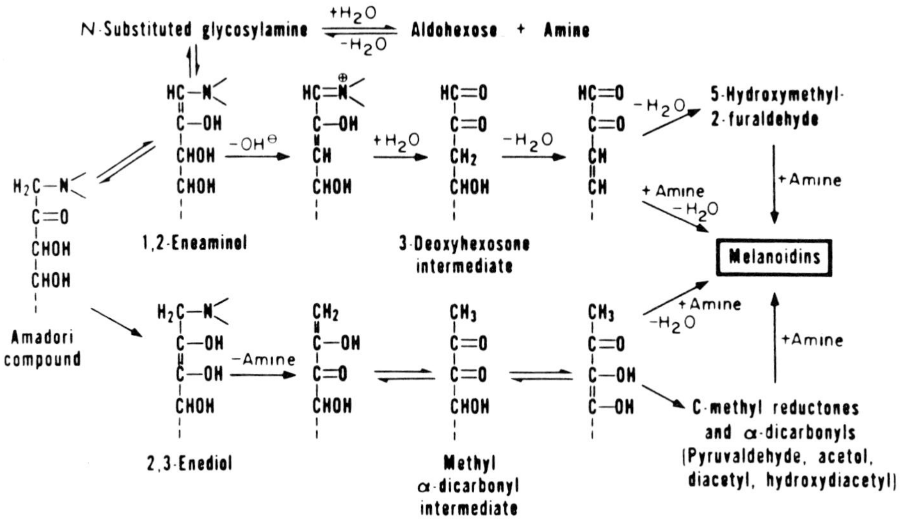

FIGURE 4.. Sugar-amine (Maillard) reactions. (From Hodge, J. E. and Osman, E. M., *Principles of Food Science,* Fennema, O. R., Ed., Marcel Dekker, New York, 1976, 84. With permission.)

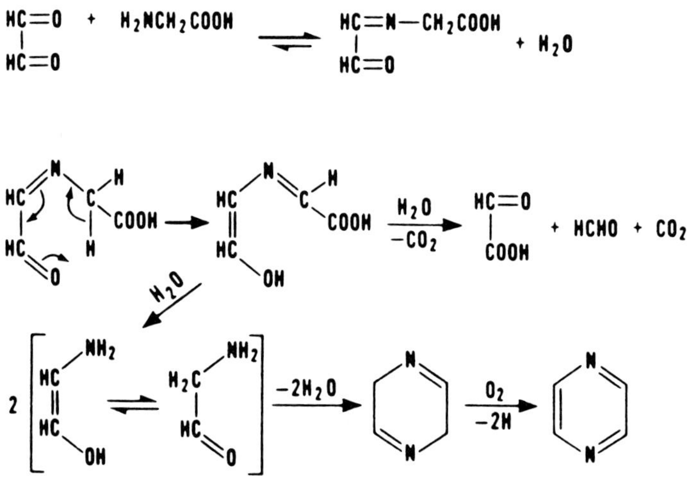

FIGURE 5. Strecker degradation of amino acids. (From Hodge, J. E. and Osman, E.M., *Principles of Food Science,* Fennema, O. R., Ed., Marcel Dekker, New York, 1976, 86. With permission.)

III. MUTAGENIC ACTIVITY OF HEATED FOODS

Mutagens and carcinogens have been detected in heated foods. These compounds may be present as a result of smoke condensation on the product during heat treatment or of thermal degradation of food constituents. Lijinsky and Shubik[17] were the first to report the presence of benzo[*a*]pyrene (BaP) and related PAH in charcoal-broiled meat. The crust of beef steak was found to contain between 8 and 50 μg of BaP/kg, that was present in the deposited smoke condensate. According to Lintas et al.,[18] the BaP content in cooked hamburger (heated in a pan on an electric grill) was only 0.05

μg/kg. Smoked foods possess a variety of PAH.[19] In wood-smoked fish, the level of BaP was estimated to be about 37 μg/kg.[20] Five PAH in smoked fish were found to have mutagenic activity when tested with several strains of *Salmonella typhimurium* after treatment of the compounds with S9 mix.[21] Nagao et al.[22] estimated that the mutagenicity of the charred surface of 190 g beef steak was equivalent to that of 860 μg BaP. These results indicate that the major mutagen in the charred surface of muscle is not BaP.

Nagao et al.[22] showed that the smoke condensates from some pyrolyzed proteins possessed strong mutagenic activities (TA98 strain with metabolic activation), but the smoke condensates from pyrolyzed DNA and RNA displayed weak mutagenicity, and that from starch and vegetable oil did not contain mutagens. Tars from many pyrolyzed amino acids were found to be highly mutagenic. Since free amino acids are present in muscle foods, decomposition of these acids could occur during broiling to the extent of producing mutagens. Certainly, it would be expected that at temperatures of 250°C and higher, pyrolysis of amino acids would occur, together with Maillard reactions in carbonyl-amino acid systems. Carbonyl compounds such as ribose and ribose phosphate are commonly present in meat.

Upon the pyrolysis of proteins and amino acids, compounds which become potent mutagens after metabolic activation with S9 mix are formed. *S. typhimurium* TA98, a frameshift mutant, was more sensitive to these pyrolysates than TA100, a base-pair substitution mutant. Sugimura[23] reported that the tryptophan pyrolysate contained two potent mutagens, 3-amino-1,4-dimethyl-5*H*-pyrido-[4,3-*b*] indole (Trp-P-1) and 3-amino-1-methyl-5*H*-pyridol[4,3-*b*] indole (Trp-P-2). When glutamic acid was pyrolyzed, 2-amino-6-methyl-dipyrido[1,2-*a*:3',2'-*d*]-imidazole (Glu-P-1) and 2-amino-dipyrido[1,2-*a*:3',2'-*d*]amidazole (Glu-P-2) were formed as mutagens.[24] Wakabayashi et al.[25] isolated a mutagen from lysine pyrolysate and identified it as 3,4-cyclopentenopyrido[3,2-*a*]carbazole (Lys-P-1). According to Sugimura,[23] Trp-P-1 and Trp-P-2 have much higher mutagenic activities than that of aflatoxin B_1. Matsukura et al.[26] found that Trp-P-1 and Trp-P-2, when administered orally to mice at a concentration of 200 ppm in the diet, induced hepatic tumors.

Yamaizumi et al.[27] found that dried sardines broiled over a high flame of a gas burner for about 5 min possessed 13.3 ng/g Trp-P-1 and 13.1 ng/g Trp-P-2. Kasai et al.[28] identified two potent mutagens in a methanol extract of high-flame broiled sardines. The compounds were 2-amino-3-methyl-imidazo[4,5-*f*] quinoline (IQ) and 2-amino-3,4-dimethylimidazo[4,5-*f*] quinoline (MeIQ). These compounds required metabolic activation by S9 mix for mutagenic activity with *S. typhimurium* TA98 and TA100.

During the pyrolysis of soybean globulin, moderately strong mutagens called amino-α-carbolines are formed. The pyrolytic products of casein, albumin, and gluten included two comutagens called harman and norharman.[29] These compounds, called β-carbolines, have been found in tobacco tars. Recently, Matsumoto et al.[30] found two amino-α-carbolines in grilled (gas-flame broiled) beef, chicken, and mushrooms. The two compounds were identified as 2-amino-9*H*-pyrido[2,3 *b*] indole (AC) and 2-amino-3-methyl-9*H*-pyrido[2,3 *b*] indole (MeAC). The AC and MeAC content in 1 g of grilled beef was 650.8 ng and 63.5 ng, respectively. In the same study, the smoke condensate of a blended cigarette was reported to contain 79.7 ng of AC and 6.2 ng of MeAC.

In North America, muscle foods are generally cooked at temperatures below 250°C, temperatures much lower than those used in the Japanese studies involving gas-flame and charcoal broiling. Commoner et al.[31] fried lean ground beef in a hamburger cooking appliance at 200°C for 1.5, 3.0, and 5.5 min, and after homogenization, the hamburger was extracted with methylene chloride. After taking up the methylene chloride residues in DMSO, the samples were treated against *S. typhimurium* TA1538 (sensitive to frameshift mutagens). As the cooking time progressed, the mutagenicity values increased, but only when an S9 mixture was used. These investigators found that the mutagens in cooked meat and beef extract (concentrated beef broth) are chromatographically distinguishable from both BaP and mutagens formed by amino acid and protein pyrolysis.

Spingarn and Weisburger[32] reported that mutagens, detectable by *S. typhimurium* TA98 (sensitive to frameshift mutagens) after activation with S9 mix, were formed during broiling, frying, and boiling of ground beef, but no mutagens were present in the raw products. After a frying time of 5 min, the amount of mutagens produced in the crust of beef patties increased rapidly. With the broiling of patties, the induction time was about 35 min and thereafter the mutagenic activity in the beef crust rose very rapidly. During the induction period in meat frying and broiling, the moisture content is reduced in the surface area, the reactants in the surface region are concentrated, and eventually the surface temperature rises above 100°C. Spingarn and Weisburger[32] noted that the surface temperature of a well-done broiled hamburger reached 130°C. At such a temperature, the Maillard reactions would be progressing at a rapid rate.[5]

It would be expected that the frying pan temperature would have a profound influence on the rate of mutagen formation. Pariza et al.[33] reported that, at a pan surface temperature of 143°C, mutagenic activity in fried hamburgers remained at a low level between frying times of 4 to 20 min. On the other hand, with a pan surface temperature of 191°C, and after an induction period of 4 min, the mutagenic activity of fried hamburger increased with time up to 10 min of frying time. Similar results were obtained with a temperature of 210°C.

When Nader et al.[34] cooked slices (1.5 cm thick) of beef in a microwave oven for 12 min (3 times the normal cooking time), no mutagenic activity was detected in the dichloromethane extract of alkaline dispersion of meat (2 mm deep surface samples) when *S. typhimurium* TA98 and TA100 were used, with and without S9 mix. The authors suggest that lack of mutagen formation with microwave treatment of beef muscle may be attributed to uniform generation of heat energy at relatively low temperatures, even though the 12 min microwave samples were markedly brown and hard.

The fat content in a ground beef patty has an influence on the amount of mutagens formed during frying. Spingarn et al.[35] have stated that with an increase in added fat content of ground beef to 10%, the mutagenicity of fried patties rose to a maximum; thereafter it dropped with higher fat levels. Fat in the frying pan can act as an effective heat transfer medium for ground meat to enhance moisture loss in the patty surface and to expedite thermal degradation of reactants.

According to Rappaport et al.,[36] the amount of mutagenic activity lost by volatilization during the cooking of lean ground beef was 83 to 99%. In contrast, Felton et al.[37] recovered more than 20 times more mutagens in cooked ground beef than in the pan fat and vapor.

It is known that the mutagenic compounds are extractable from a basic aqueous meat dispersion with an organic solvent such as methylene chloride.[32,33,37] Felton et al.[37] found no mutagenic activity in neutral and acidic methylene chloride extracts when assessed with five strains of *S. typhimurium*. The basic fraction was mutagenic only to *Salmonella* strains TA1537, TA1538 and TA98, all sensitive to frameshift mutagens. Strains sensitive to base-pair substitution mutagens showed no response to this basic fraction. It should be noted that metabolic activation was a requirement for mutagenic response to these frameshift compounds. A heterocyclic amine, 2-amino-3-methylimidazo[4,5-*d*]quinoline, has been identified as a potent mutagen in fried beef.[38] Heterocyclic compounds have been identified as reaction products in cooked meat and are undoubtedly formed through the Maillard reactions.[5]

With the idea that free amino acids may decompose during the frying of meat to form mutagens, Ashoor et al.[39] added 20 common amino acids individually to ground beef prior to frying at 191°C for 4 min. Proline was the only amino acid that increased the mutagenic activity of the fried meat. The reducing sugars, being at a very low concentration level in meat, may have been the limiting factor in mutagen production if the Maillard reactions are involved. It has been reported that cysteine and ribose completely disappear with the cooking of muscle.[5]

Pariza et al.[33] observed a mutagenic inhibitor in both raw and fried beef hamburger. In this study, the authors tested the degree of mutagenicity of a mixture of the basic methylene chloride fraction from beef extract (DIFCO) and that from either raw or fried beef hamburger. This approach to investigating the mutagenic inhibitory activity of raw and fried beef hamburger was based on the fact that liver S9 from rats not treated with Aroclor 1254 can activate the beef extract mutagens but not the fried beef mutagens. Thus any reduction in mutagenicity of the beef extract fraction by the addition of the beef muscle fraction would indicate the presence of a mutagen inhibitor. The Ames test using *S. typhimurium* TA1538 was employed for assessing mutagenicity.

Spingarn et al.[40] found that when foods with high starch contents were fried, toasted, and baked mutagens detected by *S. typhimurium* TA98 and TA100 after S9 activation were produced. These mutagens were detected in the basic fractions that were used in these mutagenesis assays. With an increase in the heating time (no oil added to the pan at 280°C) of white potatoes, the revertants per unit initial surface area rose. The uncooked potatoes had no detectable mutagenic activity. Spingarn et al.[40] reported that the mutagen formation in white bread increased with toasting time up to 6 min, whereas with brown bread the mutagenic activity continued to rise up to 12 min of toasting. A pancake fried at 230°C attained a high level of mutagenicity (about 8 times higher than that for a slice of toasted white bread) within 4 min. To account for the mutagen formation in these heated, starch-containing foods, Spingarn et al.[40] have postulated that amino acids and other amines are decomposed during heat treatment or participate in the Maillard browning reaction to form heteroaromatic compounds.

When the mutagenicity of heated, starch-containing food was compared by Spingarn et al.[40] to that of fried beef on the basis of a per square meter of surface area, the value for fried beef was about 12 times higher than that for the heated starchy foods. Moreover, the mutagens in the fried meat are more strain-specific for TA98 over TA100 than is the case for heated, starch-containing foods.

IV. MUTAGENIC ACTIVITY OF NONENZYMIC BROWNING REACTION PRODUCTS IN MODEL SYSTEMS

For the most part, mutagens formed in heated foods have not been identified. In addition, the reaction pathways for their creation have not been constructed. Presumably, nonenzymic browning reactions are involved in the chemical events towards mutagen development in thermally treated foods. The assessment of the mutagenic activity of heated sugar and sugar-amino acid model systems should be helpful in the elucidation of the reaction mechanisms involved in mutagen formation in heated complex food products that contain specific amino acids and sugars.

Spingarn and Garvie,[41] who refluxed solutions containing reducing sugars and ammonium hydroxide (NH_4OH) as the source of amine, found that all of these solutions possessed mutagenic activity when assayed with strain TA98 after activation with S9. Preliminary studies with refluxed rhamnose-NH_4OH solution revealed that *S. typhimurium* TA98 was more responsive than TA100, and that both strains were more responsive when S9 was added to the solution. The sugars included arabinose, 2-deoxy-glucose, galactose, glucose, rhamnose and xylose. Studies with refluxed glucose-NH_4OH solution showed a parallelism between mutagen formation and pyrazine development over an 80 min time period. The investigators hypothesized that the reactions that form pyrazines also form the mutagens. However, it should be pointed out that pyrazines had no mutagenic activity in the mutagenesis assay. With the refluxed glucose-NH_4OH system, propyl gallate acted as an inhibitor for mutagen formation, whereas tryptophan enhanced the formation.

According to Mihara and Shibamoto,[42] several fractions isolated from a heated (100°C/2 hr) cysteamine-glucose model system possessed mutagenic activity assessed by the Ames *Salmonella* mutagenesis test using strains TA98 and TA100 with and without microsomal S9 activation. Cysteamine is a decarboxylated compound of cysteine and presumably occurs after the formation of Schiff's base during the Maillard reaction in food products. In a methylene chloride extract of the heated cysteamine-NH_4OH system, thiazolidine and 2-methylthiazolidine were identified and found to possess mutagenic activity. The aqueous nonvolatile fraction possessed a fairly strong mutagen identified as 2-(1,2,3,4,5′-pentahydroxy)-*n*-pentylthiazolidine. Some unidentified compounds in the heated cysteamine-glucose system suppress mutagenic activity.

Shinohara et al.[43] presented a dose-response curve for the mutagenic activity of a lysine-glucose solution heated for 10 hr when the Ames test (TA100 strain) was used without S9 mix. The revertant number increased to about 750 per plate over a narrow dose range and the doses were lethal to the bacteria afterwards. It is of interest to note that the revertant number increased with a rise in brown color intensity of the reaction solution.

Powrie et al.[44] employed three short-term tests to assess the clastogenic and point mutagenic activities of Maillard reaction sugar-amino acid solutions at initial pH values of 7 and 10. The amino acids used were arginine, lysine, histidine, serine, glutamine, and glutamic acid, in combination with either glucose or fructose. All of the Maillard reaction solutions without S9 mix induced significant increases in chromosome aberrations in Chinese hamster ovary (CHO) cells. Mitotic recombination and mutation occurred in *Saccharomyces cerevisiae* strain D5 cells when exposed to all Maillard reaction solutions. When *Salmonella typhimurium* TA100 cells were treated with Maillard reaction solutions without S9 activation, induction of revertants to histidine prototrophy occurred. Mutagen concentration in the browning reaction solution was dependent on initial pH.

Caramel, the brown food colorant, is prepared by heating (120 to 160°C) a concentrated sugar solution containing ammonia or ammonium salts with or without catalytic inorganic salts.[45] Melanoidins, the brown-colored pigments in caramel, are formed by the polymerization of some Maillard reaction products. Stich et al.[46] subjected CHO cells, *Saccharomyces cerevisiae* strain D7, and *Salmonella typhimurium* TA98 and TA100, with and without S9 mix, to a commercial caramel colorant. Mutagenic activity was detected with both tester strains when caramel was assessed with the *S. typhimurium* mutagenicity test. In addition, the caramel colorant was a potent inducer of gene conversion in the D7 yeast strain. Of particular interest was the observation that caramel caused a high frequency of chromosome aberrations in CHO cells.

The decomposition of proteins in stored reducing sugar-protein systems occurs through the Maillard browning reaction, and thus mutagen formation would be expected. However, Pintauro et al.[47] found no mutagenic activity in lipid- and water-soluble fractions of an egg albumin-glucose mixture stored at 37°C for up to 40 days. The samples were assessed by the Ames *Salmonella*/microsome mutagenicity test.

Studies by Bjeldanes and Chew[48] showed that several 1,2-dicarbonyl compounds such as maltol and diacetyl, found in Maillard reaction systems, displayed weak mutagenic activity with the *S. typhimurium* TA100 mutagenicity assay.

Three short-term assays were used by Stich et al.[49] to examine pyrazine and four of its alkyl derivatives (2-methylpyrazine, 2-ethylpyrazine, 2,5-dimethylpyrazine and 2,6-dimethylpyrazine) for the presence of mutagenic activity. Exposure of *Salmonella* cultures to these compounds in an agar overlay did not result in the induction of revertants to histidine prototrophy. However, cultures of stationary phase *Saccharomyces cerevisiae* strain D5 showed an increase in aberrant colonies (but not mitotic recombinants) after exposure to each of the pyrazine compounds. Pyrazine and its derivatives induced a significant (7- to 57-fold) increase in the frequency of chromosome aberrations (breaks and exchanges) in CHO cells.

The caramelization reactions of sugars occur during the heat treatment of foods. According to Okamoto et al.,[50] carbohydrate pyrolysates produced by heating mono- and disaccharides and cellulose at 700°C enhanced the mutagenicities of 2-acetylaminofluorene (2-AAF), Trp-P, BaP and several other mutagenic agents on *Salmonella typhimurium* TA98 and TA100 strains in the presence of S9. Pyrolysis of carbohydrates at temperatures lower than 600°C resulted in progressively lower mutagenic enhancement. Sucrose pyrolysates had the highest enhancing effect. The numbers of his^+ revertants induced by 2-AAF were increased 10-fold in the presence of sucrose pyrolysate heated at 700°C. By itself, sucrose pyrolysate has a low mutagenic activity.

Furan and furan derivatives have been identified as compounded in heated sugars. Stich et al.[51] investigated the clastogenic activity of furans and some derivatives. Prior to chromosome aberration analysis, cultures of CHO cells were exposed for 3 hr to 2-methyl furan, 2,5-dimethyl furan, furfural, 5-methyl furfural, furfuryl alcohol, and 2-furyl methyl ketone. These furans have been found in numerous food products, including meat and poultry products, canned beef, milk products, various nuts and fruits, tea, coffee, and cocoa. For furan derivatives with no S9 treatment, the frequency of metaphase plates with chromosome aberrations increased with a rise in concentration. Furans without S9 treatment had no clastogenic activity. Furan derivatives respond differently with added S9 mix. The clastogenic activities of 5-methyl furfural, 2-furyl methyl ketone, furfuryl alcohol and furfural increased, while those of 2-methyl furan and 2,5-dimethyl furan decreased.

V. SUMMARY

Food systems contain hundreds of chemical compounds, some of which may be mutagenic, some antimutagenic, and others wih comutagenic activity. Recent studies have indicated that foods exposed to drying, frying, roasting, baking and broiling conditions possess net mutagenic activity as assessed by the *Salmonella*/microsome mutagenicity test and the chromosome aberration assay with CHO cells. The nonenzymic browning reactions are probably involved in the development of mutagens in heated foods.

Caramelization and Maillard reactions are two important pathways in the nonenzymic browning of food. Enolization, dehydration, fragmentation, and polymerization are reaction segments for the creation of volatile aromatic compounds, intermediate nonvolatile compounds, and brown pigments called melanoidins.

Heated sugar-amino acids have been shown to have mutagenic activity with short-term bioassays. Purified Maillard and caramelization reaction products such as reductones, dicarbonyls, pyrazines, and furan derivatives exhibit mutagenicity and clastogenicity.

REFERENCES

1. **Bengtsson, N. E., Jakobsson, B., and Dagerskog, M.,** Cooking of beef by oven roasting: a study of heat and mass transfer, *J. Food Sci.,* 41, 1047, 1976.
2. **Godsalve, E. W., Davis, E. A., Gordon, J., and Davis, H. T.,** Water loss rates and temperature profiles of dry cooked bovine muscle, *J. Food Sci.,* 42, 1038, 1977.
3. **Dagerskog, M.,** Pan frying of meat patties. I. A study of heat and mass transfer, *Lebensm. Wiss. + Technol.,* 12, 217, 1979.
4. **Skjoldebrand, C. and Olsson, H.,** Crust formation during frying of minced meat products, *Lebensm. Wiss. + Technol,* 13, 148, 1980.
5. **MacLeod, G. and Seyyedaia-Ardebili, M.,** Natural and simulated meat flavors, in *CRC Crit. Rev. Food Sci. Nutr.,* Vol. 14, Furia, T. E., Ed., CRC Press, Boca Raton, Fla., 1981, 309.
6. **Dagerskog, M.,** Pan frying of meat patties. II. Influence of processing conditions on heat transfer, crust color formation, cooking losses and sensory quality, *Lebensm. Wiss. + Technol.,* 12, 225, 1979.
7. **Feeney, R. E., Blankenhorn, G., and Dixon, H. B. F.,** Carbonyl-amine reaction in protein chemistry, in *Advances in Protein Chemistry,* Vol. 29, Alfinsen, C. B., Edsall, J. T., and Richards, F. M., Eds., Academic Press, New York, 1975, 135.
8. **Labuza, T. P., Warren, R. M. and Warmbiar, H. C.,** The physical aspects with respect to water and non-enzymic browning, in *Protein Crosslinking, Nutritional and Medical Consequences,* Friedman, M., Ed., Plenum Press, New York, 1977, 379.
9. **Labuza, T. P. and Saltmarch, M.,** The non-enzymatic browning reaction as affected by water in foods, in *Water Activity: Influences on Food Quality,* Rockland, L. B. and Stewart, G. F., Eds., Academic Press, New York, 1981, 605.
10. **Powrie, W. D. and Nakai, S.,** Processing effects on protein systems, in *Utilization of Protein Resources,* Stanley, D. W., Murray, E. D., and Lees, D. H., Eds., Food and Nutrition Press, Westport, Conn., 1981, 73.
11. **Burton, H. S. and McWeen D. J.,** Non-enzymatic browning: consideration of sugar stability, *Nature (London),* 197, 266, 1963.
12. **Spark, A. A.,** Role of amino acids in non-enzymic browning, *J. Sci. Food Agric.,* 20, 308, 1969.
13. **Schallenberger, R. S. and Birch, G. G.,** *Sugar Chemistry,* AVI Publishing, Westport, Conn., 1975, 89.
14. **Hodge, J. E. and Osman, E. M.,** Carbohydrates, in *Principles of Food Science, Part 1, Food Chemistry,* Fennema, O., Ed., Marcel Dekker, New York, 1976, 41.
15. **Hodge, J. E.,** Origin of flavor in foods — nonenzymatic browning reactions, in *Chemistry and Physiology of Flavors,* Schultz, H. W., Day, E. A., and Libbey, L. M., Eds., AVI Publishing, Westport, Conn., 1967, 465.
16. **Takeoka, G. R., Coughlin, J. R., and Russell, G. F.,** High pressure liquid chromatographic separation of Amadori compounds in model Maillard browning systems, in *Liquid Chromatographic Analysis of Food and Beverages,* Charalamkous, G., Ed., Academic Press, New York, 1979, 179
17. **Lijinsky, W. and Shubik, P.,** Benzo(a)pyrene and other polynuclear hydrocarbons in charcoal-broiled meat, *Science,* 145, 53, 1964.
18. **Lintas, C., De Mattheis, M. C., and Merli, F.,** Determination of benzo(a)-pyrene in smoked, cooked and roasted food products, *Food Cosmet. Toxicol.,* 17, 325, 1979.
19. **Howard, J. W., Tesque, R. T., White, R. H., and Fry, B. E.,** Extraction and estimation of polycyclic aromatic hydrocarbons in smoked foods. I. General method, *J. Am. Oil Chem. Soc.,* 49, 596, 1966.
20. **Masuda, Y. and Kuratsune, M.,** Polycyclic aromatic hydrocarbons in smoked fish "katsuobushi", *Gann,* 62, 27, 1971.
21. **Mossanda, K., Poncelet, F., Fouassin, A., and Mercier, M.,** Detection of mutagenic polycyclic aromatic hydrocarbons in African smoked fish, *Food Cosmet. Toxicol.,* 17, 141, 1979.
22. **Nagao, M., Yahagi, T., Kawachi, T., Seino, Y., Honda, M. Matsukura, N., Sugimura, T., Wakabayashi, K., Tsuji, K., and Kosuge, T.,** Mutagens in foods, and especially pyrolysis products of protein, in *Progress in Genetic Toxicology,* Scott, D., Bridges, B. A., and Sobels, F. H., Eds., Elsevier/North-Holland, Amsterdam, 1977, 259.
23. **Sugimura, T.,** Naturally occurring genotoxic carcinogens in *Naturally Occurring Carcinogens - Mutagens and Modulators of Carcinogenesis,* Miller, E. C., Miller, J. A., Hirono, I., Sugimura, T., and Takayama, S., Eds., University Park Press, Baltimore, 1979, 241.
24. **Yamamoto, T., Tsuji, K., Kosu, T., Okamoto, T., Shudo, K., Takeda, K., Iitaka, Y., Yamagushi, K., Seino, Y., Yahagi, T., Nagao, M., and Sugimura, T.,** Isolation and structure determination of mutagenic substances in L-glutamic acid pyrolysate, *Proc. Jpn. Acad.,* 54B, 248, 1978.

25. **Wakabayashi, K., Tsuji, K., Kosuge, T., Takeda, K., Yamaguchi, K., Shudo, K., Iitaka, Y., Okamoto, T., Yahagi, T., Nagao, M., and Sugimura, T.,** Isolation and structure determination of a mutagenic substance in L-lysine pyrolysate, *Proc. Jpn. Acad.,* 54B, 569, 1978.
26. **Matsukura, N., Kawashi, T., Morino, K., Ohgaki, H.,and Sugimura, T.,** Cacinogenicity in mice of mutagenic compounds from a tryptophan pyrolysate, *Science,* 213, 346, 1981.
27. **Yamaizumi, Z., Shiomi, T., Kasai, H., Nishimura, S., Takahashi, Y., Nagao, M., and Sugimura, T.,** Detection of potent mutagens, Trp-P-1 and Trp-P-2, in broiled fish, *Cancer Lett.,* 9, 75, 1980.
28. **Kasai, N., Yamaizumi, Z., Wakabayashi, K., Nagao, M., Sugimura, T., Nishimura, S., Yokoyama, S., Miyazawa, T., Spingarn, N. E., and Weisburger, J. H.,** Isolation, structure determination and synthesis of new mutagens in broiled fish, in *23rd Symp. Chemistry Natural Products,* Japanese Chemical Society, Japan Scientific Society Press, Tokyo, 1980, 679.
29. **Nishigata, H., Yoshida, D., and Matsumoto, T.,** Determination of the yield of norharman and harman in the pyrolytic products of protein, *Agric. Biol. Chem.,* 44, 209, 1980.
30. **Matsumoto, T., Yoshida, D., and Tomita, H.,** Determination of mutagens, amino-α-carbolines, in grilled foods and cigarette smoke condensate, *Cancer Lett.,* 12, 105, 1981.
31. **Commoner, B., Vithayathil, A. J., Dolara, P., Nair, S., Madyastha, P., and Cuca, G. C.,** Formation of mutagens in beef and beef extract during cooking, *Science,* 201, 913, 1978.
32. **Spingarn, N. E. and Weisburger, J. H.,** Formation of mutagens in cooked foods. I. Beef, *Cancer Lett.,* 7, 259, 1979.
33. **Pariza, M. W., Ashoor, S. H., Chu, F. S., and Lund, D.B.,** Effect of temperature and time on mutagen formation in pan fried hamburger, *Cancer Lett.,* 7, 63, 1979.
34. **Nader, C. J., Spencer, L. K., and Weller, R. A.,** Mutagen production during pan-broiling compared with microwave irradiation of beef, *Cancer Lett.,* 13, 147, 1981.
35. **Spingarn, N. E., Garvie-Gould, C., Vuolo, L. L., and Weisburger, J. H.,** Formation of mutagens in cooked foods. IV. Effect of fat content in fried beef patties, *Cancer Lett.,* 12, 93, 1981.
36. **Rappaport, S. M., McCartney, M. C., and Wei,. T.,** Volatilization of mutagens from beef during cooking, *Cancer Lett.,* 8, 139, 1979.
37. **Felton, J. S., Healy, S., Stuermer, D., Berry, C., Timourian, H., Hatch, F. T., Mors, M., and Bjeldanes, L. F.,** Mutagens from the cooking of food. I. Improved extraction and characterization of mutagenic fractions from cooked ground beef, *Mutat. Res.,* 88, 33, 1981.
38. **Spingarn, N. E., Kasai, H., Vuolo, L. L., Nishimura, S., Yamaizumi, Z., Sugimura, T., Matsushima, T., and Weisburger, J. H.,** Formation of mutagens in cooked foods. III. Isolation of a potent mutagen from beef, *Cancer Lett.,* 9, 177, 1980.
39. **Ashoor, S. H., Dietrich, R. A., Chu, F. S., and Pariza, M. W.,** Proline enhances mutagen formation in ground beef during frying, *Life Sci.,* 26, 1801, 1980.
40. **Spingarn, N. E., Slocum, L. A., and Weisburger, J. H.,** Formation of mutagens in cooked foods. II. Foods with high starch content, *Cancer Lett.,* 9, 7, 1980.
41. **Spingarn, N. E. and Garvie, C. T.,** Formation of mutagens in sugar-ammonia model systems, *J. Agric. Food Chem.,* 27, 1319, 1979.
42. **Mihara, S. and Shibamoto, T.,** Mutagenicity of products obtained from cysteamine-glucose browning model systems, *J. Agric. Food Chem.,* 28, 62, 1980.
43. **Shinohara K., Wu, R.-T., Jahan, N., Tanaka, M., Morinaga, N., Murakami H., and Omura, H.,** Mutagenicity of the browning mixtures of amino-carbonyl reactions on *Salmonella typhimurium* TA100, *Agric. Biol. Chem.,* 44, 671, 1980.
44. **Powrie, W. D., Wu, C. H., Rosin, M. P., and Stich, H. F.,** Clastogenic and mutagenic activities of Maillard reaction model systems, *J. Food Sci.,* 46, 1433, 1981.
45. **Greenshields, R. N.,** Caramel. II. Manufacture, composition and properties, *Process Biochem.,* 8, 17, 1973.
46. **Stich, H. F., Stich, W., Rosin, M. P., and Powrie, W. D.,** Clastogenic activity of caramel and caramelized sugars, *Mutat. Res.,* 91, 129, 1981.
47. **Pintauro, S. J., Page, G. V., Solberg, M., Lee, F. C., and Chichester, C. O.,** Absence of mutagenic response from extracts of Maillard browned egg albumin, *J. Food Sci.,* 45, 1442, 1980.
48. **Bjeldanes, L. F. and Chew, H.,** Mutagenicity of 1,2-dicarbonyl compounds: maltol, kojic acid, diacetyl, and related substances, *Mutat. Res.,* 67, 367, 1979.
49. **Stich, H. F., Stich, W., Rosin, M. P., and Powrie, W. D.,** Mutagenic activity of pyrazine derivatives: a comparative study with *Salmonella typhimurium, Saccharomyces cerevisiae* and Chinese hamster ovary cells, *Food Cosmet Toxicol.,* 18, 581, 1980.
50. **Okamoto, H., Mizusaki, S., Yoshida, D., and Matsumoto, T.,** Enhancing effect of carbohydrate pyrolysates on mutagenesis in *Salmonella typhimurium, Agric. Biol. Chem.,* 32, 1433, 1979.
51. **Stich, H. F., Rosin, M. P., Wu, C. H., and Powrie, W. D.,** Clastogenicity of furans found in food, *Cancer Lett.,* 13, 89, 1981.

Chapter 11

PLANT PHENOLICS AS GENOTOXIC AGENTS AND AS MODULATORS FOR THE MUTAGENICITY OF OTHER FOOD COMPONENTS

H. F. Stich and W. D. Powrie

TABLE OF CONTENTS

I. INTRODUCTION

It is now generally assumed that approximately 5% of all chemicals are genotoxic and, by implication, should have a carcinogenic capacity. If this notion is correct, then a considerable number of mutagenic, clastogenic, and recombinogenic chemicals should be present in food products, since they usually consist of thousands of individual compounds. Furthermore, since the largest number of chemicals enters man through the ingestion of food and beverages, one should expect to see the diet as the main source of genotoxic and carcinogenic agents to which man is exposed. Epidemiological evidence supports the idea that dietary and lifestyle factors are involved in the etiology of a large proportion of human cancers.[1,2] Surprisingly, the detection and identification of mutagens and carcinogens in food products has received only scarce attention. The situation, however, is rapidly changing. The introduction of gas chromatography (GC) and high performance liquid chromatography (HPLC) combined with mass spectrometry (MS) permits a thorough analysis of the chemical composition of foods and beverages. In addition, the application of the newly designed short-term tests for genotoxic agents can readily uncover the mutagenic, clastogenic, or recombinogenic potential of chemicals or complex mixtures.[3] A combination of refined chemical separation procedures with the rapid detection of genotoxicity appears to be the method of choice to gain information about the quantities of mutagens in food products and to assess their potential health hazard.

II. METHODS, MATERIALS, AND RESULTS

The authors have chosen simple and complex phenolics to demonstrate the involvement of common plant compounds in a multitude of reactions that may conceivably lead to genotoxic or carcinogenic effects. An understanding of the contribution of phenolics to the total genotoxic burden of a diet appears to be important, considering their wide distribution among plants of many taxonomic groups and their relatively high concentrations in vegetables, fruits, and beverages. The quantities of simple phenols (C6), phenolic acids (C6-C1) and cinnamic acids (C6-C3) in a few plant products are given in Table 1. The consumption of another group of phenolics, the flavonoids, is also by no means negligible. A rough estimate places the ingested levels of flavonoids at approximately 1g per day per individual.[4] In addition, man is exposed to phenolics through cigarette smoke or smoke or burned rubbish. The mainstream of smoke of commercial cigarettes contains catechol levels ranging from 89 to 284 μg.[5] Considering the extensive exposure of human populations to a variety of phenolics, greater attention should be given to the toxicology, genotoxicity, and possible carcinogenicity of this important group of food components.

As part of the assessment of the role of phenolics in the overall genotoxicity of mixtures, their capacity to induce genetic lesions was examined.[6] The clastogenic activities of several simple phenolic compounds, monohydroxylated, dihydroxylated, and trihydroxylated phenolic acids, one phenylacetic acid (C6-C2), and monohydroxylated and dihydroxylated cinnamic acids are shown in Table 2. The clastogenic activity is represented as a percentage of metaphase plates of CHO cells with at least one chromatid break or chromatid exchange, and as the average number of chromatid breaks and exchanges per metaphase plate. The concentration of each compound that is presented is half the dose that induces mitotic inhibition, defined as one metaphase plate or less in 6000 CHO cells. The strong clastogenic effects of phenolics in mammalian cells (Table 2) and plant cells[7] are not matched by a strong mutagenic activity in the *Salmonella typhimurium* assay. Several of the examined phenolics, including guaia-

Table 1
THE OCCURRENCE OF PHENOLICS IN FOOD PRODUCTS AND BEVERAGES

Compound	Source	Content (mg/100 g)	Ref.
Cinnamic acid	Plums, cherries, apples, pears, grapes	90—182	65
Caffeic acid	Green, red and white cabbage, Brussel sprouts, radishes	1—30	66
	Aubergine	36—44	66
	Carrot, celery, lettuce, chicory	2—90	67
Chlorogenic acid	Dry tea shoots	559—674	68
	Apples	89	69
	Crab apples	46—205	70
	Cup of coffee	250—260	36
Vanillic acid	Barley seeds	211	71
p-Coumaric acid	Green, red and white cabbage, radish	2—15	66
	Spinach	7—13	67
Ferulic acid	Brussel sprouts	1—2.9	72
	Barley seed	1—480	71
Gallic acid	Instant tea	64	73
	100 mℓ black tea (from tea bag)	3—5	74
	Red wine (mg/ℓ)	35,000	75
	White wine (mg/ℓ)	2,000	75
Syringic acid	Wine (mg/ℓ)	5,500	75
	Instant tea	20	73
Protocatechuic acid	Stuart pecans	0.7	73

col,[8] caffeic acid,[9] and eugenol,[10] did not induce a significant number of his$^+$ revertants in the standard plate test using several strains of *S.typhimurium*. This apparent discrepancy between different well-established test systems for genotoxicity is causing considerable concern. Do the negative mutagenicity results represent "false negatives" or do the positive clastogenicity data give "false positives"? Discussions of this issue can be fruitless for two simple reasons. Reliable carcinogenicity data on most phenolic compounds are lacking. Thus no comparison between genotoxicity data and carcinogenicity data can be made. Secondly, it is very difficult to prove or disprove negative results. Some chemicals belonging to the large groups of phenolics were found to lack mutagenic activity, whereas others seem to possess mutagenic properties, including catechol,[11] pyrogallol,[12] and its oxidative derivative, purpurogallin.[13] Nevertheless, different outcomes of different genotoxic test systems are worrisome. The authors addressed this topic by comparing the three most widely used tests: the Ames *S. typhimurium* assay for mutagenicity,[14] the chromosome aberration test on CHO cells,[15] and the gene conversion assay on strain D7 of *Saccharomyces cerevisiae.*[16] Chlorogenic acid and one of its components, caffeic acid, were chosen as test subjects. The results are summarized in Table 3. Chlorogenic acid lacked mutagenic activity for strains TA98 and TA100 of *Salmonella typhimurium* when examined at two dose levels (0.16 and 1.60 μmol per plate) with or without an S9 mixture from Aroclor 1254-treated rats.[9] Even the application of higher concentrations of chlorogenic acid in a more sensitive preincubation assay[17] did not uncover a significant increase in the frequency of his$^+$ revertants (Table 3). However, a two- to three-fold increase of his$^+$ revertants over background levels occurred when TA98 or TA100 strains were exposed to a mixture of the transition metal manganese and chlorogenic acid. Manganese exerts an enhancing effect at concentrations that occur in many plant products.[18,19] The results

Table 2
CLASTOGENIC ACTIVITY OF PHENOLIC COMPOUNDS IN CHO CELLS

Phenolic group	Compound	Concentration (mg/mℓ)	Activation	Metaphases with chromosome aberrations (%)	Chromatid breaks /cell	Chromatid exchanges /cell
Simple phenols (C6)	Catechol	0.05	−S9	23.5	0.17	0.67
			+S9	2.0	0.00	0.03
	4-Methyl catechol	0.01	−S9	20.0	0.10	0.73
			+S9	0.0	0.00	0.00
	Resorcinol	1.6	−S9	14.3	0.08	0.95
			+S9	13.5	0.05	0.52
	Phloroglucinol	3.0	−S9	10.1	0.03	0.21
			+S9	1.5	0.01	0.02
	Pyrogallol	0.1	−S9	22.6	0.14	1.08
			+S9	18.0	0.06	0.84
Phenolic acids (C6-C1)	*p*-Hydroxybenzoic acid	25.0	−S9	0.0	0.00	0.00
			+S9	6.7	0.09	0.01
	Protocatechuic acid	3.0	−S9	20.5	0.04	0.87
			+S9	2.5	0.01	0.05
	Vanillic acid	25.0	−S9	4.8	0.02	0.07
			+S9	15.7	0.04	0.34
	Gallic acid	0.05	−S9	24.1	0.03	0.70
			+S9	2.0	0.00	0.00
	Syringic acid	3.0	−S9	38.6	0.17	1.44
			+S9	3.0	0.01	0.01
	Salicylic acid	25.0	−S9	1.5	0.02	0.00
			+S9	1.0	0.01	0.00
Phenylacetic acid (C6-C2)	3,4-Dihydroxyphenylacetic acid	0.1	−S9	18.5	0.01	1.23
			+S9	1.5	0.00	0.02
Cinnamic acids (C6-C3)	*p*-Coumaric acid	6.0	−S9	3.0	0.01	0.08
			+S9	17.8	0.08	1.17
	Caffeic acid	0.2	−S9	33.3	0.08	3.08
			+S9	1.0	0.01	0.00
	Ferulic acid	25.0	−S9	7.8	0.12	0.15
			+S9	0.5	0.01	0.00
Others	Eugenol	0.4	−S9	21.5	0.08	0.20
			+S9	M.I.	—	—
		0.05	−S9	0.5	0.01	0.00
			+S9	13.9	0.17	0.24

point to the importance of using different endpoints, different tester cells, and different modulating factors when assessing the genotoxicity of a compound. Considering the enhancing effect of transition metals (Mn^{++}, Cu^{++} and Fe^{++}) on clastogenicity, mutagenicity, and gene conversion of diverse groups of environmental chemicals,[6,15,20,21] the question must be raised as to the necessity of testing compounds in conjunction with several transition metals prior to claiming a lack of mutagenicity.

A second set of experiments was designed to study the behavior of phenolics in complex mixtures. In spite of the difficulty of handling mixtures, some effort must be made to elucidate the possible interactions of two or more compounds in the induction of genotoxicity. Cost and logistic reasons severely restrict the usefulness of the classical carcinogenicity test on rodents in unraveling the interactions of mixtures. Thus one is forced to rely on the economic and rapid in vitro tests for genotoxicity to predict carcinogenicity of mixtures to which man is daily exposed. As a first step in under-

Table 3
GENOTOXICITY OF CHLOROGENIC ACID DETECTED BY THREE DIFFERENT TEST SYSTEMS

		Increase above background[a]		
		S. typhimurium plate test, TA100	Yeast, gene conversion	CHO, chromosome aberrations
Chlorogenic acid	$-Mn^{++}$	1.2	2.2	27.7
	$+Mn^{++}$	2.3	4.3	50.6[b]
Caffeic acid	$-Mn^{++}$	1.0	3.2	41.6
	$+Mn^{++}$	2.3	25.5	68.1[b]
Quinic acid	$-Mn^{++}$	1.0	1.0	24.0
	$+Mn^{++}$	1.0	2.2	—

[a] Actual frequencies of mutations, gene conversions, and chromosome aberrations are given in Stich et al.[76] Figures represent ratios of genotoxicity in treated cells/controls.

[b] The concentration of chlorogenic acid and caffeic acid that induces this increase in chromatid breaks and exchanges is 1/40 of that used without the addition of Mn^{++}.

Table 4
ENHANCING CLASTOGENIC ACTION OF THE COMBINED APPLICATION OF TANNIC ACID AND QUERCETIN SHOWING CHROMOSOME ABERRATIONS

Tannic acid (μg/mℓ)	Quercetin (μg/mℓ) —	10	5
—	0.8 (0.1)	0.0 (0.0)	0.5 (0.0)
5	2.6 (0.1)	50.0 (1.8)	25.4 (1.7)
2.5	0.7 (0.0)	30.2 (0.8)	7.5 (0.7)
1.2	0.8 (0.1)	18.2 (0.5)	5.2 (0.1)
0.6	0.8 (0.0)	11.8 (0.2)	3.1 (0.1)

Note: The frequency of chromosome aberrations is expressed as (a) the percentage of metaphase plates with at least one chromatid break and one chromatid exchange, and (b) the average number of chromatid breaks and exchanges/cell (figures in parentheses).

standing the behavior of complex mixtures, the authors estimated the chromosome-breaking capacity of two compounds, tannic acid and quercetin, when applied to CHO cells individually or in combination. The results clearly show that the clastogenic effects of the combined application of these two common compounds was greater than the sum of each individual action (Table 4). Similar potentiation effects have been observed when CHO cells were exposed to various combinations of phenolics, including eugenol, chlorogenic acid, quercetin, and the alkaloid arecoline.[22] These combinations were chosen since they occur when betel quid ("Paan") is chewed, which leads to a high incidence of oral carcinomas.[23,24] Potentiation effects can be detected by using various test systems, including point mutations,[25-27] chromosome aberrations,[15,28,29] unscheduled DNA synthesis,[21] and neoplastic transformation in vitro.[30]

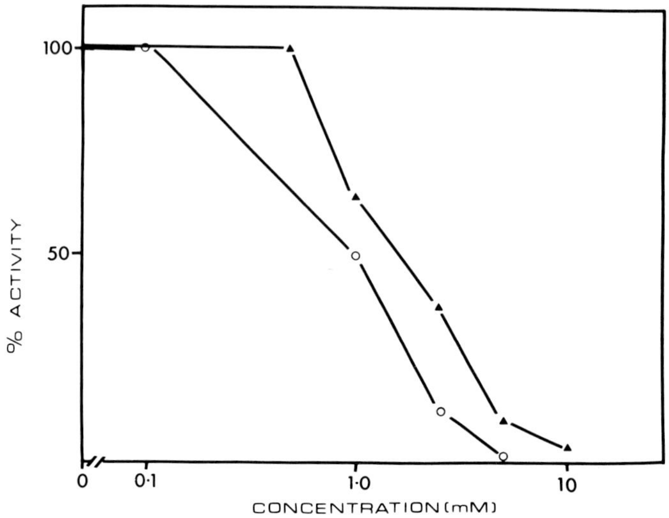

FIGURE 1. Inhibitory effect of gallic acid (▲) and ascorbic acid (o) on the mutagenicity of *Salmonella typhimurium* TA1535 exposed to a nitrosation mixture of nitrite and methylurea at pH 3.6.

Since potentiation effects appear to be quite common events, the question must be raised as to the validity of basing "no-effect", "acceptable", or "tolerable" levels of a chemical on genotoxicity results obtained by testing single compounds.[22]

III. CONCLUSIONS

In the past, emphasis was placed on a search for initiators of carcinogenesis. However, the importance of modulating factors in carcinogenesis and mutagenesis were probably underestimated. Phenolics seem to affect carcinogenesis as modulators rather than as initiators of neoplastic transformation. For example, gallic acid, pyrogallol, vanillin, and hydroquinone inhibited the formation of several nitrosamines when added to amine and sodium nitrite mixtures.[31,32] Guaiacol, 4-methyl catechol, and pyrrole suppressed the formation of *N*-nitrosomorpholine when added to an equimolar aqueous system of morpholine and nitrite at pH 3.[33] The reaction of phenolics with nitrite seems to explain their inhibitory effect on *N*-nitrosamine formation.[34,35] However, under certain conditions, phenolics, including 4-methyl catechol and chlorogenic acid,[36] gallic acid,[37] *p*-cresol,[38] and nitrosophenol,[38,39] seem to act as catalysts for nitrosation reactions. The biological significance of an increased or decreased rate of nitrosation cannot be deduced from these chemical data. To yield information on this question the action of phenolics on the mutagenicity of nitrosation products was examined. The nitrosation of methylurea was chosen as a model since the nitrosation products of methylurea are direct-acting mutagens that do not require S9-mediated activation. When several phenolics, including gallic acid, catechol, caffeic acid, chlorogenic acid, and catechin were added to the reaction mixture (nitrite:methylurea at pH 3.6), the mutagenic activity, as detected by the *S. typhimurium* assay, was strongly inhibited. An example is shown in Figure 1.

The tannins, polymers of phenolics, seem to be of particular concern because of their possible involvement in human carcinogenesis. Tannin-rich herbal teas seem to be associated with a high frequency of human EC.[40-42] In rodents, tannic acid and

Table 5
CLASTOGENIC ACTIVITY OF TANNINS FROM VARIOUS SOURCES

Source	Concentration (μg/mℓ)	Metaphase plates with chromosome aberrations (%)
Apple juice	450	20.0
Grape juice	450	48.0
Wine	220	36.3
Betel nut	10 mg[a]	7.8
Commercial (for wine making)	10	42.9
Commercial[b]	20	53.8

Note: CHO cells were exposed to the tannins for 3 hr and sampled 20 hr later for chromosome analysis.

[a] Dose expressed as mg total nut equivalent.
[b] Sigma Chemical Co., St. Louis, Mo.

Table 6
EFFECT OF TRANSITION METALS AND S9 MIXTURE ON THE CLASTOGENIC ACTIVITY OF TANNIC ACID[a]

Treatment	Metaphase plates with chromosome aberrations
Tannic acid (10 μg/mℓ)	20.1
Tannic acid + Cu^{++} (10^{-4} *M*)	29.8
Tannic acid + Mn^{++} (10^{-4} *M*)	48.4
Tannic acid + Fe^{++} (10^{-4} *M*)	0.7
Tannic acid + S9 mixture	8.5

[a] See Table 4 for exposure pattern.

tannin-containing plant extracts induce cancers at various sites.[43-49] When short-term tests for genotoxicity were applied to the tannins, they were found to be strongly clastogenic (Table 5) but lacked the capacity to induce mutations in the *S. typhimurium* assay. This observation again shows the necessity of using a battery of different genotoxic test systems to avoid misleading predictions of carcinogenicity. Of particular interest is the observation that the "nonmutagenic" tannic acid can be changed into a "mutagen" by the addition of Mn^{++}.[50] However, the addition of transition metals does not always enhance genotoxicity. For example, Fe^{++} virtually abolishes the clastogenic activity of tannic acid (Table 6) that is probably due to its well-known complex formation capacity.

The aforementioned observations on phenolics seem to exemplify the complexity of actions that are encountered when examining complex food mixtures. Some phenolics, including those belonging in the categories C6, C6-C1, C6-C2, C6-C3 (cinnamic acids), C6-C3 (coumarins), and flavonoids[4] are potent genotoxic agents, but they can also inhibit or enhance mutagenesis of other compounds in a mixture.[51] Once we leave the simplicity of studying the reactions of single compounds, we must be prepared to find this kaleidoscopic pattern. Apparently contradictory results were also seen with ascorbic acid, which shows a strong genotoxic activity,[29,52-54] a co-carcinogenic effect,[53,55] an inhibition of mutagenic nitrosation products,[56] and a preventive effect on carcinogenesis following application of nitrate/nitrite and secondary amines.[57-59] Considering this multitude of in vitro actions, it should not come as a surprise to find a similar complexity when phenolics are studied in in vivo systems. Several phenolics can act as tumor promoters,[60,61] co-carcinogens,[62] strong anticarcinogens,[63] or carcinogens, as in the case of quercetin and kaempferol[64] or tannins.[44,49]

The wide distribution of phenolics in vegetables, fruits, nonalcoholic as well as alcoholic beverages, and smoke, and their involvement in genotoxicity and carcinogenicity indicates that more attention is warranted for this group of chemicals. The relatively large number of in vitro results indicating a potential health hazard, and the lack of any epidemiological evidence for the carcinogenicity in human populations creates a dilemma for those responsible for regulatory decisions.

ACKNOWLEDGMENTS

This work was supported by the Natural Sciences and Engineering Research Council, Canada, and the National Cancer Institute of Canada.

REFERENCES

1. **Wynder, E. L., Reddy, B. S., and McCoy, G. D. S.,** Diet and cancer of the gastrointestinal tract, *Adv. Intern. Med.,* 22, 397, 1977.
2. **Hirayama, T.,** A large scale cohort study on the relationship between diet and selected cancers of digestive organs, in *Gastrointestinal Cancer: Endogenous Factors,* Bruce, W. R., Correa, P., Lipkin, M., Tannenbaum, S. R., and Wilkins, T. D., Eds., Banbury Rep. 7, Cold Spring Harbor Laboratory, Cold Spring Harbor, N.Y., 1981, 409.
3. **Stich, H. F. and San, R. H. C., Eds.,** *Short-Term Tests for Chemical Carcinogens,* Springer-Verlag, New York, 1981.
4. **Brown, J. P.,** A review of the genetic effects of naturally occurring flavonoids, anthraquinones and related compounds, *Mutat. Res.,* 75, 243, 1980.
5. **Brunnemann, K. D., Lee, H. C., and Hoffmann, D.,** Chemical studies on tobacco smoke. XLVII. On the quantitative analysis of catechols and their reduction, *Anal. Lett.,* 9, 939, 1976.
6. **Stich, H. F., Rosin, M. P., Wu, C. H., and Powrie, W. D.,** The action of transition metals on the genotoxicity of simple phenols, phenolic acids, and cinnamic acids, *Cancer Lett.,* 14, 251, 1981.
7. **Levan, A. and Tjio, J. H.,** Induction of chromosome fragmentation by phenols, *Hereditas,* 34, 453, 1948.
8. **Nestmann, E. R., Lee, E.G-H., Matula, T. I., Douglas, G. R., and Mueller, J. C.,** Mutagenicity of constituents identified in pulp and paper mill effluents using the *Salmonella*/mammalian-microsome assay, *Mutat. Res.,* 79, 302, 1980.
9. **MacGregor, J. T. and Jurd, L.,** Mutagenicity of plant flavonoids: structural requirements for mutagenic activity in *Salmonella typhimurium, Mutat. Res.,* 54, 297, 1978.
10. **Swanson, A. B., Chambliss, D. D., Blomquist, J. C., Miller, E. C., and Miller, J. A.,** The mutagenicities of safrole, estragole, eugenol, *trans*-anethole, and some of their known or possible metabolites for *Salmonella typhimurium* mutants, *Mutat. Res.,* 60, 143, 1979.
11. **Kunz, B. A., Hannan, M. A., and Haynes, R. H.,** Effect of tumor promoters on ultraviolet light-induced mutation and mitotic recombination in *Saccharomyces cerevisiae, Cancer Res.,* 40, 2323, 1980.
12. **Bilimoria, M. H.,** The detection of mutagenic activity of chemicals and tobacco smoke in a bacterial system, *Mutat. Res.,* 31, 328, 1975.
13. **Ben-Gurion, R.,** Mutagenic and colicine-inducing activity of two antioxidants: pyrogallol and purpurogallin, *Mutat. Res.,* 68, 201, 1979.

14. **Ames, B. N., McCann, J., and Yamasaki, E.,** Methods for detecting carcinogens and mutagens with the *Salmonella*/mammalian-microsome mutagenicity test, *Mutat. Res.,* 31, 347, 1975.
15. **Stich, H. F., Wei, L., and Whiting, R. F.,** Enhancement of the chromosome-damaging action of ascorbate by transition metals, *Cancer Res.,* 39, 4145, 1979.
16. **Zimmermann, F. K.,** Procedure used in the induction of mitotic recombination and mutation in the yeast *Saccharomyces cerevisiae*, in *Handbook of Mutagenicity Test Procedures,* Kilbey, B. J., Legator, M., Nichols, W., and Ramel, C., Eds., Elsevier/North-Holland, Amsterdam, 1977, 119.
17. **Nagao, M., Yahagi, T., Seino, Y., Sugimura, T., and Ito, N.,** Mutagenicity of quinoline and its derivatives, *Mutat. Res.,* 42, 335, 1977.
18. **Prasad, M. and Spiers, M.,** Comparative study of ashing techniques for the digestion of horticultural plant samples, *J. Agric. Food Chem.,* 26, 824, 1978.
19. **Enkelmann, R.,** Mangangehalte in Rebblättern, Most und Wein, *Z. Lebensm. Unters. Forsch.,* 167, 171, 1978.
20. **MacRae, W. D. and Stich, H. F.,** Induction of sister-chromatid exchanges in Chinese hamster ovary cells by thiol and hydrazine compounds, *Mutat. Res.,* 68, 352, 1979.
21. **Whiting, R. F., Wei, L., and Stich, H. F.,** Enhancement by transition metals of unscheduled DNA synthesis induced by isoniazid and related hydrazines in cultured normal and Xeroderma pigmentosum human cells, *Mutat. Res.,* 62, 505, 1979.
22. **Stich, H. F., Stich, W., and Lam, P. P. S.,** Potentiation of genotoxicity by concurrent application of compounds found in betel quid: arecoline, eugenol, quercetin, chlorogenic acid and Mn^{++}, *Mutat. Res.,* 90, 355, 1981.
23. **Muir, C. S. and Kirk, R.,** Betel, tobacco, and cancer of the mouth, *Br. J. Cancer,* 14, 597, 1960.
24. **Jussawalla, D. J.,** The problem of cancer in India: an epidemiological assessment, *Gann,* 18, 265, 1976.
25. **Ashby, J. and Styles, J. A.,** Comutagenicity, competitive enzyme substrates, and in vitro carcinogenicity assays, *Mutat. Res.,* 54, 105, 1978.
26. **Nagao, M., Yahagi, T., and Sugimura, T.,** Differences in effects of norharman with various classes of chemical mutagens and amount of S9, *Biochem. Biophys. Res. Commun.,* 83, 373, 1978.
27. **Saffiotti, U., Rice, J. N., and Donovan, P.,** Interactions of multiple carcinogens at low levels of exposure: preliminary mutagenicity studies using the Ames *Salmonella* system, in *Short-Term Tests for Prescreening of Potential Carcinogens,* Santi, L. and Parodi, S., Eds., Istituto Scientifico per lo Studio e la Cura dei Tumori, Genoa, 1979, 99.
28. **Stich, H. F., Wei, L., and Lam, P.,** The need for a mammalian test system for mutagens: action of some reducing agents, *Cancer Lett.,* 5, 199, 1978.
29. **Stich, H. F., Wei, L., and Whiting, R. F.,** Chromosome aberrations in mammalian cells exposed to vitamin C and multiple vitamin pills, *Food Cosmet. Toxicol.,* 18, 497, 1980.
30. **Borek, C. and Ong, A.,** The interaction of ionizing radiation and food pyrolysis products in producing oncogenic transformation in vitro, *Cancer Lett.,* 12, 61, 1981.
31. **Gray, J. I. and Dugan, L. R.,** Inhibition of *N*-nitrosamine formation in model food systems, *J. Food Sci.,* 40, 981, 1975.
32. **Kawabata, T., Ohshima, H., Vibu, J., Nakamura, M., Matsui, M., and Hamano, M.,** Occurrence, formation and precursors of *N*-nitroso compounds in Japanese diet, in *Naturally Occurring Carcinogens-Mutagens and Modulators of Carcinogenesis,* Miller, E. C., Miller, J. A., Sugimura, T., Takayama, S., and Hirono, I., Eds., University Park Press, Baltimore, 1979, 195.
33. **Groenen, P. J.,** A new type of *N*-nitrosation inhibitor, in *Proc. 2nd Int. Symp. Nitrite in Meat Products,* Tinbergen, B. J. and Krol, B., Eds., Pudoc, Wageningen, Netherlands, 1977, 171.
34. **Bogovski, P., Castegnaro, M., Pignatelli, B., and Walker, E. A.,** The inhibiting effect of tannins on the formation of nitrosamines, in *N-Nitroso Compounds, Analysis and Formation,* Bogovski, P., Preussmann, R., and Walker, E. A., Eds., IARC Sci. Publ. No. 3, International Agency for Research on Cancer, Lyon, 1972, 127.
35. **Challis, B. C.,** Rapid nitrosation of phenols and its implications for health hazards from dietary nitrites, *Nature (London),* 244, 466, 1973.
36. **Challis, B. C. and Bartlett, C. D.,** Possible cocarcinogenic effects of coffee constituents, *Nature (London),* 254, 532, 1975.
37. **Walker, E. A., Pignatelli, B., and Castegnaro, M.,** Effects of gallic acid on nitrosamine formation, *Nature (London),* 258, 176, 1975.
38. **Davies, R., Dennis, M. J., Massey, R. C., and McWeeny, D. J.,** Some effects of phenol and thiol nitrosation reactions on *N*-nitrosamine formation, in *Environmental Aspects of N-Nitroso Compounds,* Walker, E. A., Castegnaro, M., Griciute L., and Lyle, R. E., Eds., IARC Sci. Publ. 19, International Agency for Research on Cancer, Lyon, 1978, 183.
39. **Walker, E. A., Pignatelli, B., and Castegnaro, M.,** Catalytic effect of *p*-nitrosophenol on the nitrosation of diethylamine, *J. Agric. Food Chem.,* 27, 393, 1979.

40. **Morton, J. F.**, Plants associated with esophageal cancer in Curacao, *Cancer Res.*, 28, 2268, 1968.
41. **Morton, J. F.**, Tentative correlations of plant usage and esophageal cancer zones, *Econ. Bot.*, 24, 217, 1970.
42. **Morton, J. F.** Plant products and occupational materials ingested by esophageal cancer victims in South Carolina, *Q. J. Crude Drug Res.*, 13, 2005, 1973.
43. **Korpassy, B. and Mosonyi, M.**, The carcinogenic activity of tannic acid. Liver tumours induced in rats by prolonged subcutaneous administration of tannic acid solutions, *Br. J. Cancer*, 4, 411, 1950.
44. **Kirby, K. S.**, Induction of tumours by tannin extracts, *Br. J. Cancer*, 14, 147, 1960.
45. **Kapadia, G. J., Paul, B. D., Chung, E. B., Ghosh, B., and Pradhan, S. N.**, Carcinogenicity of *Camellia sinensis* (tea) and some tannin-containing folk medicinal herbs administered subcutaneously in rats, *J. Natl. Cancer Inst.*, 57, 207, 1976.
46. **Kapadia, G. J., Chung, E. B., Ghosh, B., Shukla, Y. N., Basak, S. P., Morton, J. F., and Pradhan, S. N.**, Carcinogenicity of some folk medicinal herbs in rats, *J. Natl. Cancer Inst.*, 60, 683, 1978.
47. **Pradhan, S. N., Chung, E. B., Ghosh, B., Paul, B. D., and Kapadia, G. J.**, Potential carcinogens. I. Carcinogenicity of some plant extracts and their tannin-containing fractions in rats, *J. Natl. Cancer Inst.*, 52, 1579, 1974.
48. **Wang, C. Y., Chiu, C. W., Pamukcu, A. M., and Bryan, G. T.**, Identification of carcinogenic tannin isolated from bracken fern *(Pteridium aquilinum), J. Natl. Cancer Inst.*, 56, 33, 1976.
49. **Pamukcu, A. M., Wang, C. Y., Hatcher, J., and Bryan, G. T.**, Carcinogenicity of tannin and tannin-free extracts of bracken fern *(Pteridium aquilinum)* in rats, *J. Natl. Cancer Inst.*, 65, 131, 1980.
50. **Rosin, M. P.**, Manuscript in preparation, 1982.
51. **Buening, M. K., Chang, R. L., Huang, M. T., Fortner, J. G., Wood, A. W., and Conney, A. H.**, Activation and inhibition in human liver microsomes by naturally occurring flavonoids, *Cancer Res.*, 41, 67, 1981.
52. **Stich, H. F., Karim, J., Koropatnick, D. J., and Lo, L. W.**, Mutagenic action of ascorbic acid, *Nature (London)*, 260, 722, 1976.
53. **Rosin, M. P., Peterson, A. R., and Stich, H. F.**, The effect of ascorbate on 3-methylcholanthrene-induced cell transformation in C3H/10T½ mouse embryo fibroblast cell cultures, *Mutat. Res.*, 72, 533, 1980.
54. **Galloway, S. M. and Painter, R. B.**, Vitamin C is positive in the DNA synthesis inhibition and sister-chromatid exchange tests, *Mutat. Res.*, 60, 321, 1979.
55. **Banic, S.**, Vitamin C acts as a carcinogen to methylcholanthrene in guinea-pigs, *Cancer Lett.*, 11, 239, 1981.
56. **Lo, L. W. and Stich, H. F.**, The use of short-term tests to measure the preventive action of reducing agents on formation and activation of carcinogenic nitroso compounds, *Mutat. Res.*, 57, 57, 1978.
57. **Mirvish, S. S., Arnold, S., Conrad, E., Ghadirian, P., Kommineni, V. R. C., and Sams, J.**, Formation of *N*-nitroso compounds, preparation of heptafluorobutyryl derivatives of ureas, and fate of nitrite in the rat stomach, in *N-Nitroso Compounds in the Environment*, Bogovski, P. and Walker, E. A., Eds., IARC Sci. Publ. No. 9, International Agency for Research on Cancer, Lyon, 1974, 67.
58. **Sander, J., Schweinsberg, F., LaBar, J., Bürkle, G., and Schweinsberg, E.**, Nitrite and nitrosable amino compounds in carcinogenesis, in *Recent Topics in Chemical Carcinogenesis*, Odashima, S., Takayama, S., and Sato, H., Eds., Gann Monogr. on Cancer Res. No. 17, University of Tokyo Press, Tokyo, 1975, 145.
59. **Raineri, R. and Weisburger, J. H.**, Reduction of gastric carcinogens with ascorbic acid, *Ann. N.Y. Acad. Sci.*, 258, 181, 1975.
60. **Boutwell, R. K.**, Phenolic compounds as tumor-promoting agents, in *Phenolic Compounds and Metabolic Regulation*, Ginkle, B. J. and Rueckles, V. C., Eds., Appleton-Century-Crofts, New York, 1967, 120.
61. **Boutwell, R. K. and Bosch, D. K.**, The tumor-promoting action of phenol and related compounds for mouse skin, *Cancer Res.*, 19, 413, 1959.
62. **Van Duuren, B. L. and Goldschmidt, B. M.**, Cocarcinogenic and tumor-promoting agents in tobacco carcinogenesis, *J. Natl Cancer Inst.*, 56, 1237, 1976.
63. **Wattenberg, L. W., Coccia, J. B., and Lam, L. K. T.**, Inhibitory effects of phenolic compounds on benzo(a)pyrene-induced neoplasia, *Cancer Res.*, 40, 2820, 1980.
64. **Pamukcu, A. M., Yalcimer, S., Hatcher, J. F., and Bryan, G. T.**, Quercetin, a rat intestinal and bladder carcinogen present in bracken fern *(Pteridium aquilinum), Cancer Res.*, 40, 3468, 1980.
65. **Mosel, H. D. and Herrmann, K.**, The phenolics of fruits. III. The contents of catechins and hydroxycinnamic acids in pome and stone fruits, *Z. Lebensm. Unters. Forsch.*, 154, 6, 1974.
66. **Herrmann, K.**, Übersicht über nichtessentielle Inhaltsstoffe der Gemüsearten. I. Gurken, Melonen, Kürbrisse, Gemüsepaprika, Auberginen, Erbsen, Bohnen, und Puffbohnen, *Z. Lebensm. Unters. Forsch.*, 165, 87, 1977.

67. **Herrmann, K.**, Übersicht über nichtessentielle Inhaltsstoffe der Gemüse-arten. III. Möhren, Sellerie, Pastinaken, rote Rüben, Spinat, Salat, Endiven, Trebzichorie, Rhabarber, und Artischoken, *Z. Lebensm. Unters. Forsch.*, 167, 262, 1978.
68. **Ullah, M. R. and Jain, J. C.**, Seasonal variations in the chlorogenic acid content of tea, *J. Sci. Food Agric.*, 31, 355, 1980.
69. **Walker, J. R. L.**, Flavonoid pigments in the skins of New Zealand apples, *N.Z. J. Sci.*, 7, 585, 1976.
70. **Kuusi, T. and Pajunen, E.**, Less grown apple varieties in juice production, and the influence of polyphenols and added ascorbic acid on the juice quality, *J. Sci. Agric. Soc. (Finland)*, 43, 20, 1971.
71. **Shominski, B. A.**, Phenolic acids in the meal of developing and stored barley seeds, *J. Sci. Food Agric.*, 31, 1007, 1980.
72. **Schmidtlein, H. and Herrmann, K.**, Über die Phenolsäuren des Gemüses. I. Hydroxyzimtsäuren und Hydroxybenzoesäuren der Kohlarten und anderer Cruciferen-Blätter, *Z. Lebensm. Unters. Forsch.*, 159, 139, 1975.
73. **Horvat, R. J. and Senter, S. D.**, A gas-liquid chromatographic method for analysis of phenolic acids in plants, *J. Agric. Food Chem.*, 28, 1292, 1980.
74. **Sanderson, G. W., Ranadive, A. S., Eisenberg, L. S., Farrell, F. J., Simons, R., Manley, C. H., and Coggon, P.**, Contribution of polyphenolic compounds to the taste of tea, *Am. Chem. Soc. Symp. Ser.*, 26, 14, 1976.
75. **Amerine, M. A. and Ough, C. S.**, *Methods for Analysis of Musts and Wines*, John Wiley & Sons, New York, 1980.
76. **Stich, H. F., Rosin, M. P., Wu, C. H., and Powrie, W. D.**, A comparative genotoxicity study of chlorogenic acid (3-*O*-caffeoylquinic acid), *Mutat. Res.*, 90, 201, 1981.

Chapter 12

MUTAGENS FROM THE COOKING OF FOOD: THERMIC MUTAGENS IN BEEF

F. T. Hatch, J. S. Felton, and L. F. Bjeldanes

TABLE OF CONTENTS

I. INTRODUCTION

Recently concern has arisen that the cooking of foods, under certain circumstances, may produce genotoxic substances that could contribute to the high incidence of cancer of the stomach, large intestine, or other organs.[1-3] The available information on the production of mutagens or carcinogens during the cooking of foods indicates that there are at least three potential sources of the toxic substances. Over 15 years ago Lijinsky and Shubik,[4,5] and subsequently others, detected the well-known carcinogen BaP and related PAH in meat that had been smoked or cooked in such a way that fat dripped onto open flames, and the smoke from its burning enveloped the food, depositing hydrocarbons.[4-11] In the past few years two groups of Japanese investigators, headed by Sugimura of the National Cancer Center Research Institute of Japan and Yoshida of the Japan Tobacco and Salt Public Corporation's Central Research Institute, have demonstrated that when amino acids, pure proteins, or protein foods are heated to comparatively high temperatures, sufficient to char the food surface, several mutagens that are extremely potent in *Salmonella* bioassays are formed.[12-18] More recently, Commoner and co-workers[19-22] reported that cooking foods at comparatively low temperatures, or prolonged dehydration by mild heat, also produces mutagenic substances whose chemical identity is, so far, unknown. Further research along this line has been reported.[23-36] Because the lower temperatures are commonly employed in American households and restaurants, the authors have focused attention on foods cooked to a well-done but noncharred state,[37] and have studied the major U.S. dietary sources of protein, particularly meats.[38,39] The objectives of this research are to determine the nature and mechanism of production of the mutagens formed during cooking, to identify the chemical structure of the principal mutagens, to quantify mutagen formation with a standard bacterial bioassay, and to assess the spectrum and potency of genetic toxicity of the mutagens with in vitro and in vivo short-term mammalian bioassays. Thus, a data base shall be established that can be used to estimate the risk, if any, to humans posed by the ingestion of mutagens derived from the cooking of foods. This article presents an overview of the current status of our research on mutagens in cooked beef, the major source of protein in the American diet.[40]

II. MUTAGENIC ACTIVITY OF FRIED GROUND BEEF

Lean ground beef patties (average 15% fat content) were fried on a thermostatically controlled griddle at a surface temperature of 200°C for 6 min per side. These conditions produced a well-done, browned, but not charred product that is preferred by a significant (though unknown) proportion of the U.S. population. Patties prepared under other cooking conditions were compared to this standard preparation. The exterior portions that contained all assayable mutagens were sliced off, homogenized, extracted, and assayed with the Ames *Salmonella* histidine reversion bioassay.[37] This bacterial mutagenesis bioassay was the mainstay of investigations in the U.S. and Japan because of its sensitivity, rapidity, and reasonable cost. It has also served as a guide in the early phase of this study, although the authors plan to apply a broad spectrum of genetic assays. Several of the products isolated from the pyrolysis of amino acids and proteins exhibit extraordinarily high mutagenic potency in the *Salmonella* assay,[1] and several of them have been reported positive for animal carcinogenesis.[41]

A. Biophysical Parameters of Cooking: Their Effect on Mutagenic Activity

In an extensive study of the relationships between the formation of mutagenic activ-

Table 1
FORMATION OF MUTAGENIC ACTIVITY BY DIFFERENT COOKING METHODS

Method or surface	Temp. (°C)	Time (min)	TA1538 (rev/gE[a])
Steel	200	6/side	63
Steel	300	6/side	286
Ceramic	200	6/side	9
Teflon®/steel	200	6/side	43
Broil	300	11/side	100
Roast (bake)	176	90	35
Stew/simmer	100	165	N.S.

[a] Number of revertant colonies per plate (approximately 10^8 bacteria)/dose of homogenate or extract equivalent to 1 g fresh wt of beef. Response was calculated by linear regression over the linear region of the dose-response curve. N.S. = response not significant.

ity and cooking time, temperature, and water loss from the exterior portion of the beef patty, temperature had the strongest influence.[42] Temperature was also reported to be the major factor by Yoshida et al.[43] At temperatures ⩾ 200°C on a stainless steel griddle, mutagen formation proceeded rapidly until 10 min of cooking per side, followed by a plateau. Mutagenic activity formed in 6 min per side at 300°C was about 4.5-fold that at 200°C (Table 1).

The water content of uncooked ground beef was 67 to 70%. Little mutagen formation was observed between 150 and 250°C until the water content of the exterior portion was reduced to ⩽ 55%. Below that there was a fairly linear relationship between mutagen formation and water content down to about 35% water, the approximate limit for edible product.

Reduction of the initial water content of beef by reconstitution of dehydrated meat into patties with varying proportions of water, showed that below 50% initial water content little mutagen was formed at 200°C. Therefore, the authors believe that the chemical reactions leading to mutagen formation occur within a narrow range of water content in the meat. It will be shown in a later section that these reactions involve water-soluble precursors and apparently occur in aqueous solution rather than on a solid surface or in a lipid phase.

Reduction of the initial fat content of the beef by solvent extraction of fats, followed by graded reconstitution, resulted in a moderate increase in mutagen formation per patty but essentially the same yield per gram of protein.

Frying of ground beef on a ceramic surface at 200°C resulted in a slower rate of cooking and water loss, and lower mutagen formation at short cooking times (Table 1). However, the amount of mutagenic activity eventually reached almost the same level observed with the stainless steel griddle. On the other hand, frying in a pan with a thin coating of Teflon® on steel resulted in mutagen formation that was similar to that on uncoated steel.

A survey of methods for cooking meats, other than frying, showed that substantial mutagen formation resulted from broiling meat close to the heating element. Baking or roasting for a prolonged time resulted in moderate mutagen formation. However,

Table 2
COMPARISON OF MUTAGEN EXTRACTION PROCEDURES APPLIED TO FRIED GROUND BEEF

Extraction method	*Salmonella* TA 1538 revertants/10^8 cells/gE[a]
1. Aqueous homogenate	440[b]
2. Acid-$(NH_4)_2SO_4$	38
3. Acetone	97
4. CH_2Cl_2:CH_3OH:H_2O (45:45:10, v/v/v)	112
5. Aqueous acid-XAD-2[c]	153
6. Aqueous pH 2.0-XAD-2[d]	300—400

[a] See footnote *a*, Table 1.

[b] Extrapolated from a low applied dose (0.156 gE) after correction for estimated histidine effects on background and mutant colony counts.

[c] Method described in text applied without rigorous pH control. Actual pH during extraction >2.

[d] Method applied with rigorous pH control.

methods involving much lower temperatures and preservation of water content, such as stewing, braising, simmering, and microwave cooking[22,44] resulted in minimal mutagenic activity. Thus, cooking methods that involve direct contact with a hot metal surface or radiant heating, and associated with rapid, partial dehydration of the exterior portions of the meat, lead to greater mutagen formation than methods utilizing lower temperatures and the continued presence of water and steam.

B. Total Mutagen Content of Fried Ground Beef: Comparison of Extraction Methods

Reports from Japan describing the chemical isolation of mutagens from cooked foods indicate that there are major losses of total mutagenic activity in the early steps of various fractionation schemes.[27,31-33] Accordingly, extraction methods to maximize the yield of recovered mutagenic activity were investigated. Accurate estimation of the total amount of mutagenic activity present in a sample of cooked food is difficult. Homogenization of the food in aqueous acid appears to give a maximal value. However, accurate bioassay of the homogenate with *Salmonella* is complicated because the presence of free histidine derived from the meat interferes with the mutation assay. By calibration experiments with added histidine, the authors have determined that the maximum error induced by free histidine is a two fold increase in mutation frequency. With this correction, it is estimated that the initial mutagenic activity of ground beef fried under the standard condition of 6 min per side at 200°C has a maximal value of 400 revertants per 10^8 bacteria per gram fresh weight of uncooked beef (gE).

The first reported extraction method employed low pH and $(NH_4)_2SO_4$ to precipitate proteins.[20] However, homogenization of the cooked meat in acetone and removal of solids is simpler and extracts about twice as much mutagen, as shown in Table 2.[37] A series of mixed organic solvents was tested with a maximal yield of mutagenic activity being obtained with CH_2Cl_2:CH_3OH:H_2O, 45:45:10 (v/v/v). Subsequently it was determined that greater recovery is possible by extraction with aqueous acid without salt precipitation. The histidine artifact was eliminated by trapping the mutagens at pH 7 on XAD-2 resin, followed by their elution with acetone. The improved yield was further increased to a maximum value of 400 revertants per 10^8 bacteria per gE by careful maintenance of the pH at 2.0 during the extraction step. This represents approximately a 6 fold increase above the average value of 63 revertants per 10^8 bacteria per gE that was obtained with the acetone extraction used throughout the initial phase of the au-

Table 3
MUTAGENIC ACTIVITY OF FRACTIONS OBTAINED WITH DIFFERENT INITIAL EXTRACTION METHODS

Fractionation step	Initial extraction method			
	Acid-$(NH_4)_2SO_4$ (2)[a]	Acetone (3)	Mixed solvent (4)	XAD-2 (5)
	Salmonella TA1538 revertants/10^8 cells/gE[b]			
Organic extract or base fraction	38	97	112	153
TLC				
Slow band (R_f approximately 0.1)	14	67	172	78
Fast band (R_f: method 4, 0.28—0.59; method 5, 0.54—0.72)	0	0	Broad	39[c]
HPLC				
Slow TLC band				
Early zone	7	9	26	26
Late zone	1	11[d]	14	10

[a] Number in parentheses refers to method in Table 2.
[b] See footnote *a*, Table 1.
[c] HPLC data on this fraction not yet available.
[d] Fraction chiefly consisted of two mutagenically active, sharply resolved UV peaks, giving mass spectra molecular ions at 181 with different fragments and not corresponding to available library.

thors' research.[37] The magnitude of the increase has so far been measured only with ground beef, and must be checked with the other major protein foods.

The initial extract, obtained by one of the foregoing methods, is fractionated by partitioning between water and CH_2Cl_2 phases with successive adjustment of pH to 2.0 (transferring acidic and neutral components into the organic phase) and to pH 12.0 (transferring the basic components into the organic phase). Essentially all of the detectable mutagenic activity appears in the fraction of organic bases. The conclusion from the foregoing extraction studies is that the mutagens formed in fried ground beef are polar organic bases. This is consistent with the types of heterocyclic aromatic amines that have so far been isolated from heated amino acids, proteins, and foods, identified, and synthesized.[14,32,45-53]

C. Fractionation of Thermic Mutagens in Fried Ground Beef

Exploratory fractionation of the mutagens in beef fried under standard conditions has been carried out on crude extracts or on the organic base fraction. The first step was TLC on silica gel, typically in the solvent ethyl acetate:CH_3OH:NH_4OH, 50:10:0.25 (v/v/v). Bands were removed, extracted, and assayed for mutagenicity. Table 3 shows the results. The extractions based upon acid-$(NH_4)_2SO_4$ and acetone yielded only a single active band of slow R_f. The mixed solvent method gave a broader slow band, with considerable material eluting faster without resolution. The XAD-2 method (No. 5 in Table 2) gave a similar slow band and a fairly well-resolved additional band at much higher R_f containing at least 25% of the total mutagen. Data on the more controlled XAD-2 method (No. 6) are not yet available.

The low-R_f mutagenic band of each TLC plate was removed, extracted, transferred to a HPLC C-18 column, and eluted with a gradient of CH_3OH in H_2O containing 60

Table 4
PROPERTIES OF SEVERAL THERMIC OR PYROLYTIC MUTAGEN STANDARDS

	AαC[a]	IQ[b]	MeIQ[c]	MeIQx[d]	Trp-P-2[e]
Formula	$C_{11}H_9N_3$	$C_{11}H_{10}N_4$	$C_{12}H_{12}N_4$	$C_{11}H_{11}N_5$	$C_{12}H_{11}N_3$
Mol wt	183	198	212	213	197
Mutagenicity[f]	300	433,000	18,500	663,000	104,000
HPLC elution (min)[g]	15.0	14.9	26	15.1	>70

[a] 2-Amino-9H-pyrido[2,3-*b*]indole, or 9*H*-pyrido[2,3-*b*]indol-2-amine.
[b] 2-Amino-3-methylimidazo[4,5-*f*]quinoline, or 3-methylimidazo[4,5-*f*]quinolin-2-amine.
[c] 2-Amino-3,4-dimethylimidazo[4,5-*f*]quinoline, or 3,4-dimethylimidazo[4,5-*f*]quinolin-2-amine.
[d] 2-Amino-3,8-dimethylimidazo[4,5-*f*]quinoxaline, or 3,8-dimethylimidazo[4,5-*f*]quinoxalin-2-amine.
[e] 3-Amino-1-methyl-5*H*-pyrido[4,3-*b*]indole, or 1-methyl-5*H*-pyrido[4,3-*b*]indol-3-amine.
[f] Specific mutagenic activity in *Salmonella* TA1538, no. revertants/μg/10^8 bacteria,[1,32,46,49,50,52,53]
[g] Solvent-CH_3OH:H_2O, 0.375:0.625 (v/v, with 60 ppm $(C_2H_5)_2NH$).

ppm (v/v) $(C_2H_5)_2NH$, 0.375 to 0.6 for 15 min, then to 0.60 over 15 min, finally constant at 0.60 (v/v). Timed fractions were collected and assayed for mutagenic activity (Table 3). All of the extraction methods yielded 2 zones of activity at about 12 to 18 min and 24 to 30 min. In general, mutagenic activity was associated with multiple small-to-medium sized peaks of UV absorbance, with the bulk of the UV-absorbing peaks being inactive. The peak fractions of the 12 to 18 min zone were rechromatographed in isocratic 0.35 v/v CH_3OH in H_2O containing 60 ppm v/v $(C_2H_5)_2NH$ and compared with known standards of several cooking mutagens previously characterized (Table 4).

The identification of the mutagens in fried ground beef is still in a preliminary stage. The following tentative findings appear justified at this time. The XAD-2 method No. 5 (Table 2) extracts a set of mutagens similar to those observed with solvent methods No. 3 and 4, but in higher yield. In addition, this method extracts mutagenic material that moves rapidly on silica gel TLC but that has not been further examined. From the slow TLC band transferred to HPLC, there is an early zone that coelutes approximately with the known mutagens IQ, MeIQ, and AαC, but which probably contains considerable mutagen not identical to any of these three standards. The mutagenic late zone from HPLC consists of two sharply resolved UV peaks. Two of these fractions from extraction method No. 3 gave MS molecular ions at mol wt 181, with different fragment patterns that, so far, cannot be identified. The standard MeIQ elutes near the late zone time but has a mol wt of 212 and is stated by Nishimura[54] to be undetectable in broiled beef by mass fragmentography. In summary, at this stage there are at least four separate impure fractions of basic material that are mutagenic in *Salmonella*. Only one of these fractions may have partial correspondence with one or more of three known cooking mutagens. The quantities of these agents present in fried beef are still quite uncertain. From published reports and estimates based on specific mutagenic activity, the concentrations are below 1 ppm and in some cases may be as low as 1 ppb (1 ng/gE).[33-36]

III. A MODEL SYSTEM FOR BOILED BEEF

A. Mutagen Formation

The authors wished to confirm the formation of mutagen at 100°C and to establish a simplified model system for study of the mechanism of mutagen formation. To de-

Table 5
MUTAGENIC ACTIVITY OF SUPERNATANT S_2 FRACTIONS FROM VARIOUS MEATS

Meat[a]	*Salmonella* TA1538 revertants/10^8 cells/gE[b]
Beef (muscle)	32 (200)[c]
Beef (liver)	19
Pork chops	24
Chicken (light)	30
Chicken (dark)	20
Fish (sole)	12

[a] Source of supernatant S_2 prepared as described in the text and boiled for 14 hr at pH 4 without loss of water.
[b] See footnote a, Table 1.
[c] Value after boiling for 30 hr.

termine the mutagen precursors in lean round steak, aqueous homogenates were separated by centrifugation into a soluble supernatant$_1$ (S_1) and a residue$_1$ (R_1). S_1 was then boiled for 30 min and centrifuged to remove denatured proteins, giving a soluble supernatant$_2$ (S_2) and residue$_2$ (R_2). S_2 contains 5% of the meat dry solids and $\lessdot$10% of the original protein.[55] Mutagen formation from S_2 at 98 to 100°C and constant volume increases exponentially with time, at least up to 30 hr, and displays sharp pH optima at 4.0 and 9.0. The mutagenic activity in S_2 for *Salmonella* TA1538 is equivalent to that observed when whole beef homogenate was boiled at constant volume for 4.5 to 15 hr. This signifies that all components required for mutagen formation are water-soluble and free from the meat solids. Residue R_1 yielded no significant mutagen after prolonged boiling, even when previously treated with various proteases. The exponential kinetics of mutagen formation may be consistent with autocatalytic or sequential reaction mechanisms. Table 5 shows the mutagenic activities of comparable boiled S_2 fractions from a variety of meats. The value for beef muscle boiled 14 hr is approximately ½ that observed with ground beef patties fried under standard conditions for 6 min per side at 200°C.

Proteolysis of S_1 with papain, trypsin, or chymotrypsin (± carboxypeptidase A) enhances the mutagen yield 1.7- to 4.0-fold over the S_2 control value. However, boiled proteolytic digests of R_1, soybean protein, BSA, and bovine α-lactalbumin contain much less TA1538 activity per milligram of soluble protein (0.01 to 1.5 revertants per 10^8 cells) than S_2 (7.4 revertants per 10^8 cells). Despite their lower mutagen yields, digests of R_1, soybean protein, and albumin and fish S_2 were comparable in their free amino acid compositions to beef S_2, while liver S_2 was much richer in most amino acids. The above variations in mutagen yield indicate that soluble nonamino acid substance(s) as well as amino acids are precursors for mutagen formation. A systematic search for the nature of the nonamino acid precursors is being carried out.

B. Preliminary Fractionation of Mutagens

A simple paper electrophoretic method has been applied to survey (with the *Salmonella* bioassay) the fractionation of mutagens formed during boiling of beef S_2. Preliminary results indicate that major slow and minor fast cathodal zones are formed after boiling for 30 hr at pH 4.0, the acidic pH optimum for mutagen formation in this system. The minor zone has similar mobility to the standard 2-amino-3-methylimidazo[4,5-*f*]quinoline (IQ), but is not present in S_2 boiled for only 14 hr. Trp-P-2 moves more slowly and separates from the major zone. Comparison with other standards is in progress. When beef S_2 is boiled at pH 9.0, the basic pH optimum, the two zones of mutagenic activity that are formed do not correspond to the zones formed at acidic pH, but have not yet been further identified. These products may be derived from nonenzymatic browning reactions and other reactions leading to substances responsible for the development of flavor and aroma in cooked foods. Such reactions have been studied extensively by others, both in model systems[56-60] and in cooked foods.[61-66]

IV. SURVEY OF MUTAGEN FORMATION DURING COOKING OF PROTEIN FOODS

In addition to our intensive research on mutagen formation in beef during cooking, a survey[38,39] of the major sources of cooked protein in the U.S. diet has also been completed using data bases obtained by the U.S. Department of Agriculture, 1964 to 1966[67] and by the U.S. Department of Health, Education, and Welfare (Health and Nutrition Examination Survey), 1971 to 1975.[68] In both surveys the major sources of cooked protein were beef in various forms, eggs, pork, ham, bacon, and fried chicken. Following the method of extraction with acetone, the organic base fractions were assayed for mutagenicity with *Salmonella* strain TA1538. Significant mutagen content was found in well-done beef, ham, bacon, beef sausage, pork chops, fried egg (only at high temperature), chicken, and several fish species.[38,39] Low or negligible mutagenicity was observed in fried shrimp, rock cod (pan-fried, baked, poached or broiled, except when broiled excessively with charring), tofu (cooked several ways, except when extensively deep-fried), and kidney or pinto beans (boiled, fried, or baked).

V. METABOLIC ACTIVATION OF BEEF MUTAGENS

A. Effect of Liver Preparations on Mutagenic Activity in *Salmonella*

The standard *Salmonella* activation system utilizes the S9 fraction of liver homogenates from rats pretreated with polychlorinated biphenyls (PCB), Aroclor 1254 (Monsanto Co., St. Louis, Mo.).[69,70] With the basic fraction from fried ground beef, this system gave a mean of 63 revertants per gE, averaged over 5 experiments.[32] Hamster liver preparations were explored because of reports that hamster S9 metabolizes aromatic amines (suspected constituents of the basic fraction) better than does rat S9. This turned out to be the case. Table 6 shows that hamster S9 activation exhibited approximately 50% more revertants than did rat S9 activation at similar doses of basic fraction from cooked beef.

Further studies, comparing rat, hamster, and mouse, after pretreatment with different microsomal inducers, showed that PCB-treated hamster S9 gave the highest yield of revertants. Also effective in hamster, rat, and mouse were 3-methycholanthrene (3MC) and β-naphthoflavone (BNF). Much less effective (8- to 10-fold fewer revertants) were S9 preparations from animals treated with phenobarbital, pregnenolone-16, α-carbonitrile, and corn oil (control). The increased activation with PCB-, 3MC-, and BNF-induced S9 implies association with cytochrome P-448 induction. Increased mutagenicity of the basic fraction correlates not only with cytochrome P-448 induction but with increased aryl hydrocarbon hydroxylase activity as well.

Ultimately the risk to human beings from exposure to mutagens in cooked meat will have to be assessed. One of the preliminary steps to evaluate this risk involved studying the effect of human microsomal fractions in the activation of mutagens in the basic fraction. Six human liver microsomal preparations were analyzed, each from a different autopsied individual. A comparison was made between human activation and PCB-treated and control rat activation of the basic fraction. Although the PCB-treated rat

Table 6
COMPARISON OF RAT AND HAMSTER S-9 ACTIVATION OF THE BASIC FRACTION

Sample	Addition/plate	Source of S-9 (2 mg/plate)[a] Rat	Hamster
		(TA1538 revertants/plate)	
DMSO	100 μl	35	25
2-Aminoanthracene	5 μg	3290	451
3MC	50 μg	309	188
Basic fraction	3.2 gE[b]	—	389
	8.0 gE	678	1030
	16.0 gE	1080	1488

[a] All animals were induced with 500 mg/kg Aroclor 1254 5 days before sacrifice.
[b] 1 gE = 1 gram equivalent of original wet wt of uncooked meat.

showed the highest activity, one of the human samples gave 30 revertants per gE, which was more than 50% of the induced rat activity for the basic fraction. In all cases the human liver preparations activated the basic-fraction mutagens. Uninduced (control) rat preparations were 5- to 10-fold lower than the human preparations. Although the human samples showed individual differences in metabolism, it is clear that human liver has the correct composition of hemoproteins to activate the basic-fraction mutagens. These hemoproteins appear to be naturally induced and metabolize the cooked beef mutagen(s) to produce a significant number of *Salmonella* revertants (10 to 30 per gE of ground beef). However, Dolara et al.[71] observed poor activation of beef extract mutagens with human S9.

To further understand the type of hemoprotein involved in the human metabolism of the basic fraction, α-naphthoflavone (ANF), a competitive inhibitor for cytochrome P-448, was preincubated with the human microsomes before addition of the mutagen. Inhibition was dose-dependent and reached 80% of the untreated activity at 10 μg of ANF. The Aroclor-induced rat metabolism also showed similar inhibition, but the corn oil-treated control rat, presumably not induced for cytochrome P-448, was not inhibited. These findings suggest that in the human, as well as in the rat, mouse, and hamster, cytochrome P-448 hemoproteins are metabolizing the ground beef basic fraction to active mutagen(s). These results are consistent with reports on the metabolic activation of the tryptophan pyrolysis mutagens, Trp-P-1 and Trp-P-2.[72,74]

B. Metabolic Activation by Intestinal S9

Since the intestine is the first tissue to come in contact with the food mutagen and since a possible cause and effect relationship between colon cancer and diet has been postulated, the authors explored the oral induction of intestinal microsomal proteins. The hamster was used because its liver S9 is superior for activating the basic fraction of cooked beef. Table 7 shows that BNF induced the intestinal S9 to a level about 60% higher than that induced by BaP, the next best inducer. These results are comparable to those obtained by Strobel et al.[75,76] The induction by i.p. injection was only slightly higher than with oral administration. It is significant that BNF is such a good intestinal inducer, since natural flavonoids common in our diet are structurally similar to BNF. The flavonoids eaten one day could conceivably induce the intestinal hemoproteins to metabolize promutagens derived from cooking to mutagens in the intestine two days later.

Table 7
ACTIVATION OF THE BASIC FRACTION WITH HAMSTER INTESTINAL S-9

Pretreatment	TA1538 revertants/gE/10^8 bacteria[a] Oral	i.p.
Corn oil	1.65	
Phenobarbital	1.40	
Aroclor 1254	2.05	
BaP	3.35	
BNF	5.30	5.90

[a] See footnote *a*, Table 1. gE = dose of fraction corresponding to one gram equivalent of original wet wt of uncooked meat.

Table 8
COMPARISON OF THE *SALMONELLA* STRAIN SPECIFICITY FOR THE BASIC FRACTION OF GROUND BEEF

		Histidine revertants/plate[b]				
Sample	Addition/plate[a]	TA1535	TA1537	TA1538	TA98	TA100
DMSO	100 μℓ	32	8	40	40	260
2-Aminoanthracene	5 μg	246	135	1994	2382	1567
3MC	50 μg	31	44	170	505	1466
Basic fraction	1 gE[c]	0.02[d]	2.29	18.4	25.4	1.22

[a] 2 mg/plate of Aroclor-induced rat S-9 was used.
[b] ca. 10^8 bacteria.
[c] gE = gram equivalents of original wet wt of uncooked meat.
[d] See footnote *a*, Table 1.

VI. GENETIC TOXICITY OF BEEF THERMIC MUTAGENS

A. Mutagenicity in Various *Salmonella* Strains

Table 8 compares the mutagenicity with five *Salmonella* strains of the acetone-extracted basic fraction of ground beef. The frameshift-sensitive strains TA1537, TA1538 and TA98 respond with 2.3, 18.1, and 25.4 revertants per gE, respectively. The strains sensitive to base substitutions, TA1535 and TA100, did not respond with a significant increase over the spontaneous reversion level. In fact, up to 300 gE have been tested with TA100 and no significant increase was seen. Even though TA1537 and TA1538 have different mutations in DNA, both are subject to reverse mutations correcting the wrong reading frame after exposure to the basic fraction. Table 8 shows both substitution and frameshift strain responses for 2-aminoanthracene and 3MC. It is interesting that the basic fraction from fried beef, probably composed of aromatic amines, shows different strain specificity than 2-aminoanthracene, a standard 3-ring aromatic amine.

B. Mutagenicity and DNA Repair

One way to better understand the DNA interaction with the basic fraction compo-

Table 9
DEPENDENCE OF MUTAGENICITY IN *SALMONELLA* ON uvrB REPAIR SYSTEM

Agent	Revertants/10^8 cells/gE: Strain TA1538 (rfa, uvrB)	Strain TA1978 (rfa, uvrB+)	% repaired
Fried ground beef	53.8	1.94	96
2-Aminoanthracene	2419	53	98
3MC	188	7	97

nents is to take advantage of other *Salmonella* mutants. One of these mutants, TA1978, lacks the uvrB repair defect found in the standard tester strains.[69] An endonuclease is synthesized in TA1978 that can cleave lesions involving covalent binding to DNA. It was reasoned that if the basic fraction forms such lesions, the reversion rate should be markedly reduced. As shown in Table 9, the reversion rate dropped from 53.8 per gE in TA1538 to 1.9 in TA1978. A total of 96% of the lesions appear to have been repaired by this endonuclease. These findings suggest that the reversion observed in TA1538 is due to covalent binding of mutagen to DNA, rather than to intercalations, since the endonuclease would not be expected to repair noncovalently bound lesions.

C. CHO Cells in Culture

Significant results are not yet available in mammalian bioassays with crude basic fraction from ground beef or more purified fractions. The authors' current opinion, based on fragmentary data, is that unreasonably large quantities of beef base fraction would be required to obtain positive assay results. Therefore, a more appropriate strategy appears to be that being followed in other laboratories, namely, chemical fractionation and identification of individual mutagens followed by organic synthesis in sufficient quantity. This work is in progress for mutagens from fried beef and egg.

In order to improve the sensitivity of bioassays in CHO cells, the authors have utilized DNA repair-deficient strains isolated by Thompson et al.[77,78] One of these strains, UV-5, is about ten-fold more sensitive to killing by UV light than the wild type strain. It is also more sensitive to certain chemical mutagens, as well as being more mutable at the HPRT and APRT loci. The cytotoxicity, mutagenicity, and sister chromatid exchange induction by the standard mutagens Trp-P-2 and IQ have been compared (Figure 1). Trp-P-2 is a potent genotoxic agent at all three endpoints. Furthermore, it exhibits several-fold greater potency in each assay in UV-5 than in the wild type strain, AA-8. The differential cytotoxicity provides additional evidence that DNA injury is the mechanism of action of the chemical, since DNA repair capability is the only known difference between UV-5 and AA-8. As described in the preceding section, Trp-P-2 is a very potent mutagen in *Salmonella*. It has recently been reported to be carcinogenic in rats fed the agent chronically.[41]

Whereas IQ is more mutagenic than Trp-P-2 in *Salmonella*, it was found that IQ exhibited very low, but measurable, responses in UV-5 at all three endpoints. In UV-5 the ratio of the potencies of the two agents is approximately 100-fold for mutagenesis and sister chromatid exchange. In the wild type strain, the mutagenicity of IQ was barely detectable. The unexpected behavior of IQ represents the greatest uncoupling

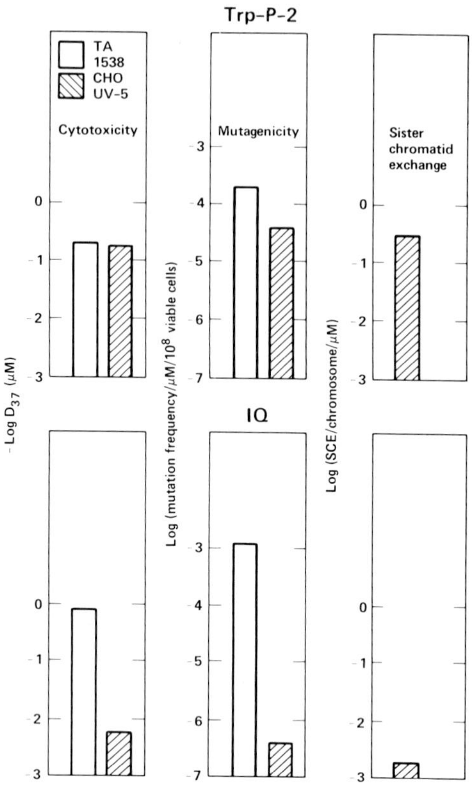

FIGURE 1. A comparison of the effects of Trp-P-2 and IQ for three test endpoints (cell killing, induced mutation, and induced sister chromatid exchange) in *Salmonella* strain TA1538, and CHO cell line UV-5 that is deficient in DNA excision repair. For comparison, all experimental values have been extrapolated to an exposure concentration of 1 μmol/ℓ.

of mutagenic potency between *Salmonella* and mammalian cellular systems so far reported. The authors are investigating the basis for these results in view of the similarity of the two agents in molecular weight, ring structure, and functional groups.

VII. SIGNIFICANCE OF THE RESEARCH

There is obviously a concern that the ingestion of genotoxic materials in the diet might contribute to the high incidence of cancer of the stomach and large intestine that afflicts the populations of many nations.[1,79-84] The known effects of the simpler aromatic amines such as 2-naphthylamine and acetylaminofluorene (AAF) are not reassuring, in that cancer of internal organs (liver, bladder) is induced by feeding these agents to animals.[85-88] However, there are many stages in the digestive process in which alteration or inactivation of particular mutagens might occur. And the doses of mutagen actually ingested by that proportion of the population that eats well-done meats must be determined and compared to the doses observed to be genotoxic in bioassay systems.

Based on the average daily intake of various sources of cooked protein,[35] the total mutagen intake for individuals cooking foods well-done is calculated to be approximately 5600 *Salmonella* revertants per day.[38,39] As reported above, improved extraction methods applied to beef have increased the yield of mutagens five- or six-fold. If similar increases also apply to other protein foods, the total daily intake would rise to approximately 30,000 revertants per day. This preliminary value must be viewed simply as a calibration point, since the quantitative genotoxic significance of consumption of bacterial mutagens is unknown. As soon as the chemical identity of the major thermic mutagens in cooked foods is known, an extensive battery of mammalian in vitro bioassays will be applied to each of the agents, including tests for cytotoxicity, drug resistance mutations, sister chromatid exchange, chromosome aberrations, and the micronucleus test — utilizing the deficient CHO cell lines for increased sensitivity. The in vitro battery will be followed by in vivo mammalian tests for cytogenetic endpoints and germ cell injury. Thus, an extensive data base in short-term bioassays shall be built to provide a foundation for assessment of human risk and for decisions concerning carcinogenesis bioassays. An indication of the particular importance of results in mammalian systems has already been provided by the testing reported here on the two previously identified cooking mutagens, Trp-P-2 and IQ. Whereas both agents proved to be extremely potent mutagens in *Salmonella,* there was a marked difference in their activities in several bioassays in CHO cells. If such variable behavior proves to be characteristic of the class of thermic mutagens, then assessment of any hazards arising from cooking the human diet will be a complex task indeed.

VIII. SUMMARY

The authors have presented an interim status report on a comprehensive investigation of the formation of mutagens during the cooking of foods under conditions prevalent in American households. The research to date has confirmed that mutagens of the frameshift type, detectable with a bacterial assay, are formed in the outer surface of many high-protein foods when cooked to a well-done state by methods involving direct heat transfer from a hot surface or by radiant heating. A study of extraction methods indicates that the mutagens are polar organic bases extracted most efficiently in an acidic aqueous medium. Preliminary fractionation experiments reveal at least four separable mutagenic components, only one of which may be identical to a series of identified mutagens derived from heated amino acids, proteins, or foods.

A model system has been devised for studying the formation of mutagens in beef at 100°C. This system shows that mutagen formation occurs in the water-soluble phase of meat and that the precursor molecules appear to be certain amino acids and an additional water-soluble component of unknown identity. The soluble system exhibits complex reaction kinetics and biphasic pH optima leading to different mutagenic products.

Metabolic activation is required for demonstration of mutagenicity in the bacterial bioassay. The activation appears to be mediated by enzymes of the cytochrome P-448 type. Both rodent and human liver S9 and microsomal preparations are effective. In the hamster intestinal S9 was effective and orally administered BNF, an analog of common dietary constituents, was the best of several inducers.

Study of two mutagens isolated from cooked foods, Trp-P-2 and IQ, in a battery of in vitro bioassays gave unexpected results. Trp-P-2 was a potent inducer of frameshift mutation in *Salmonella,* and of forward mutation to drug resistance at two loci and sister chromatid exchange in CHO cells. On the other hand, IQ, a very potent frameshift mutagen in *Salmonella* was much less potent in inducing mutation and sister chromatid exchange in CHO cells. These findings represent the most striking uncoupling of genotoxic potency between bacterial and mammalian systems so far reported. If such variable behavior proves characteristic of thermic mutagens then assessment of any human health hazard arising from cooking the human diet will require a comprehensive information base.

ACKNOWLEDGMENTS

The authors thank M. L. Mendelsohn for his continuing support of this research project and for helpful comments on the manuscript. The project overviewed in this article is a multidisciplinary team effort involving a number of staff members of the departments at Livermore and Berkeley. The following have made major contributions to the work reported here: H. Timourian for project coordination and biophysical studies; M. M. Morris on cooking methods; C. Plumlee on diet assessment; K. Grose and D. H. Stuermer, assisted by P. W. Berry, P. H. Davis, and M. G. Knize, on extraction methods and fractionation; H. Rapoport and A. Waterhouse of the Department of Chemistry at Berkeley for organic synthesis of the mutagen IQ; R. T. Taylor and V. G. Shore, assisted by E. D. Fultz, on the soluble model system; A. V. Carrano and L. H. Thompson, assisted by J. L. Minkler and E. P. Salazar on mammalian genotoxicity testing; and S. K. Healy, assisted by A. Shaieb for *Salmonella* testing.

Work performed under the auspices of the U.S. Department of Energy by the Lawrence Livermore National Laboratory under contract number W-7405-ENG-48 and supported by an Interagency Agreement (NIEHS 222Y01-ES-80038) between the National Institute of Environmental Health Sciences and the Department of Energy.

REFERENCES

1. **Sugimura, T., Kawachi, T., Nagao, M., and Yahagi, T.,** Mutagens in food as causes of cancer, *Prog. Cancer Res. Ther.*, 17, 59, 1981.
2. **Gray, J. I. and Morton, I. D.,** Some toxic compounds produced in food by cooking and processing, *J. Human Nutr.*, 35, 5, 1981.
3. **Grice, H. C., Clegg, D. J., Coffin, D. E., Lo, M.-T., Middleton, E. J., Sandi, E., Scott, P. M., Sen, N. P., Smith, B. L., and Withey, J. R.,** Carcinogens in foods, in *Carcinogens in Industry and the Environment*, Sontag, J. M., Ed., Marcel Dekker, New York, 1981, chap. 10.
4. **Lijinsky, W. and Shubik, P.,** Benzo[a]pyrene and other polynuclear hydrocarbons in charcoal-broiled meat, *Science*, 145, 53, 1964.
5. **Lijinsky, W. and Shubik, P.,** Polynuclear hydrocarbon carcinogens in cooked meat and smoked food, *Indust. Med. Surg.*, 34, 152, 1965.
6. **Lijinsky, W. and Ross, A. E.,** Production of carcinogenic polynuclear hydrocarbons in the cooking of food, *Food Cosmet. Toxicol.*, 5, 343, 1967.
7. **Masuda, Y., Mori, K., and Kuratsune, M.,** Polycyclic aromatic hydrocarbons in common Japanese foods. I. Broiled fish, roasted barley, shoyu, and caramel, *Gann*, 57, 133, 1966.
8. **Masuda, Y., Mori, K., and Kuratsune, M.,** Polycyclic aromatic hydrocarbons formed by pyrolysis of carbohydrates, amino acids, and fatty acids, *Gann*, 58, 69, 1966.
9. **Howard, J. W. and Fazio, T.,** A review of polycyclic aromatic hydrocarbons in foods, *J. Agric. Food Chem.*, 17, 527, 1969.
10. **Rhee, K. S. and Bratzler, L. J.,** Benzo[a]pyrene in smoked meat products, *J. Food Sci.*, 35, 126, 1970.
11. **Panalaks, T.,** Determination and identification of polycyclic aromatic hydrocarbons in smoked and charcoal-broiled food products by high pressure liquid chromatography and gas chromatography, *J. Environ. Sci. Health*, B11, 299, 1976.
12. **Nagao, M., Honda, M., Seino, Y., Yahagi, T., and Sugimura, T.,** Mutagenicities of smoke condensates and the charred surface of fish and meat, *Cancer Lett.*, 2, 221, 1977.
13. **Nagao, M., Honda, M., Seino, Y., Yahagi, T., Kawachi, T., and Sugimura, T.,** Mutagenicities of protein pyrolysates, *Cancer Lett.*, 2, 335, 1977.

14. **Kosuge, T., Tsuji, K., Wakabayashi, K., Okamoto, T., Shudo, K., Iitaka, Y., Itai, A., Sugimura T., Kawachi, T., Nagao, M., Yahagi, T., and Seino, Y.**, Isolation and structure studies mutagenic principles in amino acid pyrolysates, *Chem. Pharm. Bull.*, 26(2), 611, 1978.
15. **Sugimura, T., Nagao, M., Kawachi, T., Honda, M., Yahagi, T., Seino, Y., Sato, S., Matsukura, N., Matsushima, T., Shirai, A., Sawamura, M., and Matsumoto, H.**, Mutagen-carcinogens in foods, with special reference to highly mutagenic pyrolytic products in broiled foods, in *Origins of Human Cancer*, Hiatt, H. H., Watson, J. D., and Winsten, J. A., Eds., Cold Spring Harbor Laboratory, Cold Spring Harbor, New York, 1977, 1561.
16. **Matsumoto, T., Yoshida, D., Mizusaki, S., and Okamoto, H.**, Mutagenic activity of amino acid pyrolysate in *Salmonella typhimurium* TA98, *Mutat. Res.*, 48, 279, 1977.
17. **Matsumoto, T., Yoshida, D., Mizusaki, S., and Okamoto, H.**, Mutagenicities of pyrolysates of peptides and proteins, *Mutat. Res.*, 5, 281, 1978.
18. **Yoshida, D., Nishigata, H., and Matsumoto, T.**, Pyrolytic yields of 2-amino-9H-pyrido[2,3-b]indole and 3-amino-1-methyl-5H-pyrido[4,3-b]indole as mutagens from proteins, *Agric. Biol. Chem.*, 43(8), 1769, 1979.
19. **Vithayathil, A. J., Commoner, B., Nair, S., and Madyastha, P.**, Isolation of mutagens from bacterial nutrients containing beef extract, *J. Toxicol. Environ. Health*, 4, 189, 1978.
20. **Commoner, B., Vithayathil, A. J., Dolara, P., Nair, S., Madyastha, P., and Cuca, G. C.**, Formation of mutagens in beef and beef extract during cooking, *Science*, 201, 913, 1978.
21. **Commoner, B., Vithayathil, A., and Dolara, P.**, Mutagenic analysis as a means of detecting carcinogens in foods, *J. Food Prot.*, 41(12), 996, 1978.
22. **Dolara, P., Commoner, B., Vithayathil, A., Cuca, G., Tuley, E., Madyastha, P., Nair, S., and Kriebel, D.**, The effect of temperature on the formation of mutagens in heated beef stock and cooked ground beef, *Mutation Res.*, 60, 231, 1979.
23. **Pariza, M. W., Ashoor, S. H., Chu, F. S., and Lund, D. B.**, Effects of temperature and time on mutagen formation in pan-fried hamburger, *Cancer Lett.*, 7, 63, 1979.
24. **Ashoor, S. H., Dietrich, R. A., Chu, F. S., and Pariza, M. W.**, Proline enhances mutagen formation in ground beef during frying, *Life Sci.*, 26, 1801, 1980.
25. **Spingarn, N. E. and Weisburger, J. H.**, Formation of mutagens in cooked foods. I. Beef, *Cancer Lett.*, 7, 259, 1979.
26. **Spingarn, N. E., Slocum, L. A., and Weisburger, J. H.**, Formation of mutagens in cooked foods. II. Foods with high starch content, *Cancer Lett.*, 9, 7, 1980.
27. **Spingarn, N. E., Kasai, H., Vuolo, L. L., Nishimura, S., Yamaizumi, Z., Sugimura, T., Matsushima, T., and Weisburger, J. H.**, Formation of mutagens in cooked foods. III. Isolation of a potent mutagen from beef, *Cancer Lett.*, 9, 177, 1980.
28. **Spingarn, N. E., Garvie-Gould, C., Vuolo, L. L., and Weisburger, J. H.**, Formation of mutagens in cooked foods. IV. Effect of fat content in fried beef patties, *Cancer Lett.*, 12, 93, 1981.
29. **Weisburger, J. H. and Spingarn, N. E.**, Mutagens as a function of mode of cooking of meats, in *Naturally Occurring Carcinogens - Mutagens and Modulators of Carcinogenesis*, Miller, E. C., Miller, J. A., Hirono, I., Sugimura, T., and Takayama, S., Eds., University Park Press, Baltimore, 1979, 177.
30. **Uyeta, M., Kanada, T., Mazaki, M., Taue, S., and Takahashi, S.**, Assaying mutagenicity of food pyrolysis products using the Ames test, in *Naturally Occurring Carcinogens - Mutagens and Modulators of Carcinogenesis*, Miller, E. C., Miller, J. A., Hirono, I., Sugimura, T., and Takayama, S., Eds., University Park Press, Baltimore, 1979, 169.
31. **Kasai, H., Nishimura, S., Nagao, M., Takahashi, Y. and Sugimura, T.**, Fractionation of a mutagenic principle from broiled fish by high-pressure liquid chromatography, *Cancer Lett.*, 7, 343, 1979.
32. **Kasai, H., Yamaizumi, Z., Wakabayashi, K., Nagao, M., Sugimura, T., Yokoyama, S., Miyazawa, T., Spingarn, N.E., Weisburger, J. H., and Nishimura, S.**, Potent novel mutagens produced by broiling fish under normal conditions, *Proc. Jpn. Acad. Ser. B*, 56(5), 278, 1980.
33. **Yamaizumi, Z., Shiomi, T., Kasai, H., Nishimura, S., Takahashi, Y., Nagao, M., and Sugimura, T.**, Detection of potent mutagens, Trp-P-1 and Trp-P-2, in broiled fish, *Cancer Lett.*, 9, 75, 1980.
34. **Yamaguchi, K., Shudo, K., Okamoto, T., Sugimura, T., and Kosuge, T.**, Presence of 3-amino-1,4-dimethyl-5H-pyrido[4,3-b]indole in broiled beef, *Gann*, 71, 745, 1980.
35. **Yamaizumi, Z., Shiomi, T., Kasai, H., Wakabayashi, K., Nagao, M., Sugimura, T., and Nishimura, S.**, Quantitative analysis of a novel potent mutagen, 2-amino-3-methyl-imidazo(4,5-f)quinoline, present in broiled food by GC/MS, *Koenshu-Iyo Masu Kenkuyakai*, 5, 245, 1980.
36. **Matsumoto, T., Yoshida, D., and Tomita, H.**, Determination of mutagens, amino-α-carbolines, in grilled foods and cigarette smoke condensate, *Cancer Lett.*, 12, 105, 1981.
37. **Felton, J. S., Healy, S., Stuermer, D., Berry, C., Timourian, H., Hatch, F. T., Morris, M., and Bjeldanes, L. F.**, Mutagens from the cooking of food. I. Improved extraction and characterization of mutagenic fractions from cooked ground beef, *Mutat. Res.*, 88, 33, 1981.

38. **Bjeldanes, L. F., Morris, M. M., Felton, J. S., Healy, S., Stuermer, D., Berry, P., Timourian, H., and Hatch, F. T.**, Mutagens from the cooking of food. II. Survey by Ames/*Salmonella* test of mutagen formation in the major protein-rich foods of the American diet, *Food Chem. Toxicol.*, in press.
39. **Bjeldanes, L. F., Morris, M. M., Felton, J. S., Healy, S., Stuermer, D., Berry, P., Timourian, H., and Hatch, F. T.**, Mutagens from the cooking of food. III. Secondary sources of cooked dietary protein, *Food Chem. Toxicol.*, in press.
40. **Plumlee, C. Bjeldanes, L. F., and Hatch, F. T.**, Priorities assessment for studies of mutagen production in cooked foods, *J. Am. Diet. Assoc.*, 79, 446, 1981.
41. **Matsukura, N., Kawachi, T., Morino, K., Ohgaki, H., Sugimura, T., and Takayama, S.**, Carcinogenicity in mice of mutagenic compounds from a tryptophan pyrolyzate, *Science*, 213, 346, 1981.
42. **Bjeldanes, L. F., Morris, M. M., Timourian, H., and Hatch, F. T.**, Effects of meat composition and cooking conditions on mutagenicity of fried ground beef, in preparation.
43. **Yoshida, D., Matsumoto, T., and Nishigata, H.**, Effect of heating methods on mutagenic activity and yield of mutagenic compounds in pyrolysis products of protein, *Agric. Biol. Chem.*, 44(2), 253, 1980.
44. **Nader, C. J., Spencer, L. K., and Weller, R. A.**, Mutagen production during pan-broiling compared with microwave irradiation of beef, *Cancer Lett.*, 13, 147, 1981.
45. **Akimoto, H., Kawai, A., Nomura, H., Nagao, M., Kawachi, T., and Sugimura, T.**, Synthesis of potent mutagens in tryptophan pyrolysates, *Chem. Lett.*, 1061, 1977.
46. **Kasai, H., Nishimura, S., Wakabayashi, K., Nagao, M., and Sugimura, T.**, Chemical synthesis of 1-amino-3-methylimidazo-[4,5-f]quinoline (IQ), a potent mutagen isolated from broiled fish, *Proc. Jpn. Acad., Ser. B*, 56(6), 382, 1980.
47. **Yokoyama, S., Miyazawa, T., Kasai, H., Nishimura, S., Sugimura, T., and Iitaka, Y.**, Crystal and molecular structures of 2-amino-3-methylimidazo[4,5-f]quinoline, a novel potent mutagen found in broiled food, *FEBS Lett.*, 122(2), 261, 1980.
48. **Kasai, H., Yamaizumi, Z., Wakabayashi, K., Nagao, M., Sugimura, T., Yokoyama, S., Miyazawa, T., Spingarn, N. E., Weisburger, J. H., and Nishimura, S.**, Isolation and structure characterization of a potent mutagen in broiled fish. I. 2-amino-3-methylimidazo[4,5-f]quinoline, *J. Am. Chem. Soc.*, submitted.
49. **Kasai, H., Yamaizumi, Z., Wakabayashi, K., Nagao, M., Sugimura, T., Yokoyama, S., Miyazawa, T., and Nishimura, S.**, Structure and chemical synthesis of Me-IQ, a potent mutagen isolated from broiled fish, *Chem. Lett.*, submitted.
50. **Yoshida, D., Matsumoto, T., Yoshimura, R., and Matsuzaki, T.**, Mutagenicity of amino-α-carbolines in pyrolysis products of soybean globulin, *Biochem. Biophys. Res. Commun.*, 83(3), 915, 1978.
51. **Yoshida, D. and Matsumoto, T.**, Amino-α-carbolines as mutagenic agents in cigarette smoke condensate, *Cancer Lett.*, 10, 141, 1980.
52. **Kasai, H., Yamaizumi, Z., Shiomi, T., Yokoyama, S., Miyazawa, T., Wakabayashi, K., Nagao, M., Sugimura, T., and Nishimura, S.**, Structure of a potent mutagen isolated from fried beef, *Chem. Lett.*, 485, 1981.
53. **Kasai, H., Shiomi, T., Sugimura, T., and Nishimura, S.**, Synthesis of 2-amino-3,8-dimethylimidazo[4,5-f]quinoxaline (Me-IQx), a potent mutagen isolated from fried beef, *Chem. Lett.*, in press.
54. **Nishimura, S.**, personal communication, 1981.
55. **Taylor, R. T., Fultz, E., and Shore, V.**, Food mutagen formation in model boiling systems, *Environ. Mutagen.*, 3, 349, 1981.
56. **Shibamato, T. and Bernhard, R. A.**, Investigation of pyrazine formation pathways in sugar-ammonia model systems, *J. Agric. Food Chem.*, 25(3), 609, 1977.
57. **Spingarn, N. E. and Garvie, C. T.**, Formation of mutagens in sugar-ammonia model systems, *J. Agric. Food Chem.*, 27(6), 1319, 1979.
58. **Stich, H. F., Stich, W., Rosin, M. P., and Powrie, W. D.**, Mutagenic activity of pyrazine derivatives: a comparative study with *Salmonella typhimurium*, *Saccharomyces cerevisiae* and Chinese hamster ovary cells, *Food Cosmet. Toxicol.*, 18, 581, 1980.
59. **Yoshida, D. and Okamoto, H.**, Formation of mutagens by heating creatine and glucose, *Biochem. Biophys. Res. Commun.*, 96(2), 844, 1980.
60. **Shinohara, K., Wu, R.-T., Jahan, N., Tanaka, M., Morinaga, N., Murakami, H., and Omura, H.**, Mutagenicity of the browning mixtures of amino-carbonyl reactions on *Salmonella typhimurium* TA100, *Agric. Biol. Chem.* 44(3), 671, 1980.
61. **Watanabe, K. and Sato, Y.**, Some alkyl-substituted pyrazines and pyridines in the flavor components of shallow fried beef, *J. Agric. Food Chem.*, 19(5), 1017, 1971.
62. **Mussinan, C. J., Wilson, R. A., and Katz, I.**, Isolation and identification of pyrazines present in pressure-cooked beef, *J. Agric. Food Chem.*, 21, 871, 1973.

63. **Shibamoto, T.**, Heterocyclic compounds found in cooked meats, *J. Agric. Food Chem.*, 28, 237,1980.
64. **Shibamoto, T., Kamiya, Y., and Mihara, S.**, Isolation and identification of volatile compounds in cooked meat: sukiyaki, *J. Agric. Food Chem.*, 29, 57, 1981.
65. **Stich, H. F., Stich, W., Rosin, M. P., and Powrie, W. D.**, Clastogenic activity of caramel and caramelized sugars, *Mutat. Res.*, 91, 129, 1981.
66. **Powrie, W. D. and Nakai, S.**, Processing effects on protein systems, in *Utilization of Protein Resources*, Stanley, D. W., et al., Eds., Food & Nutrition Press, Westport, Conn., 1981, chap. 4.
67. **Anon.**, Food and nutrient intake of individuals in the United States, spring 1965, Rep. No. 11, U.S. Department of Agriculture, Washington, D.C., 1972.
68. **Anon.**, Plan and operation of the health and nutrition examination survey, U.S., 1971 to 1973. Vital and Health Statistics, Ser. 1(10a and 10b), DHEW Publ. 73-1310, Health Services and Mental Health Administration, U.S. Government Printing Office, Washington, D.C., 1973.
69. **Ames, B. N., Lee, F. D., and Durston, W. E.**, An improved bacterial test system for the detection and classification of mutagens and carcinogens, *Proc. Natl. Acad. Sci. U.S.A.*, 70(3), 782, 1973.
70. **Ames, B. N., Durston, W. E., Yamasaki, E., and Lee, F. D.**, Carcinogens are mutagens: a simple test system combining liver homogenates for activation and bacteria for detection, *Proc. Natl. Acad. Sci. U.S.A.*, 70(8), 2281, 1973.
71. **Dolara, P., Barale, R., Mazzoli, S., and Benetti, D.**, Activation of the mutagens of beef extract *in vitro*, and *in vivo*, *Mutat. Res.*, 79, 213, 1980.
72. **Ishii, K., Yamazoe, Y., Kamataki, T., and Kato, R.**, Metabolic activation of mutagenic tryptophan pyrolysis products by rat liver microsomes, *Cancer Res.*, 40, 2596, 1980.
73. **Ishii, K., Ando, M., Kamataki, T., Kato, R., and Nagao, M.**, Metabolic activation of mutagenic tryptophan pyrolysis products (Trp-P-1 and Trp-P-2) by a purified cytochrome P 450-dependent monooxygenase system, *Cancer Lett.*, 9, 271, 1980.
74. **Yamazoe, Y., Yamaguchi, N., Kamataki, T., and Kato, R.**, Metabolic activation of Trp-P-2, a mutagenic amine from tryptophan-pyrolysate, by liver microsomes from 3-methylcholanthrene-responsive and non-responsive mice, *Xenobiotica*, 10(7/8), 483, 1980.
75. **Fang, W.-F. and Strobel, H. W.**, Activation of carcinogens and mutagens by rat colon mucosa, *Cancer Res.*, 38, 2939, 1978.
76. **Strobel, H. W., Fang, W.-F., and Oshinsky, R. J.**, Role of colonic cytochrome P-450 in large bowel carcinogenesis, *Cancer (Philadelphia)*, 45, 1070, 1980.
77. **Brookman, K. W., Thompson, L. H., Salazar, E. P., Dillehay, L. E., and Mooney, C. L.**, Genetic complementation, DNA repair, and mutation induction in UV-sensitive Chinese hamster ovary cell mutants, *Environ. Mutagen.*, 3, 394, 1981.
78. **Thompson, L. H., Brookman, K. W., Carrano, A. V., and Dillehay, L. E.**, The role of DNA repair in mutagenesis of Chinese hamster ovary cells by 7-bromomethylbenz(a)anthracene, *Proc. Natl. Acad. Sci. U.S.A.*, 79, 534, 1982.
79. **Menck, H. R. and Henderson, B. E.**, Cancer incidence rates in the Pacific basin, *Proc. Natl. Cancer Inst. Monogr.*, 53, 119, 1979.
80. **Stemmermann, G. N., Mandel, M., and Mower, H. F.**, Colon cancer: its precursors and companions in Hawaii Japanese, *Proc. Natl. Cancer Inst. Monogr.*, 53, 175, 1979.
81. **Weisburger, J. H.**, On the etiology of gastro-intestinal tract cancers, with emphasis on dietary factors, in *Environmental Carcinogenesis: Occurrence, Risk Evaluation and Mechanisms*, Emmelot, P. and Kriek, E., Eds., Elsevier/North-Holland, Biomedical Press, Amsterdam, 1979, 215.
82. **Weisburger, J. H.** Mechanism of action of diet as a carcinogen, *Cancer (Philadelphia)*, 43, 1987, 1979.
83. **Weisburger, J. H. and Wynder, E. L.**, Nutrition and its relationship to cancer, *Adv. Cancer Res.*, 32, 237, 1980.
84. **Doll, R.**, General epidemiologic considerations in etiology of colorectal cancer, in *Colorectal Cancer: Prevention, Epidemiology, and Screening*, Winawer, S., Schottenfeld, D., and Sherlock, P., Eds., Raven Press, New York, 1980, 3.
85. **Kriek, E.**, Aromatic amines and related compounds as carcinogenic hazards to man, in *Environmental Carcinogenesis: Occurrence, Risk Evaluation and Mechanisms*, Emmelot, P. and Kriek, E., Eds., Elsevier/North-Holland, Biomedical Press, Amsterdam, 1979, 143.
86. **Thorgeirsson, S. S., Schut, H. A. J., and Wirth, P. J.**, Mutagenicity and carcinogenicity of aromatic amines, in *Molecular Basis of Environmental Toxicity*, Bhatnagar, R. S., Ed., Ann Arbor Science, Ann Arbor, Mich., 1980, chap. 11.
87. **Lotlikar, P. D.**, Nitrogenous carcinogenic industrial chemicals, in *Carcinogens in Industry and the Environment*, Sontag, J. M., Ed., Marcel Dekker, New York, 1981, chap. 9.
88. **Weisburger, E. K.**, N-2-fluorenylacetamide and derivatives, in *Carcinogens in Industry and the Environment*, Sontag, J. M., Ed., Marcel Dekker, New York, 1981, chap. 14.

Chapter 13

N-NITROSAMINES IN FOOD: OCCURRENCE AND REDUCTION OF EXPOSURE

R. Preussmann, G. Eisenbrand, and B. Spiegelhalder

TABLE OF CONTENTS

I. INTRODUCTION: CARCINOGENICITY OF *N*-NITROSO COMPOUNDS

N-nitroso compounds of the general formula $O{=}N{-}NR_1R_2$ can be subdivided into chemically stable nitrosamines where R_1 and R_2 usually are alkyl and/or aryl groups, and reactive unstable nitrosamides with R_1 an alkyl and R_2 an acyl residue. This group of compounds has been shown to possess potent carcinogenic and mutagenic activity in experimental systems.[1-4]

The known carcinogenic effects in experimental animals can be summarized as follows:

- More than 300 different *N*-nitroso compounds have been tested for carcinogenicity. About 90% of them are carcinogens.
- NDMA and NDEA have been investigated in about 30 animal species including subhuman primates. There is no animal species known to be resistant.
- The most important facet of nitrosamine carcinogenesis is the organ-specificity of this effect, which is governed mainly by the chemical structure but also by the animal species used, the dosage, and the mode of application. Malignant tumors in practically all important organs of mammals can be induced in high incidence and often selectively.
- Available low-dose experiments indicate that concentrations of between 0.03 to 1 mg/kg food of the environmentally important nitrosamines are still active.
- Although there is no direct proof of carcinogenicity of nitrosamines in man, much indirect evidence makes it highly probable that this group of known animal chemical carcinogens is also active in human beings.

II. HUMAN EXPOSURE TO *N*-NITROSO COMPOUNDS

We now have good evidence that man is exposed to *N*-nitroso compounds. This was probably first suggested by Druckrey and Preussmann,[5] however it was only after the development of highly specific and sensitive analytical methods[6,7] that exposure to volatile *N*-nitrosamines could be unequivocally proven. Our present knowledge on human exposure to *N*-nitroso compounds is summarized in Figure 1.

We must consider intake of *N*-nitroso compounds from exogenous sources such as food, tobacco, cosmetics, etc., and also occupational exposure in certain areas. In addition, and unique for this group of carcinogens, *N*-nitroso compounds may also be formed in the human body, generally in the acidic stomach milieu, from precursor amines and nitrosating agents such as nitrite or nitrous gases.[8] Both precursors occur ubiquitously in the human environment.

It is evident that human exposure to *N*-nitroso compounds is only one factor in the total burden of man.[6,9-11] This fact should always be kept in mind, together with the knowledge that exposure to nonvolatile or labile *N*-nitroso compounds, e.g., nitrosamides, cannot currently be estimated since no adequate analytical methods are available for this kind of exposure.

A summary follows of the authors' work on volatile nitrosamines in food.

III. SURVEY OF NITROSAMINES IN FOOD

A survey was performed in 1978 and 1979 on almost 3000 samples of commercially available food products. The selection of samples was based on the average consumption figures in Germany in order to give a representative picture. It has to be pointed out that not only products "under suspicion" were analyzed.

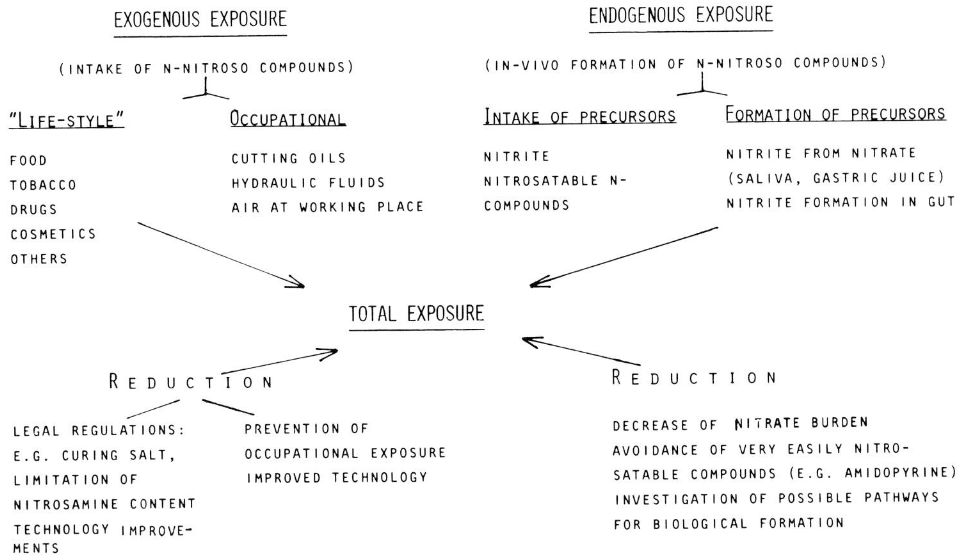

FIGURE 1. Human exposure to N-nitroso compounds.

Table 1
NITROSAMINES IN FOOD — SURVEY OF 2826 SAMPLES (169 TYPES OF FOOD) FROM THE GERMAN MARKET, 1978

ppb	NDMA (%)	NPYR (%)	NPIP (%)
<0.5	70	97	98
0.5—4.9	24	2	1
>5	6	1	1

Table 1 provides a short summary of the results from 2826 samples investigated. Only three nitrosamines, *N*-nitrosodimethylamine (NDMA), *N*-nitrosopyrrolidine (NPYR), and *N*-nitrosopiperidine (NPIP) were found more or less consistently, while all other possible volatile nitrosamines like *N*-nitrosodiethylamine (NDEA) were present only in very exceptional cases. As can be seen, NPYR and NPIP in concentrations $\geqslant$ 0.5 ppb (micrograms per kilogram) were found in only 3% and 2% of the samples, respectively. NDMA, however, was detectable in 30% of all samples, and 6% of them were found to contain more than 5 ppb.

A. Meat and Meat Products

Earlier results for this group of food products have shown that nitrite- and/or nitrate-cured products contain up to 25 ppb NDMA and up to 100 ppb NPYR and, in some cases even higher values.[12-16] The main problem in Anglo Saxon countries is fried bacon. Havery et al.,[17] however, have recently shown that analytical data from FDA investigations clearly indicate a downward trend for NPYR contamination in bacon:

Table 2
NITROSAMINES IN FOODS ON THE GERMAN MARKET, 1978, 2826 SAMPLES FROM 169 TYPES OF FOOD

Food type	*N*-nitrosamine	<0.5 ppb	0.5—4.9 ppb	>5 ppb	x_{max}ppb
Meat, sausages	NDMA	268	118	9	12
	NPYR	345	27	24	45
Cheese	NDMA	159	48	1	5

in 1971 the average value was 87 ppb and in 1977 it was 20 ppb. This decrease is the result of technological improvements, such as reduction of nitrite use and addition of sodium ascorbate. The authors' own results confirm that only nitrite- and/or nitrate-treated products are nitrosamine-contaminated, with some very rare exceptions proving the rule. A total of 395 samples of this group were investigated, and the results are shown in Table 2.

Heating of cured meat products usually does not change the concentrations of NDMA. NPYR, however, increases considerably upon heating. The authors' results confirm similar results from the literature.[18,19] NDMA is volatile and evaporates quickly during frying. The increase in NPYR probably is the result of increased nitrosamine synthesis at elevated temperatures from residual nitrite as well as from heat-induced decarboxylation of *N*-nitroso amino acids. The latter compounds (*N*-nitroso sarcosine, -proline, and -3-hydroxyproline) have been determined to be present in concentrations up to 161 ppb in cured meat products.[20] *N*-Nitroso-3-hydroxypyrrolidine, another nonvolatile nitrosamine, also has been observed[21] in 7 out of 33 samples of fried and cured meat products. It is probably formed by heat-induced decarboxylation from the corresponding *N*-nitroso-hydroxyproline.

B. Cheese

Of the 209 samples analyzed, 23% (48 samples) were found to be positive for NDMA in the range of 0.5 to 4.9 ppb; only one sample (5 ppb) was above that range. The values are shown in Table 2, and correlate well with other published data from a collaborative study.[22]

C. Other Types of Food

More than 2000 samples including all other types of food such as bread, cereals, fish, milk and milk products, potatoes, vegetables, fruit, etc., showed only sporadic occurrence of volatile nitrosamines, almost always at or near the detection limit of the analytical method (0.1 to 0.5 ppb). The results obtained were included in the evaluation of the average daily intake of volatile nitrosamines, since, on a weight basis, the consumption of these types of food is high. The "occasional" occurrence of nitrosamines should be further investigated to either confirm or refute some of these data.

D. Evaluation

The results obtained from about 3000 samples of about 170 different types of food allow a calculation of the average daily intake of volatile nitrosamines from food in Germany (Table 3). These calculations are based on the average per capita consumption for the different food products. The consumption data are based on the semiofficial "Nutritional Report 1976" ("Ernahrungsbericht 1976") and refer to adult male persons only.

Table 3
AVERAGE DAILY INTAKE OF NITROSAMINES FOR MALE ADULTS CALCULATED FROM THE AVERAGE PER CAPITA FOOD CONSUMPTION[a]

Food	Nitrosamine	μg/capita	Food consumption (g)
Meat and meat products	NDMA	0.1	210
	NPYR	0.1	
	NPIP	0.01	
Cheese	NDMA	0.01	30
Other foods	NDMA	0.2	1500
	NPYR	0.03	

[a] Average per capita food consumption. From Nutritional Report 1976, Federal Republic of Germany.

Table 4
N-NITROSODIMETHYLAMINE IN BEER: COMPARISON OF DATA BEFORE AND AFTER INTRODUCTION OF MODIFIED MALT DRYING METHODS

	1978/1979				1981			
Type of beer	No. of samples	% positive (>0.5 ppb)	Mean ppb	Max ppb	No. of samples	% positive (>0.5 ppb)	Mean ppb	Max ppb
Pilsen lager	54	65	1.2	7	169	24	0.43	6.5
Pale lager & export lager	42	67	1.2	7	179	26	0.39	2.0
Pale strong lager	25	76	1.9	8	38	26	0.42	1.6
Top-fermented pale ales	22	23	1.0	6	19	5	0.32	0.7
Top-fermented dark ales	25	76	2.7	11	21	24	0.96	7.0
Dark lager & dark strong lager	22	68	6.0	47	25	32	0.51	4.0
"Rauchbier" (from smoked malt)	9	100	18.0	68	3	100	1.50	2.0[a]
All types	199	66	2.5	68	454	24	0.44	7.0

[a] Values of 1980.

Gough et al.,[23] on the basis of 493 samples analyzed, calculated the average per capita intake from the normal U.K. diet to be about 1 μg per week for NDMA (and other dialkylnitrosamines) and 3 μg per week for NPYR and NPIP. The data compare quite well with that of the authors. Fine[24] has estimated the daily per capita intake from nitrite-preserved foods based on published data to amount to 1 μg NDMA and 5 μg NPYR "for the worst case" (i.e., 100 g of cooked bacon per day). These data seem to be too high, as compared to both Gough's and the authors' data.

IV. NDMA IN BEER

The most surprising and unexpected result of the systematic survey of commercial food samples was the finding of a regular contamination of beer with NDMA.[25] The data of 1978 and 1979 are summarized in Table 4 and show characteristic differences between different types of beer.

These data have been confirmed by several other investigators[26-32] and show that this contamination was a general situation. In view of the high consumption of beer (565 g per day for males and 255 g per day for females as a statistical average for

Table 5
NDMA CONCENTRATION (μg/kg) IN MALT, DEPENDING ON DIFFERENT DRYING TECHNIQUES

	Heating technique		
	Indirect	Direct	
Type of malt		Oil burners	Gas burners
Pale malt	<1	0.5—10	15—80
Dark malt	2—8	6—10	80—320

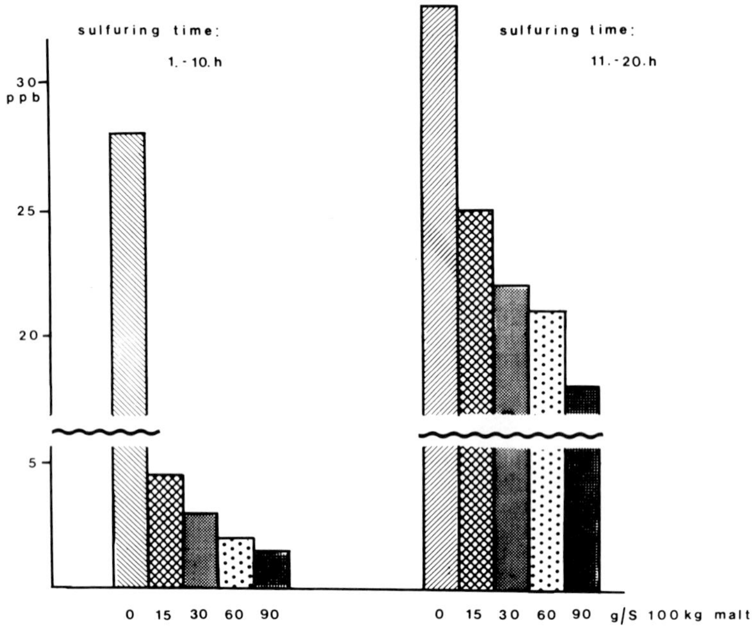

FIGURE 2. Reduction of NDMA in malt by sulfuring.

Germany, i.e., an average daily NDMA intake of 0.7 μg per male adult), further investigations on the causes and possible eliminatory steps were necessary.

A subsequent investigation on the origin of NDMA in beer was done by analyzing every stage of the brewing process together with all ingredients used. Invariably it was found that malt was practically the sole source of NDMA contamination. Therefore a wide variety of different malt samples was analyzed, and it was found that NDMA is formed during kilning (drying) of green malt. The extent of nitrosamine formation clearly depended on the type of heating technique used, as shown in Table 5.

It is obvious from Table 5 that the highest formation of NDMA occurred in malts dried in direct-fired kilns. Unfortunately, gas-fired kilns are predominant in most countries at present. The next logical step was to try and prevent this NDMA formation during kilning. In cooperation with brewing experts, it was found out that sulfur burning ("sulfuring"), i.e., SO_2-treatment during drying, significantly reduced NDMA formation. As shown in Figure 2, this method is quite effective, but only if applied during the first 10 hr when the malt is still wet. In the later stages sulfuring is much less effective.

Since NO_x is a precursor in NDMA formation, and high combustion temperatures (usually from 1500 to 1800°C) yield high reaction rates between oxygen and nitrogen, a decrease in NDMA formation can also be achieved by lowering the flame temperature. Excess air seems to be the most economic way to reduce flame temperature and NO_x synthesis. In a new type of burner developed on this principle, the resulting air had only 0.05 to 0.1 mg NO_x/m^3 as compared with 14 mg/m^3 in conventional burners. Accordingly, malt dried with such burners contains only 1.3 mg/kg NDMA, a 15- to 30-fold reduction of the NDMA concentration.

Using these improved methods for malt kilning, it is now possible to produce malt with low NDMA contamination, even with existing direct-fired kilns. A comparison of NDMA concentration in commercially available beer in 1978/1979 and in 1981 after introduction of the new technique, is given in Table 4. Similar reductions of NDMA in beer after a change of malt kilning methods have been reported recently.[33,34] Mangino et al.[33] have investigated possible main precursors of NDMA in malt. Dimethyl- and trimethylamine, and the tertiary amine alkaloids hordenine and gramine, all natural constituents of germinating green malt, have been nitrosated in model systems. It was shown that at pH 4.4 or pH 6.4 and 65°C, hordenine and gramine were both nitrosated to yield NDMA. Gramine, especially, was highly susceptible to nitrosation at pH 4.4 giving a yield of NDMA equal to that obtained from dimethylamine.

Traces of NDMA have also been found in other malt-derived beverages such as Canadian rye and Scotch whiskey,[27,29] but not in other spirits or wine.

V. CONCLUDING REMARKS

A last word on possible consequences from the data on nitrosamine content of foods. The occurrence of such potent carcinogens in the daily food supply clearly cannot be ignored. On the other hand, an inappropriate dramatization would be equally wrong, since concentrations in food and ensuing human exposures are low. On a purely scientific basis, a quantitative risk evaluation cannot be undertaken at present. This does not mean that preventive measures should be postponed. They seem necessary in view of the total exposure shown in Figure 1, especially in regard to the various open problems concerning the endogenous exposure by in vivo formation of *N*-nitroso compounds. There is much evidence for this formation, but it cannot be quantified at present. The examples of NDMA in beer and in other environmental media[34] indicate that knowledge of the causes of contamination will also lead to technological improvements that will remove, or at least decrease, the contamination and thus reduce human exposure. These preventive measures can usually be effected without great economic consequences. Similar measures in other fields of exposure may not be as simple, but they seem possible and should be tried. They are the ultimate aim of scientific work in this field.

REFERENCES

1. **Preussmann, R. and Stewart, B. W.**, *N*-nitroso carcinogens, in *Chemical Carcinogens*, Searle, C. E., Ed., Monogr. No. 173 (revised edition), American Chemical Society, Washington, D.C., in preparation.
2. **Magee, P. N., Montesano, R., and Preussmann, R.**, *N*-nitroso-compounds and related carcinogens, in *Chemical Carcinogens*, Searle, C. E., Ed., Monogr. No. 173, American Chemical Society, Washington, D.C., 1976, 491.
3. **Druckrey, H., Preussmann, R., Ivankovic, S., and Schmähl, D.**, Organotrope carcinogene Wirkung bei 65 verschiedenen *N*-Nitroso-Verbindungen an BD-Ratten, *Z. Krebsforsch.*, 69, 103, 1967.

4. International Agency for Research on Cancer, Monogr. on the Evaluation of the Carcinogenic Risk of Chemicals to Humans. Some *N*-Nitroso compounds, Vol. 17, IARC, Lyon, 1978.
5. **Druckrey, H. and Preussmann, R.**, Zur Entstehung carcinogener Nitrosamine am Beispiel des Tabakrauchs, *Naturwissenschaffen,* 49, 498, 1962.
6. **Eisenbrand, G.**, *N*-Nitrosoverbindungen in Nahrung und Umwelt, *Wissenschaften Verlagsgesellschaft, Stuttgart,* 1981.
7. **Fine, D. H. and Rounbehler, D. P.** Trace analysis of volatile *N*-nitroso compounds by combined gas chromatography and thermal energy analysis (TEA), *J. Chromatogr.,* 108, 271, 1975.
8. **Sander, J.**, Kann Nitrit in der menschlichen Nahrung Ursache einer Krebsentstehung durch Nitrosaminbildung sein? *Arch. Hyg.,* 151, 22, 1967.
9. **Fine, D. H., Rounbehler, D. P., Fan, T., and Ross, R.**, Human exposure to *N*-nitroso compounds in the environment, in *Origins of Human Cancer,* Book A, Hiatt, H. H., Watson, J. D., and Winsten, J. A., Eds., Cold Spring Harbor Laboratory, Cold Spring Harbor, N.Y., 1977, 263.
10. **Preussmann, R., Eisenbrand, G., and Spiegelhalder, B.**, Occurrence and formation of *N*-nitroso compounds in the environment and in vivo, in *Environmental Carcinogenesis: Occurrence, Risk Evaluation and Mechanisms,* Emmelot, P. and Kriek, E., Eds., Elsevier/North-Holland Biomedical Press, Amsterdam, 1979, 51.
11. **Preussmann, R., Spiegelhalder, B., and Eisenbrand, G.**, Reduction of human exposure to environmental *N*-nitroso carcinogens. Examples of possibilities for cancer prevention, in *Carcinogenesis. Fundamental Mechanisms and Environmental Effects,* Pullman, B., Ts'o, P. O. P., and Gelboin, H., Eds., D. Reidel Publishers, Boston, 1980, 273.
12. **Crosby, N. T., Foreman, J. K., Palframan, J. F., and Sawyer, R.**, Estimation of steam-volatile *N*-nitrosamines in foods at the 1 μg/kg level, *Nature (London),* 238, 342, 1972.
13. **Sen, N. P., Iyengar, J. R., Donaldson, B. A., and Panalaks, T.**, Effect of sodium nitrite concentration on formation of nitrosopyrrolidine and dimethylnitrosamine in fried bacon, *J. Agric. Food Chem.,* 22, 540, 1974.
14. **Sen, N. P., Seaman, S., and Miles, F.**, Volatile nitrosamine in various cured meat products. Effect of cooking and recent trends, *J. Agric. Food Chem.,* 27, 1354, 1979.
15. American Meat Institute Research Staff, A survey of nitrosamines in sausages and dry-cured meat products, *Food Technol. (Chicago),* 34, 45, 1980.
16. **Pensabene, W., Feinberg, J. I., Piotrowski, E. G., and Fiddler, W.**, Occurrence and determination of *N*-nitrosopyrrolidine and *N*-nitrosoproline in cured meat products, *J. Food Sci.,* 44, 1700, 1979.
17. **Havery, D. C., Fazio, T., and Howard, J. W.**, Survey of cured meat products for volatile *N*-nitrosamines, in *Environmental Aspects of N-Nitroso Compounds,* Walker E. A., Castegnaro, M., Griciute, L., and Lyle, R. E., Eds., IARC Sci. Publ. 19, International Agency for Research on Cancer, Lyon, 1978, 4.
18. **Lijinsky, W., Keefer, L., and Loo, J.**, The preparation and properties of some nitrosoamino acids, *J. Natl. Cancer Inst.,* 57, 1311, 1976.
19. **Pensabene, J. W., Fiddler, W., Gates, R. A., Fajen, J. C., and Wasserman, A. E.**, Effect of frying and other cooking conditions on nitrosopyrrolidine formation in bacon, *J. Food Sci.,* 39, 314, 1974.
20. **Janzowski, C., Eisenbrand, G., and Preussmann, R.**, Occurrence of *N*-nitrosoamino acids in cured meat products and their effect on formation of nitrosamines during heating, *Food Cosmet. Toxicol.,* 16, 343, 1978.
21. **Janzowski, C., Eisenbrand, G., and Preussmann, R.**, Occurrence and determination of *N*-nitroso-3-hydroxypyrrolidine in cured meat products, *J. Chromatogr.,* 150, 216, 1978.
22. **Elgersma, R. H. C., Sen, N. P., Stephany, R., Schuller, P. L., Webb, K. S., and Gough, T. A.**, A collaborative examination of some Dutch cheeses for the presence of volatile nitrosamines, *Neth. Milk Dairy J.,* 32, 125, 1978.
23. **Gough, T. A., Webb, K. S., and Coleman, R. F.**, Estimate of the volatile nitrosamine content of UK food, *Nature (London),* 272, 161, 1978.
24. **Fine, D. H.**, An assessment of human exposure to *N*-nitroso compounds, in *Environmental Aspects of N-Nitroso Compounds,* Walker, E. A. et al., Eds., IARC Sci. Publ. 19, International Agency for Research on Cancer, Lyon, 1978, 267.
25. **Spiegelhalder, B., Eisenbrand, G., and Preussmann, R.**, Contamination of beer with trace quantities of *N*-nitrosodimethylamine, *Food Cosmet. Toxicol.,* 17, 29, 1979.
26. **Walker, E. A., Castegnaro, M., Garren, L., Toussaint, G., and Kosalski, B.**, Intake of volatile nitrosamines from consumption of alcohols, *J. Natl. Cancer Inst.,* 63, 947, 1979.
27. **Goff, E. U. and Fine, D. H.**, Analysis of volatile *N*-nitrosamines in alcoholic beverages, *Food Cosmet. Toxicol.,* 17, 569, 1979.
28. **Scanlan, R. A., Barbour, J. F., Hotchkiss, J. H., and Libbey, L. M.**, *N*-nitrosodimethylamine in beer, *Food Cosmet. Toxicol.,* 18, 27, 1980.
29. **Sen, N. P., Seaman, S., and McPherson, M.**, Nitrosamines in alcoholic beverages, *J. Food Safety,* 2, 13, 1980.

30. **Fazio, T., Havery, D. C., and Howard, J. W.,** Determination of volatile *N*-nitrosamines in foodstuffs, in *N-Nitroso Compounds: Analysis, Formation and Occurrence,* Walker, E. A. et al., Eds., IARC Sci. Publ. 31, International Agency for Research on Cancer, Lyon, 1980, 419.
31. **Kawabata, T., Uibu, J., Oshima, H., Matsui, M., Hamano, M., and Tokiwa, H.,** Occurrence, formation, and precursors of *N*-nitroso compounds in the Japanese diet, in *N-Nitroso Compounds: Analysis, Formation and Occurrence,* Walker, E. A., et al., Eds., IARC Sci. Publ. 31, International Agency for Research on Cancer, Lyon, 1980, 481.
32. **Kann, J., Tauts, O., Kalve, R., and Bogovski, P.,** Potential formation of *N*-nitrosamines in the course of technological processing of some foodstuffs, in *N*-Nitroso Compounds: Analysis, Formation and Occurrence, Walker, E. A., et al., Eds., IARC Sci. Publ. 31, International Agency for Research on Cancer, Lyon, 1980, 319.
33. **Mangino, M. M., Scanlan, R. A., and O'Brien, T. J.,** Nitrosamines in beer, in *N-Nitroso Compounds,* Scanlan, R. A. and Tannenbaum, S. R., Eds., Am. Chem. Soc. Symp. Ser. 174, 1981, 229
34. **Preussmann, R., Spiegelhalder, B., and Eisenbrand, G.,** Reduction of human exposure to environmental *N*-nitroso compounds, in *N-Nitroso Compounds,* Scanlan, R. A. and Tannenbaum, S. R. Eds., Am. Chem. Soc. Symp. Ser. 174, 1981, 271.

Chapter 14

POLYCYCLIC AROMATIC HYDROCARBONS (PAH)

Bruce P. Dunn*

TABLE OF CONTENTS

* Dr. B. P. Dunn is a Research Scholar of the Ministry of Health and Welfare, Canada.

I. INTRODUCTION

PAH are a long-recognized group of chemical carcinogens composed chiefly of compounds consisting of multiple fused benzene rings. They are one of the major classes of carcinogens that can be isolated from cigarette tar and are a major carcinogenic constituent of air pollution particulates. They are widely distributed in the environment, and it is not surprising that they have repeatedly been found in a variety of foodstuffs. Because of their potent carcinogenicity, the presence of these compounds in food consumed by humans has been of considerable concern and has prompted a number of surveys of the levels of PAH in specific foodstuffs. The literature on levels of PAH in foods has been extensively reviewed by Lo and Sandi,[1] and Howard and Fazio.[2]

II. SOURCES OF PAH IN FOODS

There are two major sources of PAH in foods. The first is pyrolysis, in which PAH are formed by the high temperature pyrolysis or incomplete combustion of materials containing carbon and hydrogen. The mechanisms of this formation are not completely understood, although schemes for the genesis of specific hydrocarbons such as BaP have been proposed.[3] Table 1 indicates three routes by which foods may become contaminated with pyrolytically generated PAH. Air pollution particulates that contain appreciable amounts of pyrolytically generated PAH, can contaminate the surface of growing fruits, vegetables, and grain.[4] Only a small proportion of this surface contamination can be removed by washing with water,[4] presumably reflecting particulate material that has become incorporated into waxes or other materials on leaf surfaces, or that has penetrated pores in the foodstuff. A similar deposition of pyrolytically generated PAH on foods is achieved by the deliberate exposure of foods to wood smoke in order to enhance flavor or storage characteristics.[5] A third form of this contamination occurs during the charbroiling of meats or fish. The presence of PAH in such foods is dependent on grilling under conditions in which fat dripping from the meat can fall onto hot coals or metal and be pyrolyzed.[6] Foods grilled to the same degree of "doneness" by infrared heating without the possibility of flames do not show detectable levels of PAH. This, and the fact that the PAH occur at or near the surface of the meat rather than in the interior, suggest that charbroiling of meat or fish results in the deposition of soot on the food, in effect, smoking the meat as it is cooked.

The second major source of contamination of foodstuffs is through contact with either petroleum products or coal tar products. These materials are frequently rich in naturally occurring PAH that have been generated by geochemical processes. Table 2 shows some of the routes by which food may be contaminated by these sources. Low levels of some noncarcinogenic PAH have been detected in solvents used to refine vegetable oils.[7] This could potentially be a source of the PAH that have been reported to be present in vegetable oils,[8] although contamination arising from the deposition of air pollution particulates on the vegetable crop used as the source of oil is generally thought to be more important.[4] Petroleum-derived paraffin wax used for food packaging could potentially contaminate foods. This form of contamination, however, is readily controlled by applying strict standards to the purity of waxes used in food packaging.[9] Petroleum pollution and contamination of the marine environment by creosote from wharfs and docks can contaminate seafoods.[10] In this case, the contamination does not appear to be direct but to proceed via the contamination of the water column.[11] Finally, single cell protein preparations prepared from microorganisms using

Table 1
ROUTES BY WHICH FOODS MAY BECOME CONTAMINATED WITH PYROLYTICALLY GENERATED PAH

Source	Route	Contaminated food
Air pollution particulates	Deposition on growing plants	Leafy vegetables, grain, fruit
Wood smoke or smoke extracts	Smoking of foods	Smoked meats, fish
Burning fat	Deposited on grilled foods	Charbroiled steaks, fish

Table 2
ROUTES BY WHICH FOODS MAY BECOME CONTAMINATED WITH PETROLEUM AND COAL TAR PRODUCTS

Source	Route	Contaminated food
Solvents	Refining of edible oils	Vegetable oils(?)
Paraffin wax	Food packaging	Packaged foods (under control)
Crude and refined oils, coal tar products	Water pollution	Seafoods
Petroleum as substrate for microorganisms	Production of single cell protein	Single cell protein preparations

petroleum as a carbon source may contain petroleum-derived PAH.[12] This latter problem may restrict the use of such protein preparations in human foods and pose problems, such as carcinogenicity to animals and carcinogen residues in animal tissues, when single cell protein preparations are used as animal feeds.

Biosynthesis of PAH by plants, algae, or microorganisms has frequently been suggested in the past as a third source of these compounds in the environment. In a typical experiment, seeds, soil, and nutrient solutions shown to be free of PAH were used to grow plants that on analysis showed readily detectable levels of PAH. This and similar experiments with other organisms have been cited as evidence for the biosynthesis of PAH. However, a compound such as BaP, consisting of multiple fused benzene rings without any functional groups, is far from being a typical biosynthetic product, and furthermore has no obvious biological function. The experiments of Borneff et al.,[13] who grew algae with ^{14}C-acetate as a carbon source and claimed the production of radiolabeled PAH, are frequently cited as positive proof that biosynthesis of PAH can take place. A critical reading of their report, however, reveals that there is no evidence that the measured radioactivity was actually in the detected PAH.

An alternative explanation of the results is that the chromatographic spots analyzed, in addition to containing nonradioactive PAH from a nonbiosynthetic source, also contained some other undetected hydrocarbon synthesized by the algae from the radioactive precursor. This suggestion is reinforced by the fact that the specific activities of the "synthesized" PAH varied widely, and were not consistent with their production from the radioactive precursor. An alternative explanation for the apparent synthesis of PAH by organisms has been provided by Grimmer and Duvel.[14] They showed that when plants were grown in carefully filtered air, no PAH could be detected, whereas if the plants were grown in the open air, PAH could be readily demonstrated in leaves and stems. In the case of higher plants, it therefore seems that PAH are not produced by biosynthesis but arise from the deposition of air pollution particulates on the growing plant. A similar explanation probably accounts for the PAH seen in the algae experiments of Borneff et al.[13]

Seafoods present an interesting example of the various routes by which human foods may become contaminated with PAH. Bivalve molluscs, such as clams, oysters, and mussels, readily accumulate PAH from contaminated waters. Most commercial samples of these seafoods contain at least trace levels of BaP. In contrast, vertebrate fish, even living in moderately contaminated waters, do not appear to accumulate substantial levels of PAH carcinogens.[10] The most likely explanation for this difference is that, while fish readily metabolize PAH to water-soluble derivatives that are excreted, shellfish lack any appreciable ability to metabolize these compounds. PAH taken up from the water column thus have a long residence time in shellfish,[15] which can only eliminate them by partitioning them into the water column in their original, hydrophobic form.

Although vertebrate fish, when freshly caught, do not generally exhibit detectable levels of PAH, canned fish that have been processed with the addition of vegetable oil show, in some cases, appreciable levels of contamination.[10] This most probably results from contamination of the vegetable oil itself, and has nothing to do with the fish. Finally, fresh, PAH-free fish may still become highly contaminated with carcinogens if they are smoked or charbroiled.

III. ANALYTICAL PROCEDURES FOR PAH IN FOODS

The evaluation of the potential danger of PAH in foods is complicated by the difficulties of separating and measuring the large variety of PAH isomers present in foods and environmental samples. Early procedures based on column and TLC followed by UV and fluorescence spectroscopy have a limited ability to separate and measure different PAH compounds. To a large extent, the compounds chosen to be measured in various food surveys have depended more on the technical feasibility of making a measurement and the availability of standard reference compounds than on the necessary importance of the compound. There is, therefore, a large amount of data on levels of pyrene and fluoranthene in foods, since these compounds are normally present in large amounts in PAH-contaminated foods, and are easily separated and measured. These two compounds, however, are not animal carcinogens, and the extensive body of information on their occurrence in foods, although of academic interest, does not help in the evaluation of the potential human health risks of PAH in foods.

Older methods of analysis of PAH in foods have now been largely overtaken by modern GC and HPLC techniques. Table 3 indicates some of the relative merits of these procedures as an analytical tool for measuring PAH levels in foods. GC, using capillary columns, is well-established as a procedure for separating PAH.[16] The high resolving power inherent in long capillary columns and the ready availability of online MS make this technique attractive when it is desired to quantitate a large variety of PAH in food samples. The availability of relatively nonselective detectors, such as flame ionization detectors, allows ready quantitation of compounds for which no analytical reference compound is available. Capillary GC is the better technique to use when foods contain large numbers of alkylated PAH, and where the parent nonsubstituted compounds may represent only a minor fraction of the PAH present. However, for most foods, the parent nonsubstituted PAH predominate, and the enormous resolving power of capillary GC is not essential. In these cases, HPLC may be the preferable technique. HPLC is somewhat faster than GC. Although they do not have the theoretical resolving power of capillary GC columns, reversed phase HPLC columns can readily separate a number of PAH isomers that are difficult to separate by GC. Notable among these are the various benzofluoranthene isomers, some of which are carcinogenic and some of which are not. UV absorption and fluorescence detectors used in HPLC are often highly selective. This complicates quantitative analysis and

Table 3
RELATIVE MERITS OF HPLC AND GC AS ANALYTICAL TOOLS FOR MEASURING PAH LEVELS IN FOODS

HPLC	GC
Faster	Less expensive
Separates some key isomers difficult by GC	More resolving power (capillary columns)
Selective UV and fluorescence detectors aid discrimination	Nonselective detector aids quantitation
More sensitive	Easily coupled to MS

Table 4
LEVELS OF BaP FOUND IN VARIOUS FOODSTUFFS

Food	Typical level (ng/g)	
Smoked meats, fish	0—1	Lightly smoked
	1—10	Heavily smoked
Charbroiled meat, fish	1—10	Normal cooking
	10—50	Heavily charred
Leafy vegetables, grains, fruit	0—1	Grown in clean air
	1—10	Grown in polluted air
Vegetable oils	0—2	
Seafood	0—10	Grown in clean water
	10—100 +	Grown in polluted water

makes essential the availability of analytical standards for each compound to be quantitated. However, this selectivity may be of great advantage in discriminating between one compound and another, even if they are not completely resolved by the chromatographic column.[17] Finally, the extremely high sensitivity and selective nature of fluorescence detectors may ease the problems of sample preparation. Since this detector does not respond to nonfluorescent hydrocarbons present in a partially purified extract, it may be possible to simplify sample extraction and purification procedures and carry out analyses on semipurified extracts.

IV. LEVELS OF PAH CARCINOGENS IN FOODS

The most important and sometimes the only carcinogen measured in surveys for carcinogenic PAH in foods is BaP. Table 4 shows typical levels of BaP that have been found in various foodstuffs. The values shown in this table are not meant to represent absolute limits of levels that have been found in different foods, but rather as a guide to typical levels found in various surveys. It is evident that a large variety of foods are contaminated with PAH and that no one type of food predominates as a source of PAH in the diet. In view of the widespread nature of the contamination of foods by PAH, it is perhaps understandable why little or no epidemiological evidence is available linking PAH consumption to human cancer. In North America, it will probably prove impossible to clearly identify groups of people with a greater or lesser consumption of PAH. Complicating any attempts to perform an epidemiological analysis of the risks of PAH in foods is the presence of confounding factors, such as the association between consumption of charbroiled meats and consumption of animal fat, a food

Table 5
STRONG CARCINOGENS (+ + +) FOUND IN VARIOUS FOODS

Compound	Ratio relative to BaP
Dibenz[*a,h*]anthracene	0—0.2 vegetables, 0.5 lobster
Benzo[*c*]phenanthrene, Dibenzo[*a,h*]pyrene, Dibenzo[*a,i*]pyrene	Little or no data available, probably less than 0.1
DMBA, 3MC	Probably not present except in trace amounts

Note: Ratios calculated from data tabulated in Reference 1, except lobster. See Reference 10.

Table 6
STRONG CARCINOGENS (+ +) FOUND IN VARIOUS FOODS

Compound	Ratio relative to BaP
Benzo[*b*]fluoranthene	1—2 Smoked, broiled fish, 0.9 lobster
Benzo[*j*]fluoranthene	0.5—1 Smoked, broiled fish
Cholanthrene	No data

Note: Ratios calculated from data tabulated in Reference 1, except lobster. See Reference 10.

component that is an independent risk factor for some types of cancer. The only substantial study investigating a link between PAH in foods and cancer was performed by Dungal,[18] who examined the high rate of stomach cancer in Iceland in relation to the consumption of large amounts of heavily smoked foods in the traditional Icelandic diet. Although somewhat anecdotal, this study that analyzed over 2500 individual cases, strongly appears to suggest a link between stomach cancer and the consumption of smoked foods.

Tables 5, 6, and 7 relate the levels of BaP to be found in various foods to the levels of other carcinogenic PAH found in the same samples. The compounds are listed in three somewhat arbitrary groups, ranging from the most carcinogenic (+ + +) to the least carcinogenic (+), as classified by Lo and Sandi.[1] BaP is a + + + carcinogen. Among the strong carcinogens listed in Table 5, the major compound in foods according to available analytical data is dibenz[*a,h*]anthracene, having typical levels of perhaps 10 to 20% of the level of BaP. Little or no analytical data are available for the compounds benzo[*c*]phenanthrene, dibenzo[*a,i*]pyrene, and dibenzo[*a,h*]anthracene, although it is probable that the levels of these compounds are less than 10% of the levels of BaP. Although there have been occasional reports of the presence of DMBA and 3MC in foods, it seems likely that these compounds are not normally present except in trace amounts. These two compounds are not generally found in environmental samples, but are synthetic chemicals of great carcinogenic potential that were developed during investigations of the molecular basis of PAH carcinogenicity. The reported occurrence of these compounds in foods probably results from the

Table 7
WEAK CARCINOGENS (+) FOUND IN VARIOUS FOODS

Compound	Ratio relative to BaP
Benzo[*a*]anthracene	0.25—5 Smoked foods, broiled meats, vegetables, 1.1 Lobster
Benzo[*e*]pyrene	0.5—2 Smoked foods, broiled meats 1.0 Lobster
Indeno[1,2,3-*cd*]pyrene	0.25—4 Smoked foods, broiled meats 0.4 Lobster
Dibenzo[*a,j*]anthracene	No data
Dibenzo[*a,c*]anthracene	No data
Dibenzo[*e,l*]pyrene	No data
2-Methylfluorene	No data

Note: Ratios calculated from data tabulated in Reference 1, except lobster. See Reference 10.

misidentification of other PAH resulting from a combination of low resolution chromatography and the ready availability of these compounds as chromatographic standards.

Table 6 lists the levels of moderately strong PAH carcinogens in foods relative to BaP levels. Both benzo[*b*]fluoranthene and benzo[*j*]fluoranthene are present in amounts approximately comparable to BaP. Unfortunately, these compounds have not been frequently measured in foods, while the noncarcinogenic benzo[*k*]-fluoranthene is often measured. The relatively high levels of the two carcinogenic benzofluoranthene isomers in foods indicate a need in future analytical techniques to separate and measure these compounds in any survey of PAH carcinogen levels in foods.

Table 7 lists the levels of weak PAH carcinogens in foods relative to BaP levels. While this table includes all PAH listed as "+" by Lo and Sandi,[1] the carcinogenic activity of some, such as benzo[*e*]pyrene, is debatable. Benz[*a*]anthracene, chrysene, benzo[*e*]pyrene, and indeno[*1,2,3-cd*]pyrene all appear in contaminated foods at approximately the same level as BaP, while there appears to be no analytical data for dibenzo[*a,j*]anthracene, dibenzo[*e,l*]pyrene, and 2-methylfluorene.

V. SIGNIFICANCE OF PAH IN FOODS

To give some perspective to the importance of PAH in foods, it is informative to examine the occurrence of mutagens and carcinogens in cooked or charbroiled beef. As mutagens, measured by *Salmonella* mutagenesis assays, PAH as a class represent only a fraction of the total mutagenicity in this food. Hatch et al.[19] report that the total mutagenicity extractable from 100 g of cooked (not charbroiled) beef is equivalent to approximately 40,000 revertants. In comparison, 100 g of a charbroiled steak (contaminated at a level of 10 ng BaP per gram of meat) contains approximately 1 μg of BaP. This amount of carcinogen is capable of producing approximately 500 revertants in the most responsive Ames *Salmonella* strain.[20] Assuming that BaP represents approximately 10% of the total mutagenicity of the PAH in the food (as it does for soot[21]), this implies that mutagenicity from this class of compounds is limited to approximately 5000 revertants or less.

As carcinogens, however, the PAH in cooked meats (when present) may be much more important than their mutagenicity would indicate. Many food mutagens that have so far been studied, such as IQ and MeIQ, formed during normal low temperature cooking of beef, and Tryp 1 and Tryp 2, formed during the pyrolysis of protein, are extremely potent. These compounds may yield 100 to 1000 times as many revertants per microgram as does BaP. Thus the specific mutagenicity of IQ has been reported as 433,000 revertants per microgram in Ames strain TA98,[22] while BaP yields only 137 revertants per microgram in TA98 and 437 revertants per microgram in TA100.[20] Allowing for the high inherent mutagenicity of IQ, the amounts of this compound in cooked beef are exceedingly small. Even if all the mutagenicity in the basic fraction from cooked beef (approximately 2500 TA98 revertants per 100 g[19]) were accounted for by IQ, this would still represent only 6 ng of mutagen, compared with 1000 ng of BaP present if the beef were charbroiled. In addition, the importance of beef-cooking mutagens may be further diminished by low inherent carcinogenicity. The only cooking-derived mutagen so far tested for carcinogenicity, Trp-P-2, is only a weak carcinogen.[23] PAH, however, are very strong carcinogens, not only when administered by skin painting but also when administered orally. A single dose of as little as 200 μg of BaP administered orally has been reported to induce stomach tumors in 45% of mice.[24]

PAH carcinogens that have been found in foods are carcinogenic in a variety of animal species.[1,25] PAH-containing materials, such as shale oil distillates and creosote, are recognized human carcinogens, as are air pollution particulates, cigarette smoke, and coke oven emissions. Those PAH carcinogens that have been tested by oral administration to animals show activity by this route.[1,25] There are suggestive, although not conclusive, studies linking PAH-containing foods to human cancer.[18] Thus there are reasons for believing that PAH in human foods may pose a risk to human health. The main point at issue is the extent of this risk and whether it is acceptable, or should steps be taken to reduce it.

REFERENCES

1. **Lo, M. T. and Sandi, E.,** Polycyclic aromatic hydrocarbons (polynuclears) in foods, *Residue Rev.,* 69, 35, 1978.
2. **Howard, J. W. and Fazio, T.,** Review of polycyclic aromatic hydrocarbons in foods, *J. Assoc. Off. Anal. Chem.,* 63, 1077, 1980.
3. **Badger, G. M.,** Mode of formation of carcinogens in human environment, *Natl. Cancer Inst. Monogr.,* 9, 1, 1962.
4. **Tilgner, D. J.,** Food in a carcinogenic environment, *Food Manuf.,* 47, 1970.
5. **Tilgner, D. J. and Daun, H.,** PAH (polynuclears) in smoked foods, *Residue Rev.,* 24, 42, 1968.
6. **Lijinsky, W. and Ross, A. E.,** Production of carcinogenic polynuclear hydrocarbons in the cooking of food, *Food Cosmet. Toxicol.,* 5, 343, 1967.
7. **Howard, J. W., Fazio, T., and White, R. H.,** PAH in solvents used in extraction of edible oils, *J. Agric. Food Chem.,* 16, 72, 1966.
8. **Howard, J. W., Turicchi, E. W., White, R. H., and Fazio, T.,** Extraction and estimation of PAH in vegetable oils, *J. Assoc. Off. Anal. Chem.,* 49, 1236, 1966.
9. **Howard, J. W. and Haenni, E. O.,** The extraction and determination of polynuclear hydrocarbons in paraffin waxes, *J. Assoc. Off. Anal. Chem.,* 46, 933, 1963.
10. **Dunn, B. P. and Fee, J.,** Polycyclic aromatic hydrocarbon carcinogens in commercial seafoods, *J. Fish. Res. Board Can.,* 36, 1469, 1979.
11. **Dunn, B. P.,** Polycyclic aromatic hydrocarbons in marine sediments, bivalves, and seaweeds: analysis by high pressure liquid chromatography, in *Polynuclear Aromatic Hydrocarbons: Chemistry and Biological Effects,* Bjorseth, A. and Dennis, A. J., Eds., Battelle Press, Columbus, Ohio, 1980, 367.

12. **Santoro, A., Modica, R., Paglialunga, S., and Bartosek, I.,** Rapid determination of polycyclic aromatic hydrocarbons (PAH) on yeasts grown on normal paraffins and molasses, *Toxicol. Lett.*, 3, 85, 1979.
13. **Borneff, J., Selenka, F., Kunte, H., and Maximos, A.,** Experimental studies on the formation of polycyclic aromatic hydrocarbons in plants, *Environ. Res.*, 2, 22, 1968.
14. **Grimmer, G. and Duvel, D.,** Untersuchungen zur endogenen Bildung von polycyclischen Kohlenwasserstoffen in hoheren Pflanzen, *Z. Naturforsch.*, 25b, 1171, 1970.
15. **Dunn, B. P. and Stich, H. F.,** Release of the carcinogen benzo(a)pyrene from environmentally contaminated mussels, *Bull. Environ. Contam. Toxicol.*, 15, 398, 1976.
16. **Grimmer, G. and Bohnke, H.,** PAH profile analysis of high protein foods, oils, and fats by gas chromatography, *J. Assoc. Off. Anal. Chem.*, 58, 725, 1976.
17. **Dunn, B. P. and Armour, R. J.,** Sample extraction and purification for determination of polycyclic aromatic hydrocarbons by reversed-phase chromatography, *Anal. Chem.*, 52, 2027, 1980.
18. **Dungal, N.,** The special problem of stomach cancer in Iceland with particular reference to dietary factors, *JAMA*, 178, 93, 1961.
19. **Hatch, F. T., Felton, J. S., and Bjeldanes, L. F.,** Mutagens from the cooking of food: thermic mutagens in beef, in *Carcinogens and Mutagens in the Environment, Vol. 1, Food Products*, Stich, H. F., Ed., CRC Press, Boca Raton, Fla., 1983.
20. **McCann, J., Spingarn, N. E., Koboari, J., and Ames, B. N.,** Detection of carcinogens as mutagens: bacterial tester strains with R factor plasmids, *Proc. Natl. Acad. Sci. U.S.A.*, 72, 979, 1975.
21. **Kaden, D. A., Hites, R. A., and Thilly, W. G.,** Mutagenicity of soot and associated polycyclic aromatic hydrocarbons to *Salmonella typhimurium*, *Cancer Res.*, 39, 4152, 1979.
22. **Kasai, H., Yamaizumi, Z., Wakabayashi, K., Nagao, M., Sugimura, T., Yokoyama, S., Miyazawa, T., and Nishimura, S.,** Structure and chemical synthesis of Me-IQ, a potent mutagen isolated from broiled fish, *Chem. Lett.*, 1391, 1980.
23. **Matsukura, N., Kawachi, T., Morino, K., Ohgaki, H., and Sugimura, T.,** Carcinogenicity in mice of mutagenic compounds from a tryptophan pyrolysate, *Science*, 213, 346, 1981.
24. **Pierce, W. E. H.,** Tumor promotion by lime oil in the mouse forestomach, *Nature (London)*, 189, 497, 1961.
25. International Agency for Research on Cancer, Monogr. on the Evaluation of the Carcinogenic Risk of Chemicals to Man, Vol. 3, Certain Polycyclic Aromatic Hydrocarbons and Heterocyclic Compounds, IARC, Lyon, 1973.

Chapter 15

THE HEALTH SIGNIFICANCE OF MUTAGENS IN FOODS

D. R. Stoltz, B. Stavric, R. Klassen, and T. Muise

TABLE OF CONTENTS

I. INTRODUCTION

Sugimura[1] published a plea in 1978 to the mutagen research community titled "Let's Be Scientific About the Problem of Mutagens in Cooked Food." The parting line in that communication stressed that more exhaustive data on mutagens and carcinogens in foods should be accumulated. While data indicating the presence of mutagens in many foods have piled up, comparable information on nonmutagenic foods has received less attention. Furthermore, progress in determining the health significance of mutagens in food has not been dramatic.

The authors' approach to the problem of mutagens in food consists of an attempt to develop perspective by mutagenicity screening of a variety of widely consumed foods coupled with a more detailed investigation of foods exhibiting mutagenic activity.

II. SCREENING FOODS FOR MUTAGENIC ACTIVITY

A. Objectives

The major aim of this survey of foods for mutagenic activity was to develop a background against which to view new reports of mutagens in foods. Foods were to be chosen on the basis of high consumption, without singling out those suspected of containing mutagenic natural constituents or nitrosation or pyrolysis products. Bias in the mutagenicity test itself was also to be avoided.

B. Methods

1. Sample Preparation

To date, studies have included only beverages, fruits, and vegetables. In the beverage series, a total of 28 different products, all widely consumed national brands were selected for testing (Table 1). A total of 13 fresh and dried fruits and 27 raw or cooked, fresh, canned, or frozen vegetables have been screened (Table 2).

Foods were concentrated and fractionated to give 5 to 7 fractions. A 750 mℓ sample of each beverage was obtained and prepared as illustrated in Figure 1. In the case of fruits and vegetables, 250 g were blended with 500 mℓ water prior to extraction. Fractions 2 and 6 of fruits and vegetables were discarded since they contained mainly insoluble fiber.

2. Mutagenicity Assay

The *Salmonella*/microsome plate incorporation test was performed as described by Ames et al.,[2] except that plates were incubated for 3 days.[3] Food fractions were assayed for mutagenic activity with two strains of *Salmonella*, TA98 and TA100, with and without Aroclor-induced rat S9 mix. One sample of each product was subjected to the mutagenicity screen outlined in Figure 2.

C. Results

Few of the 28 beverages and 40 fruits and vegetables screened were mutagenic. Results for only those products that demonstrated appreciable mutagenic activity are given in Table 3. Three beverages, red wine, grape juice, and instant coffee showed reproducible mutagenic activity, while strawberries, raspberries, peaches, raisins, and onions showed potent mutagenic activity for *Salmonella*.

D. Discussion

A number of limitations of this strategy for screening food for mutagenic activity are recognized, relating to sampling, the extraction procedures, and the mutagenicity screen itself. Possible limitations include:

Table 1
BEVERAGES ASSESSED FOR MUTAGENIC POTENTIAL

Tap water	Pineapple	Beer and ale
	Orange	Light
Bottled carbonated		Dark
soft drinks	Vegetable juice	
Cola	Mixed	Wine
Ginger ale	Tomato	Red
Fruit-flavored		White
Root beer	Dairy products	
	Milk — high fat	Liquor
Powdered soft drinks	Milk — low fat	Whiskey
Orange	Chocolate drink	Gin
Cherry		
	Tea	Liqueur
Frozen concentrated	Brewed	
fruit juice (orange)	Instant	
Canned or bottled	Coffee	
fruit juice	Percolated	
Grape	Instant	
Apple		

Table 2
FRUIT AND VEGETABLES ASSESSED FOR MUTAGENIC POTENTIAL

Raw fresh fruit	Cooked frozen vegetables	Carrot
Apple	Bean	Corn
Banana	Corn	Pea
Cherry	Pea	Potato, boiled
Grapefruit		Rhubarb
Orange	Raw fresh vegetables	Spinach
Peach	Cabbage	Turnip
Pear	Celery	
Pineapple	Cucumber	
Plum	Lettuce	
Raspberry	Onion	
Strawberry	Pepper	
	Radish	
Dried fruit	Tomato	
Prune		
Raisin	Cooked fresh vegetables	
	Asparagus	
Cooked canned vegetables	Bean sprouts	
Bean	Broccoli	
Corn	Brussels sprouts	
Tomato	Cabbage	

1. Only one sample of each product was screened
2. No hydrolysis or simulated digestive process was employed
3. Mutagens partitioned into several fractions
4. Volatile mutagens might have been lost
5. Only two strains of *Salmonella* and one S9 concentration were used

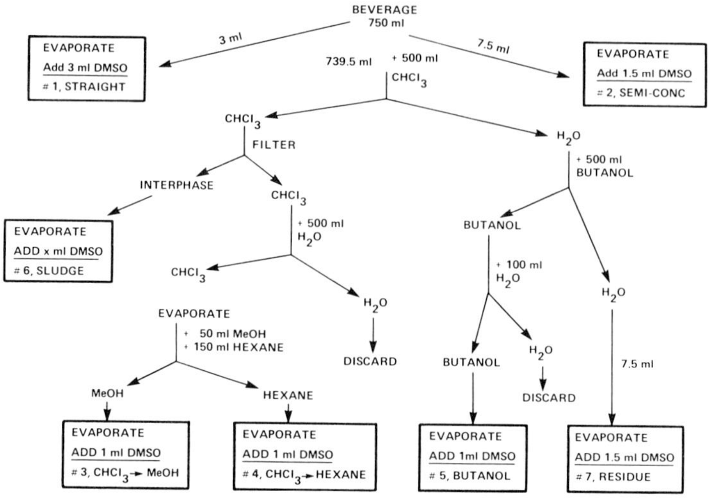

FIGURE 1. Preparation of beverage samples for mutagenicity testing. Fruit and vegetables were prepared similarly (see text).

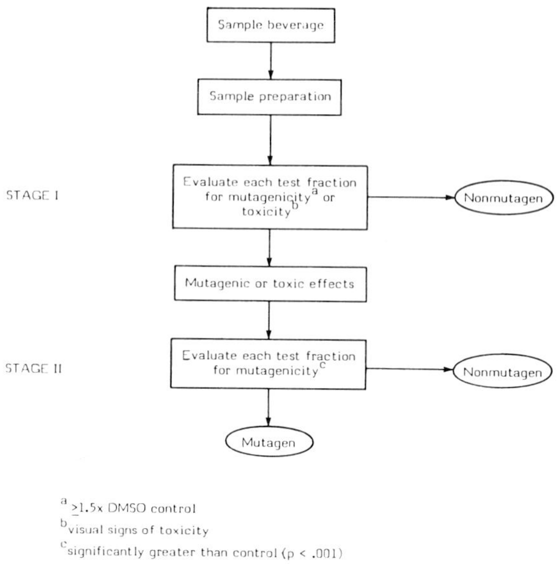

FIGURE 2. Mutagenicity test strategy. At stage I, any treatment for which the estimated mean number of revertants per plate exceeded 1.5 times that of the concurrent solvent control was repeated. In addition, any treatment resulting in visual signs of toxicity was repeated. At stage II, each of the repeated test fractions was then assessed statistically for mutagenic potential.

Table 3
RESULTS OF MUTAGENICITY SCREEN[a]

Product	Maximum treated vs. control ratio at stage II for fraction:						
	1	2	3	4	5	6	7
Beverages							
Instant coffee	—[b]	—	1.8	2.0	1.7	—	—
Grape juice	—	—	1.4	—	4.9	—	—
Red wine	—	2.2	10.6	2.1	12.3	—	—
Fruit and vegetables							
Strawberries	—	NP[c]	4.7	3.9	5.3	NP	0.9
Raspberries	—	NP	10.7	9.5	6.7	NP	1.1
Peaches	—	NP	2.1	8.8	5.8	NP	—
Raisins	1.9	NP	1.3	0.9	8.3	NP	1.8
Onions	1.1	NP	2.3	1.0	17.9	NP	—

[a] Only products showing appreciable mutagenic activity are shown.
[b] Fraction not repeated at stage II.
[c] Fraction not prepared at stage I.

Table 4
MUTAGENICITY SCREEN OF SPIKED BEVERAGES

Product	Spiking agent	ppm	Treated vs. control ratio for fraction		
			3	4	5
Cola	BaP	1.3	6.6	12.6	5.1
Ginger ale	β-Naphthylamine	6.5	6.2	6.2	1.6
Tomato juice	Quercetin	13	—[a]	—	30.4
Tomato juice	Quercetin	0.2	—	0.8	3.4
Beer	Tryptophan pyrolysate	13	12.1	4.7	1.5
Milk	4-Nitrophenylene-*o*-diamine	6.5	12.5	—	10.3

[a] Fraction not repeated at stage II.

Nevertheless, the utility of the fractionation procedure and the sensitivity of the screening strategy were established initially by assaying six beverages spiked with known mutagens at concentrations of 0.2 to 13 ppm prior to fractionation (Table 4).

When the mutagenic products (Table 3) were examined in terms of the fraction containing the activity, the differential response of TA98 and TA100, and the effect of metabolic activation, certain patterns emerged. Not surprisingly, grape juice, red wine, and raisins were similar, with TA98 more responsive than TA100 and a greatly enhanced effect with metabolic activation. Onion showed the same pattern of responses. The naturally occurring flavonol, quercetin, has been identified in these products.

Raspberries, strawberries, and peaches shared a pattern of responses, with TA100 being more responsive, and S9 decreasing the mutagenic activity. The mutagenicity of these fruits is due to the fungicide, captan.

Coffee stood alone with marginal, but reproducible mutagenic activity primarily for TA100 in the absence of liver homogenate. The origin of the weak mutagenicity observed with coffee is under investigation.

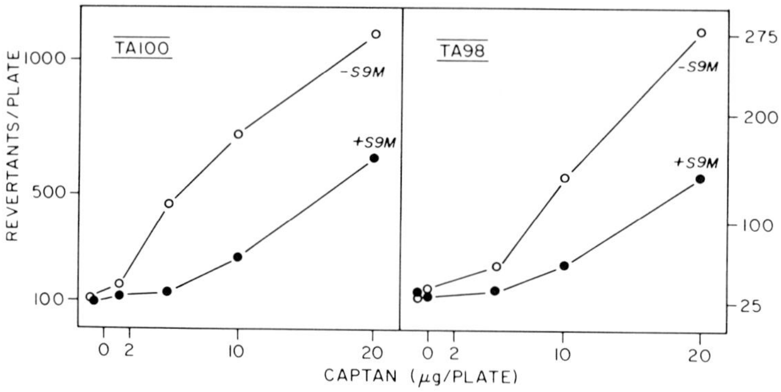

FIGURE 3. Mutagenicity of captan for *Salmonella typhimurium* TA98 and TA100. Fortified liver homogenate (S9M) was prepared from Aroclor-induced rats.

III. CAPTAN

A. Identification as the Mutagen on Strawberries, Raspberries, and Peaches

Strawberries were one of the first products to be screened in the fruit and vegetable series. The berries were purchased at a local grocery store. As they were being prepared for extraction, it was noticed that the strawberries on top were in perfect shape while those at the bottom of the package were covered with a fine fuzz of mold. Although only good berries had been selected for the mutagenicity screen, when a strong mutagenic response was observed it was assumed that the mutagen was a mold product acquired by contact with the fuzzy berries at the bottom of the package. Fortunately, these were still on hand, covered with a much denser growth of mold. When tested for mutagenic activity, however, the moldy strawberries gave a response identical to that of the good berries, lending little support to the initial assumption implicating a mold metabolite.

At this point, a colleague donated fresh home-grown strawberries produced without the help of any pesticides. When these were found to be completely nonmutagenic, the possibility of a mutagenic pesticide was considered. Examination of Division 15, Table II of the Food and Drug Regulations for mutagenic agricultural chemicals permitted on strawberries focused attention on the fungicide, captan, which has a maximum residue limit of 25 ppm on strawberries in Canada.[4]

Captan is a well-known bacterial mutagen with a pattern of responses for *Salmonella* strains identical to that observed with fractions of strawberries (Figure 3).[5] Confirmation of the presence of captan in strawberries was achieved by sequential HPLC: GC:MS (Figure 4). Captan was also identified in the mutagenic fractions of the samples of raspberries and peaches. The amount of captan on these fruits was estimated on the basis of mutagenic potency and by analytical chemistry to be about 2 to 3 ppm (assuming 100% recovery) which is well within the permitted residue limit.

B. Captan Residues on Fruit

Shortly after the identification of captan as the mutagen in the fruit samples, the safety of captan was questioned in an RPAR (Rebuttable Presumption Against Registration) issued by the Environmental Protection Agency (EPA) in the U.S., and repeatedly by the press.[6,7] A relevant question in determining the safety of captan as a food chemical is the stability of residues on food crops and the effects of normal food

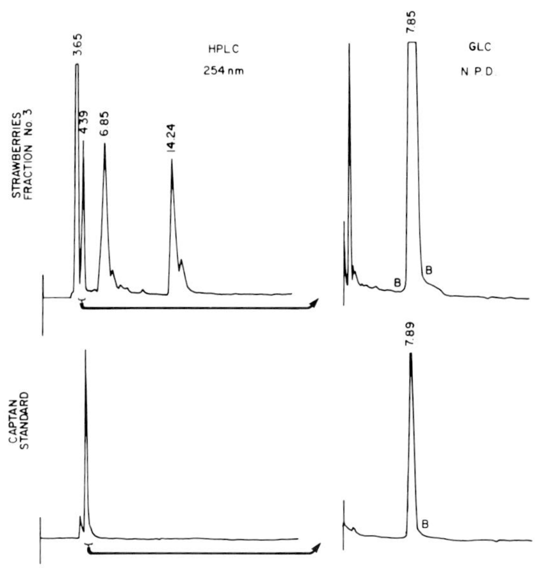

FIGURE 4. Identification of captan in strawberries. Strawberry fraction no. 3 (top) and standard (bottom) were collected from HPLC and compared by GLC (nitrogen: phosphorus detector).

processing. In particular, the question of residues surviving in preserves such as strawberry jam had arisen.

Since earlier independent studies had indicated that boiling destroyed captan, and mutagenic activity did not disappear entirely upon autoclaving, a study was initiated to monitor loss of both captan and mutagenic activity upon heating.[8,9] This coupling of a chemical detector with a biological assay has provided some interesting results. Figure 5 illustrates the fairly rapid degradation of captan in water as detected by GC, and the much slower loss of mutagenic activity. This study suggests that biologically relevant degradation products of captan might persist after disappearance of the parent compound and that estimates of exposure should take these into account.

The captan example illustrates two aspects of the authors' approach to mutagens in foods. The first involves identification of active components, a process that may also serve to exclude artifacts. Secondly, having identified the mutagen, one should not forget that it is the biological activity rather than the chemical entity that is important.

C. Carcinogenicity of Captan

The results of different cancer bioassays with captan have been far from unanimous. Nevertheless, the EPA has judged that captan might pose a cancer risk on the basis of elevated intestinal and liver tumor incidences in male mice following treatment with captan. No carcinogenic effect was observed in two rat studies.[6]

IV. QUERCETIN

A. Identification as the Mutagen in Red Wine, Grape Juice, Raisins, and Onions

Quercetin was identified amongst the butanol-solubles (fraction no. 5) of red wine, but not without difficulty. For instance, HPLC analysis of standard quercetin produced a single peak that was mutagenic. However, the MS was not in agreement with

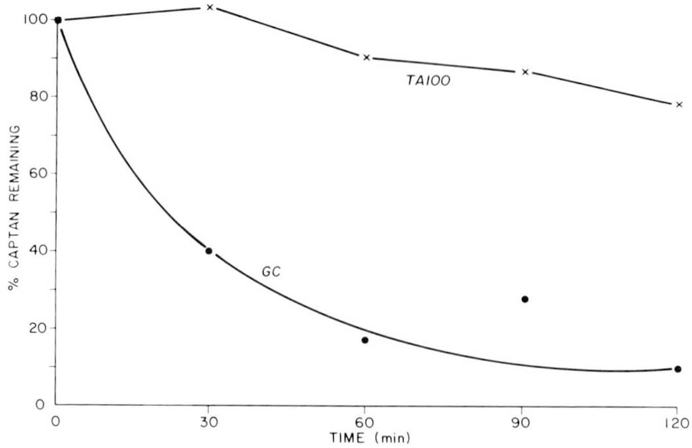

FIGURE 5. Disappearance of captan with heating. Captan suspended in water was incubated at 60°C. Ethyl acetate extracts of aliquots were assayed for captan by GLC and for mutagenic activity with TA100-S9M.

the MS of quercetin by direct probe. It was observed that a number of conditions altered the chemical characterization by MS of quercetin, with or without loss of mutagenic activity. Eventually, quercetin was confirmed in red wine, grape juice, raisins, and onions by derivatization and GC-MS. Nevertheless, experience indicates the need for more cautious assessment of the absence of quercetin in a food sample on the basis of chemical analysis alone.

Others[10,11] have reported the identification of quercetin as the mutagen in foods under test or have simply attributed the mutagenicity of their samples to quercetin without actual identification.

B. Mutagenicity of Urine of Wine Drinkers

Since a number of mutagens can be detected in urine of exposed individuals, the authors examined the possibility that the potent mutagenicity of red wine might be reflected in the urine of wine drinkers. The data for one volunteer, a cigarette smoker, showed an apparent reduction of urine mutagenic activity for TA1538 after red wine consumption (Figure 6). Several other subjects showed the same trend, but not as dramatically, perhaps because they were not smokers. Although comforting, these results are difficult to explain.

C. Enzymatic Browning of Quercetin

The discoloration of fruit and vegetable tissue injured during handling or processing is due to oxidation of phenolic compounds into *o*-quinones, mediated by polyphenoloxidase (PPO). The *o*-quinones formed from phenolic substrates usually participate in secondary reactions such as coupled oxidation of other substrates, complexing with proteins, and condensation and polymerization.[13]

Since quercetin may act as a natural substrate for PPO, it was of interest to examine the mutagenicity of the reaction products. Quercetin was reacted with PPO at pH 6.5 for up to 2 hr. Samples were assayed at various times for residual quercetin by GC and for appearance or disappearance of mutagenic activity for TA98 ± S9M. Figure 7 illustrates the disappearance of quercetin and mutagenic activity with enzymatic browning. Similar experiments employing nonmutagenic chlorogenic acid as substrate for PPO gave no evidence for the formation of mutagenic activity during enzymatic browning.

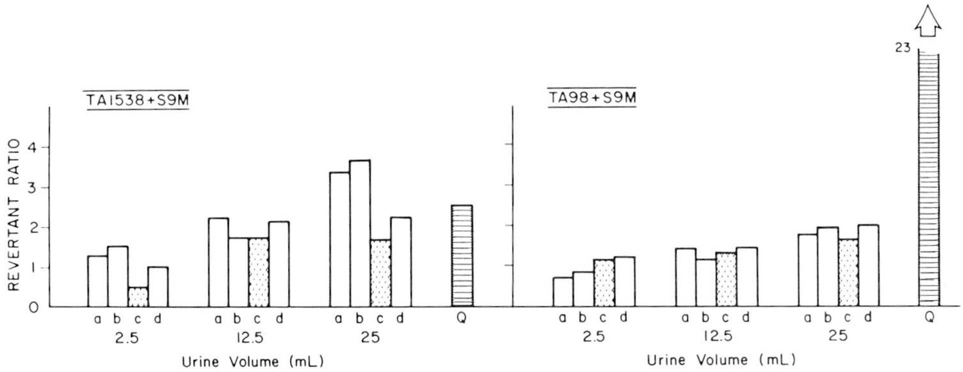

FIGURE 6. Mutagenicity of a wine drinker's urine. Urine was collected for 3 days: a) day 1, control; b) day 2, a.m., control; c) day 2, p.m., after 1.5ℓ red wine; d) day 3, control. XAD-2 column concentrates were plated at concentrations representing 2.5, 12.5, and 25 mℓ of original urine.[12] 50 μg quercetin (Q) served as a positive control. Results are expressed as revertant ratio or number of revertants per treated plate/control plate.

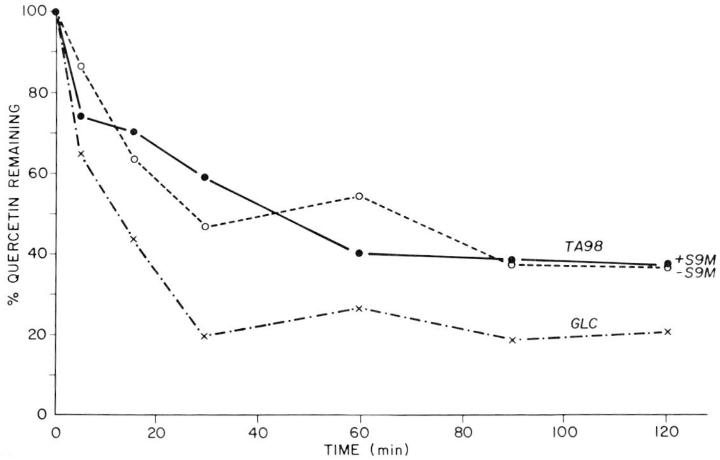

FIGURE 7. Effect of enzymatic oxidation on quercetin. Residual quercetin was estimated from standard curves of mutagenic activity and from GC of the trimethylsilyl derivative.

D. Carcinogenicity of Quercetin

Quercetin has been reported to induce intestinal and bladder cancer in rats, to be noncarcinogenic for rats, and to induce a few unusual tumor types in mice.[14-16] Certain activities of quercetin, e.g., inhibition of Na^+, K^+ ATPase and cyclic AMP phosphodiesterase, might complicate cancer bioassays.[17]

V. COFFEE

A. Mutagenicity of Coffee

The weak mutagenicity obtained with fractions of coffee (Table 3) is similar to reported observations with respect to magnitude of the response and specificity for TA100-S9M. Values of 2 to 4 times background at coffee concentrations as high as 50

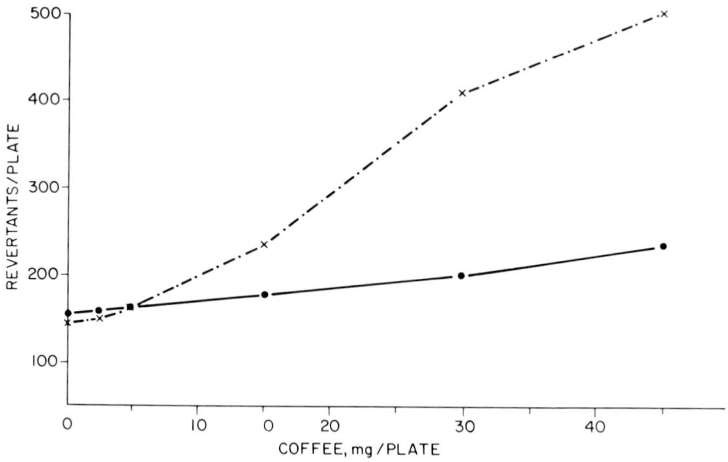

FIGURE 8. Mutagenicity of coffee for TA100-S9M by preincubation (dashed line) and standard (solid line) techniques. The coffee, a popular instant brand, was dissolved in hot water.

mg per plate are representative.[18-20] Since large quantities of coffee are required to produce an effect, it has been argued that the mutagenicity of coffee might really be an artifact due to bactericidal effects.[21] The recent epidemiological association of pancreatic cancer with coffee consumption renders the issue important.[22]

The liquid preincubation modification of the original *Salmonella*/microsome test, popularized by Japanese investigators,[23] has proved valuable for certain chemical classes. Since Japanese researchers, using their preincubation method, had reported the highest mutagenic activity for coffee, the authors directly compared techniques (Figure 8).[18] The superior sensitivity of the preincubation technique for coffee is clear. The positive control in these experiments, azaserine, was specifically chosen because it is a pancreatic carcinogen, highly mutagenic for TA100-S9M, and could conceivably occur in coffee as a result of microbial contamination. The mutagenicity of azaserine was neither enhanced in the preincubation assay nor suppressed by liver homogenate, in contrast on both counts, to the mutagenic activity of coffee.

Numerous candidate mutagens may exist in the wide range of compounds that have been identified in coffee, or in the unidentified fraction (Table 5). Indeed, certain aspects of coffee production such as fermentation, prolonged storage, and roasting are quite compatable with the development of mutagenic activity.

B. Carcinogenicity of Coffee

With its long history of use, coffee has accumulated an equally long and often controversial list of associated diseases, including cancer at six different sites.[22,25] Although reports have suggested enhancement by coffee of cycasin-induced intestinal carcinogenesis and *N*-2-fluorenylacetamide-induced mammary tumorigenesis in rats, a recent carcinogenesis bioassay in rats found no carcinogenic effect of coffee.[26-28]

VI. DETERMINING THE HEALTH SIGNIFICANCE OF MUTAGENS IN FOODS

The three situations that have emerged so far from the screening survey of foods for mutagenic activity provide ample basis for discussion. On the one hand, the potent

Table 5
SOME CONSTITUENTS OF ROASTED COFFEE[a]

Natural Components	
Caffeine	Ascorbic acid
Cellulose	Nitrate
Hemicelluloses	Nicotinic acid
Lignin	Nicotinamide
Fat	Tocopherol
Amino acids	Carboxy-5-hydroxytryptamides
Sucrose	Cycloalkapyrazines
Phenolic acids	Oxazoles

Contaminants
Microbes
Mycotoxins: aflatoxin, sterigmatocystin, ochratoxin A
Metals: lead, cobalt, cadmium
Polycyclic aromatic hydrocarbons
Residues: pesticides, methylene chloride, trichloroethylene
Misc: insects, cashew nut shells

[a] Unknown components account for about 30% on a dry wt basis.[24]

mutagens, captan and quercetin, have been isolated and identified. Theoretically, mutagens should act as initiators of carcinogenesis. Thus the quickest route to determining the health significance of potent mutagens is to subject them to sensitive assays for initiating activity. For instance, if skin is considered a potential target, then skin painting followed by phorbol myristate acetate (PMA) is an expedient method of confirming initiating activity. Other organ systems are available as initiator assays.

The case of coffee is more complex. Here, consumption of a food exhibiting weak mutagenic activity has been associated with increased risk of pancreatic cancer. There are two areas that necessitate attention. First, the mutagens in coffee, if any, should be isolated and then their health significance as carcinogens determined in assays for initiators. Second, the possible presence in coffee of agents able to enhance carcinogenesis should be assessed in tests specific for that purpose.

Figure 9 illustrates an approach to determining the health significance of mutagens and carcinogens in foods based on the presumption that genotoxic carcinogens (initiators, first order carcinogens) should be differentiated from carcinogenesis enhancers (positive modifiers, promoters, epigenetic carcinogens) and regulated separately.[29,30] In this particular approach, bacterial mutagens found in foods would be tested for initiating activity while foods suspected of carcinogenic activity would be tested thoroughly for mutagenic activity and then for initiating and/or enhancing activity. Such studies should supplement, not replace, traditional bioassay, and information generated on mechanism of action should assist risk assessment.

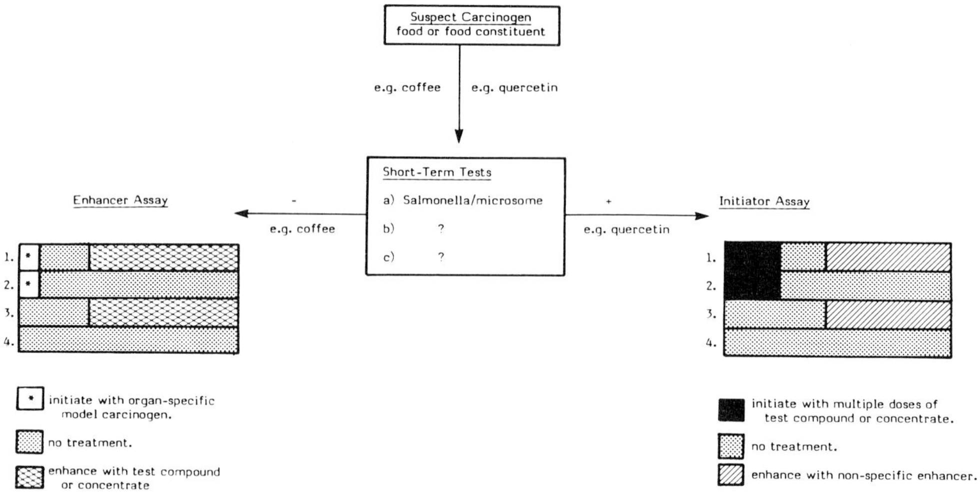

FIGURE 9. Determining the health significance of mutagens in foods. Mutagens; e.g., quercetin, should be tested in the Initiator Assay in which the mutagen is employed as an initiator followed by enhancement with PMA, dietary fat, or other agents known to enhance carcinogenesis. Foods linked to cancer but devoid of mutagenic activity, or for which there is some other reason to suspect enhancing activity; e.g., coffee, should be employed as enhancers in the Enhancer Assay following initiation with a model carcinogen.

REFERENCES

1. **Sugimura, T.**, Let's be scientific about the problem of mutagens in cooked food, *Mutat. Res.*, 55, 149, 1978.
2. **Ames, B. N., McCann, J., and Yamasaki, E.**, Methods for detecting carcinogens and mutagens with the *Salmonella*/mammalian microsome mutagenicity test, *Mutat. Res.*, 31, 347, 1975.
3. **DeSerres, F. J. and Shelby, M. D.**, The *Salmonella* mutagenicity assay: recommendations, *Science*, 203, 563, 1979.
4. Department of National Health and Welfare, Canada, The Food and Drugs Act and Regulations, 1980, 65C.
5. **Bridges, B. A.**, The mutagenicity of captan and related fungicides, *Mutat. Res.*, 32, 3, 1975.
6. U.S. Environmental Protection Agency, Pesticide Programs, Notice of rebuttable presumption against registration (RPAR) and continued registration of pesticide products containing captan, *Fed. Regist.*, 45, 54938, 1980.
7. **Cox, K.**, Pregnant women warned to stay out of strawberry fields, *The Toronto Globe and Mail*, p. 5, July 4, 1981.
8. **Koivistoinen, P., Karinpaa, A., Kononen, M., and Roine, P.**, Postharvest preservatives: magnitude and stability of captan residues in fresh and preserved plant products, *J. Agric. Food Chem.*, 13, 468, 1965.
9. **Ficsor, G. and Nii, G. M.**, The effect of temperature on the mutagenicity of captan, paper presented at 1st Ann. Meet. Environ. Mutagen Soc., Washington, D.C., March 22, 1970.
10. **Sugimura, T.**, Naturally occurring genotoxic carcinogens, in *Naturally Occurring Carcinogens-Mutagens and Modulators of Carcinogenesis*, Miller, E. C., Miller, J. A., Hirono, I., Sugimura, T., and Takayama, S. Eds., University Park Press, Baltimore, 1979, 241.
11. **Tamura, G., Gold, C., Ferro-Luzzi, A., and Ames, B. N.**, Fecalase: a model for activation of dietary glycosides to mutagens by intestinal flora, *Proc. Natl. Acad. Sci. U.S.A.*, 77, 4961, 1980.
12. **Yamasaki, E. and Ames, B. N.**, Concentration of mutagens from urine by adsorption with the non-polar resin XAD-2: cigarette smokers have mutagenic urine, *Proc. Natl. Acad. Sci. U.S.A.*, 74, 3555, 1977.

13. **Mathew, A. G. and Parpia, H. A. B.**, Food browning as a polyphenol reaction, in *Advances in Food Research,* Vol. 19, Chichester, C. O., Mrak, E. M., and Stewart, G. E., Eds., Academic Press, New York, 1971, 75.
14. **Pamukcu, A. M., Yalciner, S., Hatcher, J. F., and Bryan, G. T.**, Quercetin, a rat intestinal and bladder carcinogen present in bracken fern *(Pteridium aquilinum), Cancer Res.,* 40, 3468, 1980.
15. **Hirono, I., Ueno, I., Hosaka, S, Takanashi, H. Matsushima, T., Sugimura, T., and Natori, S.**, Carcinogenicity examination of quercetin and rutin in ACI rats, *Cancer Lett.,* 13, 15, 1981.
16. **Saito, D., Shirai, A., Matsushima, T., Sugimura, T., and Hirono, I.**, Test of carcinogenicity of quercetin, a widely distributed mutagen in food, *Teratogen. Carcinogen. Mutagen.,* 1, 213, 1980.
17. **Brown, J. P.**, A review of the genetic effects of naturally occurring flavonoids, anthraquinones, and related compounds, *Mutat. Res.,* 75, 243, 1980.
18. **Nagao, M., Takahashi, Y., Yamanaka, H., and Sugimura, T.**, Mutagens in coffee and tea, *Mutat. Res.,* 68, 101, 1979.
19. **Aeschbacher, H. U. and Wurzner, H. P.**, An evaluation of instant and regular coffee in the Ames mutagenicity test, *Toxicol. Lett.,* 5, 139, 1980.
20. **Stich, H., Rosin, M. P., Wu, R., and Powrie, W. D.**, Mammalian systems for testing for genotoxicity, paper presented at 13th Ann. Symp. Soc. Toxicol. Can., Montreal, December 2, 1980.
21. **Aeschbacher, H. U., Chappuis, C., and Wurzner, H. D.**, Mutagenicity testing of coffee: a study of problems encountered with the Ames *Salmonella* test system, *Food Cosmet. Toxicol.,* 18, 605, 1980.
22. **MacMahon, B., Yen, S., Trichopoulos, D., Warren, K., and Nardi, G.**, Coffee and cancer of the pancreas, *N. Engl. J. Med.,* 304, 630, 1981.
23. **Yahagi, T., Nagao, M., Seino, Y., Matsushima, T., Sugimura, T., and Okada, M.**, Mutagenicities of *N*-nitrosamines on *Salmonella, Mutat. Res.,* 48, 121, 1977.
24. **Feldman, J. R., Ryder, W. S., and Kung, J. T.**, Importance of nonvolatile compounds to the flavour of coffee, *J. Agric. Food Chem.,* 17, 733, 1969.
25. **Heyden, S., Tyroler, H. A., Heiss, G., Hames, C. G., and Bartel, A.**, Coffee consumption and mortality, *Arch. Intern. Med.,* 138, 1472, 1978.
26. **Mori, H. and Hirono, I.**, Effect of coffee on carcinogenicity of cycasin, *Br. J. Cancer,* 35, 369, 1977.
27. **Fujii, M., Mori, H., Nishikawa, A., and Takahashi, M.**, Effect of coffee on carcinogenicity of *N*-2-fluorenylacetamide, *Gifu Daigaku Igakubu Kiyo,* 28, 295, 1980.
28. **Wurzner, H. P., Lindstrom, E., Vuataz, L., and Luginbuhl, H.**, A 2-year feeding study of instant coffees in rats. II. Incidence and types of neoplasms, *Food Cosmet. Toxicol.,* 15, 289, 1977.
29. **Campbell, T. C.**, Chemical carcinogens and human risk assessment, *Fed. Proc. Fed. Am. Soc. Exp. Biol.,* 39, 2467, 1980.
30. **Clayson, D. B.**, Carcinogens and carcinogenesis enhancers, *Mutat. Res.,* 86, 217, 1981.

Part III

Methodological and Regulatory Issues

Chapter 16

METHODS FOR THE DETECTION OF CARCINOGENS AND MUTAGENS IN FOOD: AN INTRODUCTORY REVIEW

P. S. Elias

TABLE OF CONTENTS

I. INTRODUCTION

The presence of toxic compounds in human food has been known since antiquity. That these toxic substances include carcinogens[1-4] and mutagens[5-6] is now an accepted scientific fact. Indeed some of the biologically produced contaminants, e.g., mycotoxins, must have been associated with the human food supply since the earliest days of man's existence. However, the knowledge that processing of food, whether industrially or domestically, and more particularly, the fact that the use of heat could add to the exposure from natural sources, is comparatively recent.[7-16]

This monograph is devoted to the methodology for detecting carcinogens and mutagens in food as distinct from determining this type of toxicological activity in a clearly specified chemical entity. The toxicological methods available for detecting the carcinogenic activity of a defined chemical have been discussed in numerous publications.[17-25] An even larger number of publications has appeared on short-term screening assays for carcinogens and are based on the detection of the mutagenic activity of a substance by less cumbersome in vitro and in vivo procedures.[19,20,22,23,25-28]

Mutagenesis and carcinogenesis are similar superficially in that both involve changes in cell genotype that are heritable. Both are considered to be the result of a permanent and irreversible structural change in the critical constituent of the cell, most probably the DNA. This so-called somatic mutation theory of cancer has received considerable support during the past few years, particularly from experiments demonstrating a close correlation between carcinogenic and mutagenic activity for certain classes of carcinogens. The development of short-term screening tests, using test systems with observable mutational endpoints as indicator systems for the presence of mutagenic activity, has thus extended enormously the methodological capabilities of toxicological investigation. Consequently, it is now widely accepted that chemicals capable of inducing mutations should also be suspected of being able to cause cancer. Short-term tests and carcinogenicity bioassays each have value, and mutually increase the validity of toxicological assessments based on their outcome. However, short-term tests are not substitutes for carcinogenicity bioassays because the precise relationship of mutagenicity to carcinogenicity has not yet been elucidated.

The process of carcinogenesis comprises a number of steps that are sometimes morphologically distinguishable. In the first or initiating step the genotoxic carcinogen, in its ultimate reactive form, probably interacts chemically with the DNA of a cell. Unless the lesion is repaired, a permanent structural alteration remains in what has now become a precursor cell. At some point in time, these precursor cells begin to divide more frequently than the surrounding normal cells, thus giving rise to clones of transformed cells, some of which progress eventually to frankly malignant tumors. These steps in the further development of a cancer follow either as a result of continuing interaction of the initiating carcinogen with the precursor cells, or by "spontaneous" changes, or as a result of a previously imposed irreversible structural change.[29,30]

Such a process involving progression to the neoplastic state offers an opportunity for nongenotoxic agents that are not by themselves able to initiate a permanent structural change in cells, to enhance the activity of genotoxic carcinogens, or magnify their effect. This activity has been termed co-carcinogenesis and substances endowed with this capability are known as modifiers, co-carcinogens, or tumor promoters. Promoters, in particular, are co-carcinogens that act only after the initiating genotoxic carcinogens, while co-carcinogens may act before or together with the genotoxic carcinogen. The co-carcinogenic effect may manifest itself at a variety of functional levels, e.g., the induction of enzymes involved in the activation of carcinogens, alteration of the hormonal environment, nonspecific stimulation of cellular proliferation, alterations in the immunological surveillance, or stimulation of an oncogenic virus. Substances capable of acting as co-carcinogens may pose just as great a threat as the genotoxic agents that initiate the cancer process. However, many different mechanisms are involved, and suitable test systems are not yet available for unequivocally revealing co-carcinogenic activity.

Positive results may be obtained, therefore, in animal bioassays for carcinogenicity, not because the agent under test is an initiator of the process, but because the agent or the circumstances of its administration enhances the risk of tumor development through some other mechanism. This is likely to occur where the test agent cannot be shown to be a mutagen, where the test substance is a hormone or induces changes in the hormonal status of the test animals, or where the tumors produced in excess in exposed animals also commonly occur in control animals.

II. APPRAISAL OF ANIMAL BIOASSAYS FOR CARCINOGENICITY

The concepts on testing chemicals for carcinogenic activity in chronic experiments have not changed in the past few years despite the appearance of a host of recommendations from national and international authorities and scientific groups. Certain facts must be taken into account in both the design and interpretation of animal carcinogenicity tests. Animals of all species are susceptible to the spontaneous development of both benign and malignant neoplasms, varying widely in site and histological type. Whether this background "noise" of tumor incidence is in fact caused by environmental or endogenous factors is not known. However, the incidence of many types of tumors may be influenced by the composition of the diet of the test animals,[31] the amount they eat,[32,33] by stress, and by numerous other apparently nonspecific factors. Viruses are also involved in the causation of some commonly occurring tumors.[34,35] Consequently, experiments must be carefully designed and strictly controlled so that treated and control groups of animals are, as far as possible, not differently exposed to nonspecific factors possibly influencing tumor incidence. Otherwise, differences in tumor incidence may be attributed erroneously to treatment, or the carcinogenic effect of the test substance might be masked by a higher incidence of tumors due to nonspecific influences in the control group than is found in response to the carcinogen in the treated group.

The inception and duration of exposure and the observation period are other important aspects determining the validity of the test and the interpretation of the results.[23] Last but not least is the need to have an adequate specification of the material to be studied.[18,23,36] When the results of carcinogenicity tests are extrapolated to man, consideration has to be given to any difference in the kinetics of the test substance between the test species and man, including differences in absorption, distribution, biotransformation, and excretion. Furthermore, the receptor or target organ may differ in the two species. There is also the problem of detecting low incidence effects with a limited number of test animals. A toxic effect that occurs with an incidence of less than 1% is likely to be missed in a test design employing 50 animals per group. Little imagination is needed to appreciate the size of the cancer problem that could arise in an entire population of many millions of individuals if an undetected carcinogen were to enter the national food supply.[37]

When present concepts of the design and the methodology of carcinogen bioassays are applied to the testing of food, many major drawbacks become obvious at once. There is the impossibility of identifying the test substance precisely by physical and chemical parameters, although this is a prime requirement for conducting such tests. If an unidentified cancerogenic agent is suspected of being present in food, then its detection by traditional carcinogenicity testing is beset with many difficulties. A primary consideration is the nature of the food itself. Many foods wholesome for man are toxic for other species. Thus the incorporation of over 10% potato starch into the diet of rats reduces growth, raises mortality, and leads to cecal enlargement, while cereal starch causes no such effects.[38] If chicken and green beans are fed to dogs, or green beans alone to rats, these animals develop pulmonary lesions.[39] Yet no such effects are experienced by people consuming these foods. Several extensive feeding studies on irradiated onions have yielded results that are difficult to evaluate because of the appearance of hemolysis and anemia in both control and test groups, attributable to active constituents in the onion per se, regardless of any radiation-induced changes.[40]

Secondly, even if the food to be tested is well-tolerated by animals, it is usually necessary to feed it as a high proportion of the animal diet in order to achieve an adequate level of exposure to the carcinogen presumed to be present in the test food. It is therefore impossible to achieve an exposure of the experimental animals of at least 100 times the likely human exposure, as is normally aimed at in toxicological testing. This makes it very unlikely to detect the presence of any carcinogen unless it has considerable potency or induces an unusual type of tumor in the test species. The methodological need for the use of this high level of administration can create problems of nutritional imbalance and possible consequential deficiencies that show up only after many months of administration. Supplementation of the basic diets with additional micronutrients in order to avoid these problems has not always proved satisfactory. An additional difficulty is created by the inability to design a test procedure incorporating the determination of a distinct dose-response relationship. In particular, the demonstration of a "no-effect" situation is of rather limited value as evidence of safety, unless it can be related to the dosage of the potential toxicants.[41] As has been aptly stated, "both aspects — dose as well as response — may be difficult to interpret when dealing with food components (or food *per se*)".[42]

A different problem, arising in the case of many foods, is that of furnishing a valid control diet. Strictly speaking, the control groups must be fed a diet identical to that of the test group except that the test food in question should not contain the presumed carcinogen. If the agent is a natural constituent of the food, some means may have to be devised to remove it from the food to be fed to the control animals. The complications and possible sources of error introduced by such removal treatment make the interpretation of the eventual results of the test difficult. Alternatively, if the presumed carcinogen is formed in the food to be tested as a result of the technological processing to which the food has been subjected, some alternative processing to achieve the same technological result may have to be used on the control food, if such an alternative processing method exists. Even then, it is likely that purists would disagree with this suggestion.

Thirdly, to allow meaningful biological and statistical evaluation of the results a sufficient number of animals should be used in each group. However, the low exposure likely to occur from the presence of a suspect carcinogen in food, if a logistically acceptable test is to be performed under existing protocols, would require such a large number of animals per experimental group as to preclude the performance of the test except under the most extraordinary circumstances. In these circumstances, to obtain statistical advice on the suitability of the design of the study and on the methods to be used for analyzing the results may be no more than a cosmetic exercise. Golberg[42] stated some time ago that "the solution attempted for the problem of irradiated foods has proved a colossal and costly failure: feeding irradiated fruit salad to dogs has not established safety, nor has the whole vast exercise provided an advance in fundamental knowledge to serve as a springboard for further work''. The same sentiments could be equally voiced against the many feeding tests performed to establish the ''safety'' of novel proteins, processed foods, processing enzymes, bioproteins for animal feeds, and the so-called relay toxicity tests, because the carrying out of such carcinogenicity tests has been made a requirement in official guidelines on the testing of these materials. Clearly an alternative approach is necessary.

III. SHORT-TERM SCREENING TESTS

The close correlation between carcinogenic and mutagenic activity of certain classes of carcinogens has brought to the fore the use of screening procedures for detecting mutagenic activity as a credible alternative to carcinogen bioassays in animals. The scientific basis for this approach has already been discussed earlier. The changes in the structure, properties and behavior of DNA brought about by the action of mutagens are amenable to investigation by in vitro and in vivo procedures. With a few exceptions, these methods are less cumbersome, more rapid, less costly, and save experimental animals. However, just as in the case of animal bioassays, the problem of detecting mutagenic activity in defined or clearly specified chemicals or their simple mixtures, differs considerably from detecting such mutagenic activity in a complex food. The in vitro screening tests are particularly difficult to apply directly to food unless some means can be developed to incorporate the food to be examined into the test system without interference with its proper performance. The situation is far simpler for in vivo screening tests because animals can be fed a diet containing only the food to be examined for comparatively short periods of a few days, thus achieving the maximum possible exposure. The subsequent examination of various target tissues or body fluids for evidence of mutagenic effects, or the presence of mutagenic metabolites incorporates the advantage of the passage of the suspected toxicant in the food through the intact mammalian metabolic machinery.

The in vitro screening tests for mutagenic activity were developed for use with pure chemical substances and often involve the treatment of cultured organisms such as bacteria or mammalian cells.[43,44] In most cases, therefore, there is the problem of being able to employ these systems for detecting mutagenic activity in foodstuffs because of its physical nature and composition. The problem is compounded if the nature and properties of the mutagens presumably present are unknown, whether these are naturally derived or are induced in the test food by the chemical changes accompanying its processing.

In the testing of foodstuffs this problem has, until recently, been circumvented by the use of extracts and juices. Although solvent extraction is a simple and valuable technique that has led to the identification of a number of mutagens, for example in heat-treated foodstuffs, the method is open to a number of criticisms.

The complexity of the reactions induced in foodstuffs of widely varying composition by physical treatments produces a large number of compounds that have not all been identified. It is therefore difficult to devise an extraction method encompassing all possible toxic compounds from all foods. Chemically altered macromolecules that remain biologically inactive until broken down into small molecules, may be formed during processing. Active mutagens may be detectable in extracts, but may be destroyed under normal conditions of digestion. Any mutagenic activity found in an extract may therefore have little relevance to the in vivo situation. Conversely, digestion could activate otherwise innocuous substances. Reactions between food constituents and solvents are possible, and, in addition, solvent residues may persist in the samples, both events leading to erroneous conclusions.

IV. SAMPLE PREPARATION BY IN VITRO DIGESTION

In an attempt to answer some of these criticisms, the possibility was investigated in the author's laboratory of using in vitro digestion to prepare samples for testing in bacterial and cell culture systems.[45] The digestion procedure chosen mimicked as closely as possible the mammalian intestinal digestion, including as initial step a low-pH pepsin digestion, followed by an alkaline pancreatin digestion. The treated food homogenate was then sterilized by ultrafiltration and its osmolality adjusted to correspond to that of the normal culture media for cells or bacteria. No adverse effects were observed when the appropriate cell lines were incubated in the presence of digests, and it was found to be possible eventually to grow the mutagenicity indicators directly in the food digest, supplemented appropriately for maximum growth. In this way maximum exposure of the indicator organisms could be achieved. This method has been applied successfully to a large number of foods preserved by irradiation as part of the wholesomeness investigation of this processing technology.[46,47] This approach may be useful for studying processed foods, but it may also be a useful adjunct to the study of the effects of food additives and contaminants where an indirect effect through reaction with food components is possible, or where digestion might affect the mutagenic activity of the toxicant.

The method described above is obviously not entirely comparable with in vivo digestion because certain elements of this process are absent, such as the action of the enzymes of the intestinal mucosa and the gut flora. The satisfactory mimicking of fat digestion and absorption has not yet been discovered and further development of the methodology is needed to encompass this aspect. On the other hand, it is necessary to avoid too complex a procedure if routine use is contemplated. Nevertheless, this methodology achieves the reduction of the bulk of solid foods to manageable solutions of materials of comparatively low molecular weight. It also takes account of the possible interactions of food constituents with one another and with the digestive secretions, and of the modifying effects of digestion on ingested toxic substances. The latter aspect has, in the past, been rather neglected. It can be argued, however, that the similarity between the in vitro and in vivo digestion, in its main features, allows the conclusion that the results of tests using this methodology are more, rather than less, meaningful, and more realistically reflect the eventual human hazard than tests carried out under standard simplified conditions.

It is clearly not the task of this monograph to critically review the multitude of available short-term tests for mutagenic activity as predictors of carcinogenic activity generally, and of mutagenic or carcinogenic hazard for man. It is generally accepted that none of these tests is sufficiently reliable to be used on its own as a screen for genetic toxicity, but that batteries of complementary techniques that determine different genetic endpoints, should be used. For this reason an arbitrary choice of short-term tests has to be made that may then be applied simultaneously to food. In the author's experience the following methods have proved advantageous:

1. The *Salmonella typhimurium*/mammalian microsome mutagenicity test of Ames
2. Sister chromatid exchange and point mutation assays in cultured mammalian cells
3. The micronucleus, sister chromatid exchange and germ-cell cytogenetic tests in mammals
4. BM cytogenetics and DNA repair induction in mammals
5. The sex-linked recessive lethal mutation test in *Drosophila melanogaster*

This program also includes the investigation of the utility of cell transformation assays and of assays for promoting activity.

V. APPRAISAL OF SHORT-TERM TESTS

Short-term tests in bacteria measure the ability of an agent to damage DNA. They are particularly well-suited for inclusion in screening batteries and for detecting and identifying mutagenic components in complex mixtures. Their predictive value is still being established empirically by comparison of the agreement of the results from short-term tests with those of animal carcinogenicity tests.

Several experimental systems have been developed to measure specific locus mutations in mammalian cells in culture. A correlation has been demonstrated between carcinogenicity and mutagenicity in mammalian cells for a variety of chemicals. These assays, therefore, provide valuable information when they are part of a battery of screening tests for chemical carcinogens.

Mammalian cells have the additional advantage that the same system can be used to simultaneously determine other genetic endpoints such as chromosomal damage, DNA damage and repair, as well as neoplastic transformation. A substantial proportion of carcinogenic substances are potent inducers of chromosome damage that may be detected cytologically. The finding of chromosome aberrations, or damage resulting in sister chromatid exchanges, or micronucleus formation in in vitro and in vivo studies, is therefore relevant for the identification of carcinogens. However, different in vivo systems have different sensitivities and are probably less sensitive than the in vitro systems, the latter having the advantage of permitting the direct application of high concentrations of the test substance. However, correlation between in vivo and in vitro activity is limited, and validation with respect to bacterial tests and animal bioassays has not yet reached the stage where a definite judgment can be made.[25]

DNA damage and, consequently, DNA repair may be sensitive indicators for the carcinogenic activity of a substance. A variety of procedures have been developed measuring breaks in DNA or unscheduled DNA synthesis in exposed cells. The in vitro systems have been evaluated and appear to be promising. However, carcinogens that operate by mechanisms not involving DNA damage and repair are not likely to be detected by this methodology.[25]

The use of whole mammals provides the closest practical approach to the human situation. While determinations of mutagenicity in some lower systems are used primarily as indicators of the possible existence of carcinogenic activity, determinations in whole mammals also answer the questions about mutagenesis per se. Transmitted effects in mammals allow the study of point mutations and chromosome breakage as well as enabling quantitative assessment of relative risk from various kinds of genetic damage. For the assessment of mutagenicity the whole mammal tests are irreplaceable, but as indicators of carcinogenicity they are of more limited utility.[25]

Even when using a battery of short-term tests there is insufficient evidence to suggest that the efficiency of detecting chemical carcinogens improves by the continued addition to the number of assays selected. Moreover, the inadequacy of existing data bases effectively limits the selection of the ideal combination of tests, and some agents will be misjudged or missed by any array of tests now available.[25]

VI. CONCLUSIONS

Hopefully, this review has highlighted a number of difficult problems associated with demonstrating the safety of foodstuffs suspected of containing unknown mutagens or carcinogens, either naturally or as a result of processing. In relation to assessing the hazards to human health, the suggested approaches appear to represent the best that can be achieved at present with the available test systems.

REFERENCES

1. **Miller, J. A. and Miller, E. C.**, Carcinogens occurring naturally in foods, *Fed. Proc. Fed. Am. Soc. Exp. Biol.*, 35(6), 1316, 1976.
2. **Rodricks, J. V.**, Food hazards of natural origin, *Fed. Proc. Fed. Am. Soc. Exp. Biol.*, 37(12), 2587, 1978.
3. **Anon.**, *Toxicants Occurring Naturally in Foods*, National Academy of Sciences, U.S. Government Printing Office, Washington, D.C., 1973.
4. **Ayres, J. C. and Kirschman, J. C., Eds.**, *Impact of Toxicology on Food Processing*, AVI Publishing Co., Westport, Conn., 1981.
5. **Rockwell, P. and Raw, I.**, A mutagenic screening of various herbs, spices, and food additives, *Nutr. Cancer*, 1(4), 11, 1979.
6. **Brown, J. P.**, A review of the genetic effects of naturally occurring flavonoids, anthraquinones, and related compounds, *Mutat. Res.*, 75, 243, 1980.
7. **Commoner, B., Vithayathil, A. J., Dolara, P., Nair, S., Madyastha, P., and Cuca, G. C.**, Formation of mutagens in beef and beef extract during cooking, *Science*, 201, 913, 1978.
8. **Sugimura, T. and Nagao, M.**, Mutagenic factors in cooked foods, *CRC Crit. Rev. Toxicol.*, 6, 189, 1979.
9. **Pariza, M. W., Ashoor, S. H., and Chu, F. S.**, Mutagens in heat-processed meat, bakery and cereal products, *Food Cosmet. Toxicol.*, 17, 429, 1979.
10. **Spingarn, N. E. and Garvie, C. T.**, Formation of mutagens in sugar-ammonia model systems, *J. Agric. Food Chem.*, 27(6), 1319, 1979.
11. **Spingarn, N. E. and Weisburger, J. H.**, Formation of mutagens in cooked foods. I. Beef, *Cancer Lett.*, 7, 259, 1979.
12. **Spingarn, N. E., Slocum, L. A., and Weisburger, J. H.**, Formation of mutagens in cooked foods. II. Foods with high starch content, *Cancer Lett.*, 9, 7, 1980.
13. **Shelby, M. D. and Matsushima, T.**, Mutagens and carcinogens in the diet and digestive tract, *Mutat. Res.*, 85, 177, 1981.
14. **Uyeta, M., Taue, S., and Mazaki, M.**, Mutagenicity of hydrolysates of tea infusions, *Mutat. Res.*, 88, 233, 1981.
15. **Stich, H. F., Rosin, M. P., Wu, C. H., and Powrie, W. D.**, Clastogenic activity of dried fruits, *Cancer Lett.*, 12, 1, 1981.
16. **Nader, C. J., Spencer, L. K., and Weller, R. A.**, Mutagen production during pan-broiling compared with microwave irradiation of beef, *Cancer Lett.*, 13, 147, 1981.
17. WHO, Evaluation of the carcinogenic hazards of food additives, Tech. Rep. Ser. No. 220, World Health Organization, Geneva, 1961.
18. WHO, Principles for the testing and evaluation of drugs for carcinogenicity, Tech. Rep. Ser. No. 426, World Health Organization, Geneva, 1969.
19. **Anon.**, *The Testing of Chemicals for Carcinogenicity, Mutagenicity, Teratogenicity*, Ministry of Health and Welfare, Ottawa, 1973.
20. WHO, Assessment of the carcinogenicity and mutagenicity of chemicals, Tech. Rep. Ser. No. 546, World Health Organization, Geneva, 1974.
21. National Cancer Institute, Guidelines for carcinogen bioassay in small rodents, NCI-CG-TR-1, U.S. Department of Health and Welfare, Washington, D.C., 1976.
22. WHO, *Environmental Health Criteria, No. 6: Principles and Methods for Evaluating the Toxicity of Chemicals, Part I.*, World Health Organization, Geneva, 1978.
23. **Anon.**, *Proposed System for Food Safety Assessment*, Food Safety Council, U.S. Government Printing Office, Washington, D.C., 1980.
24. Health Council of the Netherlands, *The Evaluation of the Carcinogenicity of Chemical Substances*, Gov. Publ. Office, The Hague, 1980.
25. International Agency for Research on Cancer, *Long-term and Short-term Screening Assays for Carcinogens: A Critical Appraisal*, IARC Monogr., Suppl. 2, IARC, Lyon, 1980.

26. WHO, Evaluation and testing of drugs for mutagenicity, Tech. Rep. Ser. No. 482, World Health Organization, Geneva, 1971.
27. International Agency for Research on Cancer, *Screening Tests in Chemical Carcinogenesis,* IARC Sci. Publ. 12, IARC, Lyon, 1976.
28. Department of Health and Social Security, Rep. on Health and Social Subjects, No. 24, Guidelines for the testing of chemicals for mutagenicity, Her Majesty's Stationery Office, London, 1981.
29. **Ames, B. N.,** Identifying environmental chemicals causing mutations and cancer, *Science,* 204, 587, 1979.
30. **Stoltz, D. R., Poirier, L. A., Irving, C. C., Stich, H. F., Weisburger, J. H., and Grice, H. C.,** Evaluation of short-term tests for carcinogenicity, *Toxicol. Appl. Pharmacol.,* 29, 157, 1974.
31. **Elias, P. S.,** The effect of diet on toxicological parameters, paper presented at 3rd World Cong. Animal Feeding, Madrid, October 23, 1978, 515.
32. **Roe, F. J. C.,** Are nutritionists worried about the epidemic of tumors in laboratory animals? *Proc. Nutr. Soc.,* 40, 57, 1981.
33. **Clayson, D. B.,** Nutrition and experimental carcinogenesis: a review, *Cancer Res.,* 35, 3292, 1975.
34. **Temin, H. N.,** Do viruses have a role in causing human cancer? *Trends Biochem. Sci.,* 3, N80, 1978.
35. **Mare, J. L.,** RNA tumor viruses: getting a handle on transformation, *Science,* 199, 161, 1978.
36. **Munro, I. C.,** Considerations in chronic toxicity testing: the chemical, the dose, the design, *J. Environ. Pathol. Toxicol.,* 1, 183, 1977.
37. **Doull, J.,** Assessment of food safety, *Fed. Proc. Fed. Am. Soc. Exp. Biol.,* 37(12), 2594, 1978.
38. **El-Harith, A., Dickerson, J. W. T., and Walker, R.,** On the nutritive value of various starches for the albino rat, *J. Sci. Food Agric.,* 27, 521, 1976.
39. **Palecek, F.,** Pulmonary damage in rats fed by beans *(Phaseolus vulgaris), Experientia,* 25, 285, 1969.
40. IAEA/FAO/WHO, Wholesomeness of irradiated food. Summaries of data considered by the Joint FAO/IAEA/WHO Expert Comm. Wholesomeness of Irradiated Food, World Health Organization, Geneva, 1980.
41. **Elias, P. S.,** The wholesomeness of irradiated food, *Ecotoxicol. Environ. Safety,* 4, 175, 1980.
42. **Golberg, L.,** Chemical and biochemical implications of human and animal exposure to toxic substances in food, *Pure Appl. Chem.,* 21, 309, 1970.
43. **Kilbey, B. J., Legator, M., Nichols, W., and Ramel, C., Eds.,** *Handbook of Mutagenicity Test Procedures,* Elsevier, Amsterdam, 1977.
44. **Norpoth, K. H. and Garner, R. C., Eds.,** *Short-term Test Systems for Detecting Carcinogens,* Springer-Verlag, Basel, 1980.
45. **Phillips, B. J. and Elias, P. S.,** A new approach to investigating the genetic toxicity of processed foods, *Food Cosmet. Toxicol.,* 16, 509, 1978.
46. **Phillips, B. J., Kranz, E., Elias, P. S., and Münzner, R.,** An investigation of the genetic toxicology of irradiated foodstuffs using short-term test systems, Part I., *Food Cosmet. Toxicol.,* 18, 371, 1980.
47. **Phillips, B. J., Kranz, E., and Elias, P. S.,** An investigation of the genetic toxicology of irradiated foodstuffs using short-term test systems, Part II., *Food Cosmet. Toxicol.,* 18, 471, 1980.

Chapter 17

PROBLEMS ASSOCIATED WITH THE EXTRACTION OF MUTAGENS FROM FOODS

Wayne T. Iwaoka and Cheryl A. Krone

TABLE OF CONTENTS

I. INTRODUCTION

The development of the Ames and other short-term mutagenicity tests has resulted in an avalanche of information in the literature reporting on the presence or absence of bacterial or mammalian cell mutagens or clastogens in all types of foods, including many common ones such as cooked beef,[1-6] grilled fish,[7-9] regular and decaffeinated coffee,[10] dried fruits,[11] and pickles,[12] among others.

While all the mutagens found in various foods have not yet been shown to be mammalian carcinogens, this type of work takes on added significance since the Japanese have recently reported on the carcinogenicity in mice of two orally administered mutagenic compounds (Trp-P-1 and Trp-P-2) commonly found in broiled foods.[13]

In testing various foods for the presence of mutagenic compounds, there are invariably fewer problems if portions of the whole food can be incorporated into the mutagenicity test. However, this operation is usually difficult to perform and many investigators resort to extracting the food in question to isolate and concentrate the potential mutagens.

Several different extraction procedures that isolate and concentrate specific classes of compounds have been used, and range from simple solvent extraction and centrifugation systems to those that involve protein precipitation, adjustment of pH, and extraction with different organic solvents. The more simple extraction systems have been extensively used by the Japanese in their study of mutagens found on the charred surfaces of meat and broiled fish.[7] A more complex extraction procedure involving the use of acid, base, salts, and organic solvents was developed by Commoner and coworkers[1,2] to study mutagen formation in fried hamburger. This procedure, or minor modifications of it, has been used by several other investigators to study mutagen formation in a number of other foods.[14,15]

In the process of developing and modifying mutagen extraction procedures, many diverse problems arise. They range from the incomplete extraction of the mutagenic compounds to chemical reactions occurring among food components with chemicals used to extract the food sample. While there are specific problems associated with different types of extractions, this paper will address four general problem areas of mutagen extraction:

1. Artifactual mutagenicity
2. The effect of pH and the use of different organic solvents on the extraction of mutagens
3. The difficulties that arise when extracting foods with components known to participate in the Maillard browning reaction
4. The limitations of a food extraction procedure in accurately assessing its mutagenic potential.

II. ARTIFACTUAL MUTAGENICITY

The extraction procedure developed by Commoner et al.[1] involved precipitation of proteins with ammonium sulfate and low pH, extraction of the acidified aqueous portion with an organic solvent to extract acidic and neutral compounds, basification with ammonium hydroxide, and reextraction with an organic solvent to isolate basic compounds. An abbreviated schematic diagram of this extraction procedure is shown in Figure 1.

In the authors' studies dealing with the formation of mutagens in foods, the Commoner[1] ammonia extraction procedure has been used with a number of different foods,

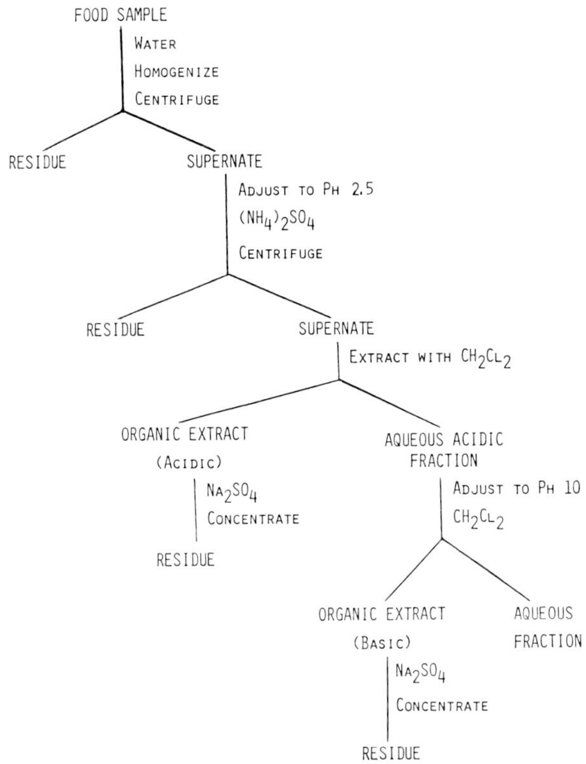

FIGURE 1. General extraction scheme for isolating and concentrating mutagens from cooked ground beef and other foods.

including fried ground beef and bakery products, and found that artifactual mutagenicity can be produced under certain conditions.[16]

These mutagens apparently are formed in large quantities only when certain foods and the ammonia extraction procedure are used. These foods appear to be those that generate, or have the potential to generate, large amounts of carbonyl compounds upon processing or cooking. Bakery products have this ability, and in testing a number of different types of biscuits it has been found that mutagens are produced in varying amounts when using the ammonia extraction procedure, but none at all are produced when using a salt other than ammonia for precipitating proteins.[16]

In work with a biscuit product the authors tested sodium sulfate and sodium hydroxide as well as ammonium sulfate and ammonium hydroxide for precipitating proteins and basifying the acidic extracts, respectively. In separate extractions, the basic aqueous fractions from the sodium and ammonium procedures were extracted twice a day over a four-day period, the extracts were kept separate, the organic solvent was evaporated from each fraction, and the individual residues were tested in the Ames assay. The results of this experiment are shown in Figure 2. High levels of mutagenicity are obtained when ammonium sulfate and ammonium hydroxide are used in the extraction procedure (upper curve), but virtually no mutagenicity is produced when sodium sulfate or sodium hydroxide are used (lower curve). Mutagenicity is produced in the sodium sulfate/hydroxide procedure only when ammonium ions (10% of the total) are added to the aqueous fraction after the fourth extraction.

This study shows that mutagens can be generated during the extraction of a food

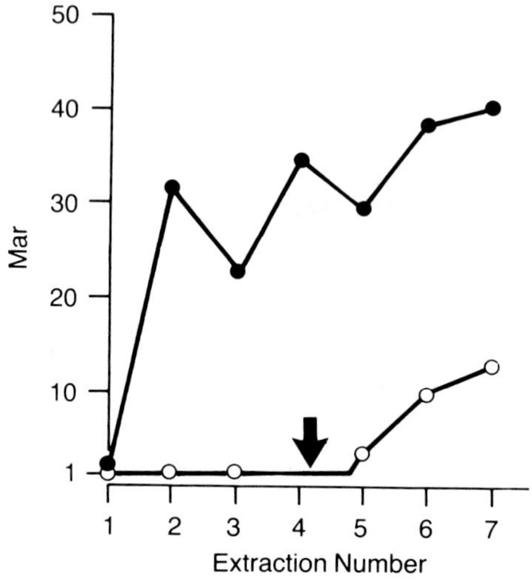

FIGURE 2. Differences in mutagenic activity towards strain TA1538 obtained in the basic organic extracts of biscuits. Symbols • and o represent $(NH_4)_2SO_4/NH_4OH$ and $Na_2SO_4/NaOH$ extraction procedures, respectively. Concentrations of extracts used in extraction No. 1 were 12 mg and 13 mg for the ammonia and sodium sulfate procedures, respectively. The other extracts (2 to 7) for both ammonium and sodium salt-treated solutions contained between 1 and 8 mg of sample. Arrow indicates where 1/10 the volume of concentrated ammonium hydroxide was added to both aqueous fractions.

product, and in this case was dependent upon the use of ammonia-containing compounds for extraction. The fact that mutagens were not obtained in the acidic or basic sodium sulfate extracts did not mean that there were no mutagens in biscuits or other bakery products, and this point will be dealt with later.

Because a large quantity of mutagens was detected in biscuits by using the ammonia procedure, the next step was to find out whether the mutagens other investigators were detecting in fried hamburger, using the same ammonia procedure, were also artifacts.[1,2,6] Consequently, ground beef was fried and subjected to the same type of extraction procedure as used for the biscuits and the basic extracts were tested in the Ames mutagenicity assay (Table 1). The mutagenicity for the sodium sulfate/hydroxide procedure appeared to be slightly higher, but there were no significant differences. A more detailed description of this work can be found elsewhere.[4]

These results show that artifactual mutagenicity can be produced in large quantities in one food product, while apparently very little or none at all is produced in another type of food. Still, it would probably not be advisable to use the ammonia procedure when studying mutagen formation in foods. The point that mutagens did not appear to form in hamburger when using the ammonia procedure is irrelevant, since conditions may change so the possibilities for formation always exist.

Table 1
EFFECT OF THE USE OF AMMONIUM AND SODIUM SULFATE AND HYDROXIDES ON APPARENT MUTAGENICITY IN FRIED HAMBURGER EXTRACTS

Basic extract (mg/plate)[a]	MAR (NH_4^+)[b]	MAR (Na^+)[b]
0.033	1.7	1.8
0.33	8.0	11.0
3.3	48.0	55.0

[a] No mutagenic activity was observed in acidic extracts (pH 2.5).

[b] Mutagenic activity ratio (MAR) for pH 10 organic extracts, from either the $(NH_4)_2SO_4/NH_4OH$ or $Na_2SO_4/NaOH$ procedure, on *Salmonella typhimurium* strain TA98 with S9 mix.

III. EFFECTS OF pH AND ORGANIC SOLVENTS ON MUTAGEN EXTRACTION

Studies have been conducted on mutagen formation in a high protein food other than ground beef (fish) and have found that mutagens are produced in this food during frying.[17] Using the extraction procedure shown in Figure 3, mutagens were found in the basic organic extracts from fried but not steamed, raw, or broiled (in an electric oven) salmon, sole, snapper, or turbot. The mutagens from the different species of fried fish all required metabolic activation by rat liver homogenates and were specific for frameshift strains of *Salmonella typhimurium.* Table 2 shows the important strains and MARs obtained in this study.

Because a precipitating salt (ammonium sulfate) was responsible for the artifactual mutagenicity in one particular class of food, but not in another, it was decided that the effects of other conditions of extraction should be examined as well. The pH of an extraction system plays a major role in the interactions, solubilities, and reaction rates of various chemicals in food. Studies were carried out to examine the effect of pH and the use of different organic solvents on the extraction of mutagens from the species of fish (sole) that appeared to produce the highest levels of mutagenicity.

Since fish proteins can be precipitated by salts at a pH more neutral than those used previously (pH 2.5 in Figure 3), the precipitation steps of the extraction were performed at both pH 4.5 and pH 6.8, in addition to pH 2.5. The remainder of the extraction procedure was then carried out in the same manner as described earlier but the filtered solids were retained for further study. An almost fivefold decrease in total mutagenicity (revertants) was observed when the pH of protein precipitation was increased from pH 2.5 to pH 6.8 (Table 3).

This phenomenon could have been due to a number of reasons. The most obvious ones were that compounds interfering with the assay were being extracted at pH 4.5 and 6.8, decreasing the total number of revertants, or that mutagens were associating with the precipitated proteins.

To test the first possibility, mixing experiments were carried out in which the pH 6.8, pH 4.5, and pH 2.5 organic extracts were combined with other extracts known to contain mutagens and these mixtures were tested in the Ames assay. The mixture con-

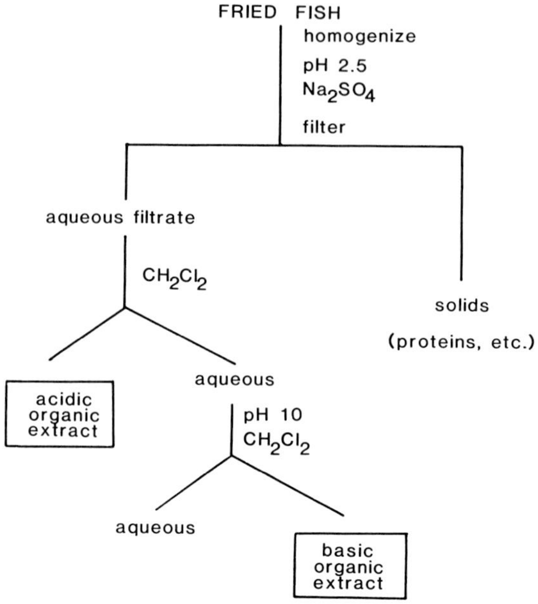

FIGURE 3. Abbreviated schematic diagram of the extraction procedure used for concentrating mutagens from cooked fish (precipitation of soluble proteins occurs in first step).

Table 2
MUTAGENICITY OF BASIC EXTRACTS FROM FRIED FISH

Salmonella typhimurium	MAR[a]/20 g fish			
	Sole	Snapper	Salmon	Turbot
TA1538	15.7	14.8	11.3	3.3
TA98	14.7	12.2	7.1	2.8
TA1537	3.4	3.0	6.0	1.0

[a] Mutagenic activity ratio = no. of revertants on plate with fish extract/no. of spontaneous revertants.

Table 3
EFFECT OF pH OF PROTEIN PRECIPITATION ON THE EXTRACTION OF MUTAGENS FROM FRIED SOLE

pH of protein precipitation	Revertants[a] produced by first (acidic) extract[b]	Revertants produced by basic extract[b]	Totals
2.5	7	421	428
4.5	63	143	206
6.8	62	28	90

[a] *Salmonella typhimurium* strain TA1538 with 80 $\mu\ell$ S9/plate. The spontaneous revertants (24/plate) have been subtracted.

[b] The organic extract equivalent to 10 g fried sole/plate was used in all cases.

taining the pH 6.8, but not the pH 4.5 or pH 2.5 extracts, produced fewer revertant colonies than expected, and it appeared that the pH 6.8 extracts reduced the mutagenicity of known mutagens by about 40%. It seemed that compounds interfering with mutagen expression were being extracted out at this more neutral pH.

By reextracting the precipitated fish proteins (from the above experiments) at different pHs, it was shown that the mutagen was binding to, or associating with these proteins to a greater extent as the pH was increased. The association may have been due to ionic interactions between the basic, positively charged mutagen and negatively charged groups on the proteins.

Thus, both the binding of mutagens to proteins and the presence of substances that interfere with mutagen testing appeared to be responsible for the differences in mutagenicity that were encountered upon varying the pH of extraction. Although in this experiment the low pH did not seem to be producing mutagenic substances, this may not be the case with other foods.

The information obtained from this study also indicated that the procedure used for the survey of mutagenicity in cooked fish was not particularly effective in concentrating mutagens from fried fish, and probably would not be effective in isolating mutagens from any other high protein food.

At this point, it was apparent that some method other than an aqueous extraction was needed to improve the recovery of mutagens from fried fish. The use of water-miscible organic solvents is a common method for precipitation of proteins, and since the mutagen appeared to be more soluble in organic solvents than water, the use of solvents such as acetone, methanol, or ethanol should be more effective than water in extracting mutagens. An extraction procedure that used these three solvents is shown in Figure 4. Briefly, the fish is ground with one of the solvents, the solids are filtered, and the solvent is removed by rotary evaporation. The residue is taken up in water, the pH adjusted to near neutrality, and the remainder of the procedure is the same type of neutral/base partitioning as has been described earlier. The results of using this type of procedure are shown in Table 4. All the organic solvents were able to extract more mutagens than any of the aqueous procedures, where at best compounds producing only 400 revertants were extracted (extracting at pH 2.5 and 10).

More importantly, if components in the food sample could react at pH extremes, this procedure would allow an investigator to extract the mutagens at near neutral pH and prevent unwanted side reactions from occurring. Felton and co-workers[3] reported earlier this year that the use of acetone or isopropanol also doubled the mutagen yield from fried ground beef, and that the use of these solvents appeared to be an acceptable method for ground beef.

A criticism for using acetone in an extraction procedure is that acetone is a carbonyl compound and, while fairly unreactive, still presents the possibility of interaction with amines found in foods. The authors investigated this possibility and refluxed an aqueous acetone/ammonia solution overnight, extracted the mixture with methylene chloride, and upon testing it in the Ames assay found that mutagenic compounds had formed. Granted, during the extraction of foods with acetone, the short time and low temperatures would probably preclude mutagen formation during extraction. However, it should be emphasized that the possibility for reaction does exist.

IV. MAILLARD OR BROWNING REACTIONS IN FOODS

It has been shown by a number of investigators, particularly Shibamoto et al., that certain Maillard browning reaction products from model systems are mutagenic in the Ames test.[18-20] In foods susceptible to the browning reaction, it has been noticed that

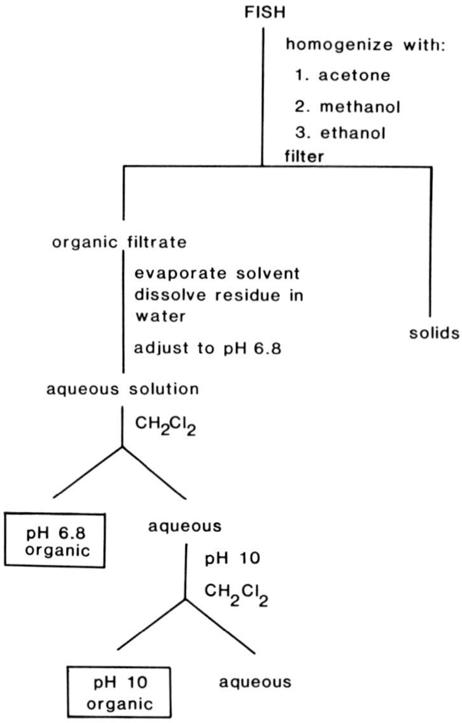

FIGURE 4. Extraction procedure using different organic solvents for the concentration of mutagens from cooked fish.

Table 4
REVERTANT COLONIES PRODUCED BY METHYLENE CHLORIDE EXTRACTS OF FRIED SOLE USING ORGANIC SOLVENTS AS PROTEIN PRECIPITANTS

	Revertants[a] from CH_2Cl_2 extracts[b]		
Solvent	pH 6.8	pH 10	Total
Acetone	613	135	748
Ethanol	500	253	753
Methanol	474	195	669
Water	62	28	90

[a] *Salmonella typhimurium* strain TA1538 with 80 μl S9/plate. The spontaneous reversion rate (24/plate) has been subtracted.

[b] The extract equivalent to 10 g fried sole was applied to each plate.

upon basifying the aqueous solutions of these foods even slightly (pH 8 to 9), the solution turns darker. In some instances, this darkening may very well be due to unstable pigments, but in most cases it is a result of the browning reaction. This reaction also seems to take place when an organic solvent extract is concentrated to near dryness. When the residue is resuspended in the same solvent, the resulting solution is noticeably darker than before. Similar darkening also occurs if extracts of foods susceptible to the browning reaction are allowed to stand at room or freezing temperatures for any length of time before they are assayed. Thus, there may be a problem of testing and working with a compound in the food extract that may not have been initially present in the food and, again, investigators may be dealing with a mutagen artifact. Because of the number and complexity of the compounds formed as a result of the Maillard reaction, it would be questionable to use the conventional extraction techniques with foods that undergo this reaction.

Perhaps the simple aqueous extraction procedure developed by Stich and co-workers,[11] and used to test for the presence of clastogenic compounds in dried fruits, would be the type of extraction that is needed for dealing with foods undergoing the browning reaction. However, this method can only be applied to a limited number of foods, and to prove that the mutagenic compound was present in the original food product may be extremely difficult, especially if the mutagen was a Maillard reaction intermediate.

V. LIMITATIONS OF CURRENT EXTRACTION TECHNIQUES

Thus far, many of the mutagens extracted and identified from foods have been relatively small molecular weight compounds. With the multitude of reactions taking place in foods during processing or cooking, it would be fair to assume that a number of reactions are occurring between reactive groups of small molecules and those of macromolecules, or between reactive groups on macromolecules (such as proteins and polysaccharides). For example, a carbonyl compound (any reducing sugar or compound formed from oxidative rancidity) could react with a protein-bound amino group, and any product formed as a result of this and further reactions probably would not be extracted with the conventional extraction systems because there was no cleavage of this new compound from the macromolecule. Earlier it was mentioned that no mutagens were found in the acidic or basic extracts of biscuits using the sodium sulfate/hydroxide extraction procedure. This did not mean that there were no mutagens present, but a reason for not seeing any is that perhaps the precipitated proteins and polysaccharides from the biscuits were discarded, and there may have been a potential mutagen attached to the protein.

However, if people were to consume this food, their digestive enzymes or enzymes from the gut bacteria might hydrolyze these macromolecules and release the small molecular weight compounds, and our present extraction system may be completely missing another set of potential mutagens.

Thus to improve the mutagen testing of foods, further work should be conducted to develop procedures that will allow an investigator to extract and test all the compounds that will eventually be exposed to and absorbed by the body. For example, the incubation of a food with certain digestive enzymes (either mammalian or bacterial, or both) or with homogenates of the digestive system prior to extraction, may aid in the release of mutagenic compounds bound to macro- or other molecules in the food.

VI. SUMMARY

Some of the problems associated with different chemicals and conditions used for mutagen extraction from foods have been highlighted. It was shown that the use of ammonium salts in the extraction procedure caused artifactual mutagenicity to occur in some foods, but not in others. Experiments have demonstrated that the pH of the aqueous fraction during extraction affects the solubility and, therefore, the partitioning of various mutagens into organic solvents. It was also pointed out that compounds that prevent the expression of mutagenicity, i.e., interfering compounds, can be extracted in greater or lesser quantities depending on the pH. The organic solvent extraction procedure described in the text and outlined in Figure 3, while not the most efficient, appears to be the most suitable for the extraction of mutagens from certain foods, primarily because pH extremes are not used.

A different type of potential artifact problem arises when mutagen formation is studied in foods having components that participate in the browning reaction. The Maillard reaction has been shown to produce mutagenic compounds, and browning is clearly observed when an aqueous extract is made slightly basic or when organic solvents are evaporated.

The last problem discussed dealt with the inadequacies of the food extraction techniques currently in use because they do not address the digestive processes functioning in mammalian systems. Potential mutagens may be covalently bound to macromolecules, and while they may be released by digestive and/or bacterial enzymes, current extraction procedures do not have provisions for dealing with this aspect of mutagen release.

REFERENCES

1. **Commoner, B., Vithayathil, A. J., Dolara, P., Nair, S., Madyastha, P., and Cuca, G. C.,** Formation of mutagens in beef and beef extract during cooking, *Science,* 201, 913, 1978.
2. **Dolara, P., Commoner, B., Vithayathil, A., Cuca, G. C., Tuley, E., Madyastha, P., Nair, S., and Kriebel, D.,** The effect of temperature on the formation of mutagens in heated beef stock and cooked ground beef, *Mutat. Res.,* 60, 231, 1979.
3. **Felton, J. S., Healy, S., Stuermer, D., Berry, C., Timourian, H., Hatch, F. T., Morris, M., and Bjeldanes, L. F.,** Mutagens from the cooking of food. I. Improved extraction and characterization of mutagenic fractions from cooked ground beef, *Mutat. Res.,* 88, 33, 1981.
4. **Iwaoka, W. T., Krone, C. A., Sullivan, J. J., and Johnson, C. A.,** Effect of pH and ammonium ions on mutagenic activity in cooked beef, *Cancer Lett.,* 12, 335, 1981.
5. **Pariza, M., Ashoor, S., Chu, F. S., and Lund, D. B.,** Effects of temperature and time on mutagen formation in pan-fried hamburgers, *Cancer Lett.,* 7, 63, 1979.
6. **Spingarn, N. and Weisburger, J.,** Formation of mutagens in cooked foods. I. Beef, *Cancer Lett.,* 7, 259, 1979.
7. **Nagao, M., Honda, M., Seino, Y., Yahagi, T., and Sugimura, T.,** Mutagenicities of smoke condensates and the charred surface of fish and meat, *Cancer Lett.,* 2, 221, 1977.
8. **Yamaguchi, K., Shudo, K., Okamoto, T., Sugimura, T., and Kosuge, T.,** Presence of 2-aminodipyrido[1,2-a:3′,2′-*d*]imidazole in broiled cuttlefish, *Gann,* 71, 743, 1980.
9. **Yamaizumi, Z., Shiomi, T., Kasai, H., Nishimuri, S., Takahashi, T., Nagao, M., and Sugimura, T.,** Detection of potent mutagens, Trp-P-1 and Trp-P-2, in broiled fish, *Cancer Lett.,* 9, 75, 1980.
10. **Nagao, M., Takahashi, H., Yamanaka, H., and Sugimura, T.,** Mutagens in coffee and tea, *Mutat. Res.,* 68, 101, 1979.
11. **Stich, H. F., Rosin, M. P., Wu, C. H., and Powrie, W. D.,** Clastogenic activity of dried fruits, *Cancer Lett.,* 12, 1, 1981.
12. **Takahashi, Y., Nagao, M., Fujino, T., Yamaizumi, Z., and Sugimura, T.,** Mutagens in Japanese pickle identified as flavonoids, *Mutat. Res.,* 68, 117, 1979.
13. **Matsukura, N., Kawachi, T., Morino, K., Ohgaki, H., Sugimura, T., and Takayama, S.,** Carcinogenicity in mice of mutagenic compounds from a tryptophan pyrolysate, *Science,* 213, 346, 1981.

14. **Pariza, M. W., Ashoor, S. H., and Chu, F. S.,** Mutagens in heat-processed meat, bakery and cereal products, *Food Cosmet. Toxicol.,* 17, 429, 1979.
15. **Spingarn, N., Slocum, L., and Weisburger, J.,** Formation of mutagens in cooked foods. II. Foods with high starch content, *Cancer Lett.,* 9, 7, 1980.
16. **Iwaoka, W. T., Krone, C. A., Sullivan, J. J., Meaker, E. H., Johnson, C. A., and Miyasato, L. S.,** A source of error in mutagen testing of foods, *Cancer Lett.,* 11, 225, 1981.
17. **Krone, C. A. and Iwaoka, W. T.,** Mutagen formation during the cooking of fish, *Cancer Lett.,* 14, 93, 1981.
18. **Mihara, S. and Shibamoto, T.,** Mutagenicity of products obtained from cysteamine-glucose browning model systems, *J. Agric. Food Chem.,* 28, 62, 1980.
19. **Shibamoto, T., Nishimura, O., and Mihara, S.,** Mutagenicity of products obtained from a maltol-ammonia browning model system, *J. Agric. Food Chem.,* 29, 643, 1981.
20. **Spingarn, N. and Garvie, C.,** Formation of mutagens in sugar-ammonia model systems, *J. Agric. Food Chem.,* 27, 1319, 1979.

Chapter 18

A NEW CHROMATOGRAPHIC METHOD FOR SEPARATING MUTAGENS FROM COMMERCIAL BEEF EXTRACT AND FRIED GROUND BEEF

W. A. Hargraves, R. Dietrich, and M. W. Pariza

TABLE OF CONTENTS

I. INTRODUCTION

It is now well-established that mutagenic heterocycles are present among the myriad of products that result from the complex chemical reactions initiated in meat by cooking at moderate temperatures.[1-6] These minor products are temperature-dependent and appear to be maximized at approximately 190°C.[3,7] Pyrolysis of proteins and amino acids also results in the formation of mutagenic products, but these reactions do not occur until temperatures of 250 to 600°C are reached.[8-10] The nonenzymatic browning reactions resulting from dehydration and cyclization of monosaccharides during heating, as well as Maillard reactions,[11,12] have often been implicated in the pathways for formation of the mutagens formed at 160 to 190°C.[13-15]

The effects of increased length of time of cooking are not well-defined, but at lower temperatures (e.g., 143°C) detectable mutagenic activity can be found after 20 min of cooking.[3]

It appears that even with different methods of cooking (grilling, broiling, microwave) the temperature that is reached at the surface of the meat is the most important parameter in determining the extent of mutagen formation.[3,7] The surface of the meat has been demonstrated both directly[6,7] and indirectly[16] to be the site of generation of the mutagenic products, but this is probably because the interior does not reach the critical temperature. Most likely, the differences in mutagenic activity found in hamburgers cooked at several commercial restaurants[3] were due to the methods of cooking. However, another important variable could be differences in composition or processing of the meat. Several different sources of proteinaceous foods have been surveyed for their mutagenic products after cooking,[17] and because variable amounts of mutagenic activities were found the composition of the meat may be another important factor in determining the number and amount of mutagens after cooking.

Although the temperature dependence of mutagen formation has been observed for crude extracts from meat[3] and other food products,[4] the effect of temperature on the formation of individual mutagenic compounds is not known. It is known that more than one mutagenic product occurs in beef extract and smoked sardines,[18-20] and three mutagens have been identified as 2-amino-3,4 methylimidazolo[4,5-*f*] quinoline, 2-amino-3-methylimidazolo[4,5-*f*] 7-methylquinoline, and 2-amino-3,8 methylimidazolo[4,5-*f*] 3-methylquinoxaline.

Because temperature, time, food source, percentage of water, cooking method, etc., can influence the formation of mutagenic products in cooked meat, methods must be devised that can quantitatively evaluate the effect of each of these parameters on each individual mutagen before more can be learned about the reaction pathways. Experimental procedures are constantly being improved in order to do this,[17,21] and in this work methods are described that will reliably compare these factors.

II. EXPERIMENTAL METHODS

A. Reagents

The beef extract used in these experiments was purchased from Difco Laboratories Inc., Detroit, Mich., in 1 lb cans labeled as Difco Certified Reagent. The lot numbers varied from can to can. Sodium chloride, sodium hydroxide, and chloroform for column chromatography were USP grade from Mallinckrodt Chemical Co., St. Louis, Mo. The methylene chloride used for extraction, as well as the methanol, were from Fisher Scientific Co., Pittsburgh, Pa., and were certified ACS grade reagents. The Adsobosil-5 silica gel was purchased from Anspec Co. as powder for TLC. Ground beef used in this work was purchased as processed frozen 115 g patties or fresh ground beef from a local supermarket. Both contained approximately 19% fat. The L-proline

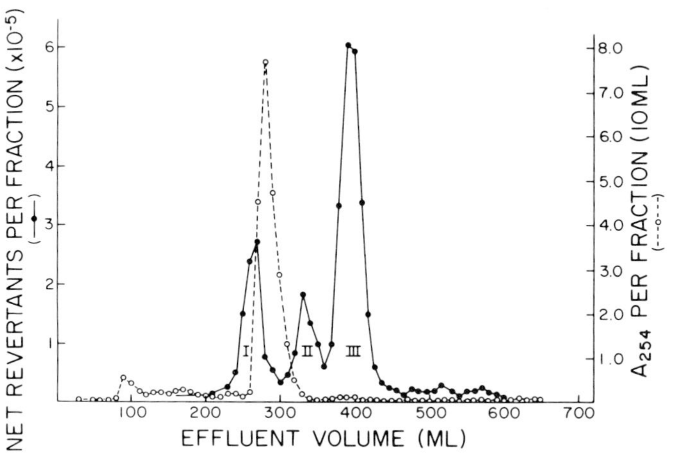

FIGURE 1. Fractionation of Difco beef extract on Adsorbsil-5 with 5% $MeOHCl_3$. (Corrected for total activity of sample.) Straight line refers to mutagenicity (peaks I, II, III) and broken line refers to absorbance at 254 nm.

was Sigma grade 3x crystallized. Authentic IQ was the kind gift of Drs. Kawachi and Sugimura.[20]

B. Procedures

The extraction procedures have been previously described by Pariza et al.[3] The silica gel column was slurry packed in 5% methanol in chloroform and the fines were removed. The column was washed with 20% MeOH in $CHCl_3$ and equilibrated overnight, again in 5% MeOH:$CHCl_3$, before application of the sample. Following equilibration, the column settled to the dimensions of 32 × 2.5 cm and it was capped with a 0.5 cm layer of anhydrous Na_2SO_4. The basic beef or hamburger extract was evaporated to dryness, redissolved in 1 mℓ of 20% MeOH:$CHCl_3$ and carefully layered on the top of the column. The mutagenic activity was eluted from the column with 5% MeOH:$CHCl_3$ in 10 mℓ fractions at a flow rate of approximately 10 mℓ/hr. A total volume of 600 to 700 mℓ of eluate was generally collected, followed by a wash of 250 mℓ of 20% MeOH:$CHCl_3$. The absorbance at 254 nm of each fraction was assayed, and the mutagenic activity determined by the standard Ames procedure using *Salmonella typhimurium* TA98.[22]

III. RESULTS AND DISCUSSION

Two major peaks of mutagenic activity were always observed in the eluate from the silica gel column, as was reported by Commoner et al.[23] and Spingarn et al.,[19] but frequently three peaks were found depending upon the lot of beef extracted as shown in Figure 1. On the average, the recovery of mutagenic activity was 105 + 6%. Purified synthetic IQ was also run on this column and the recovery was found to be 109%. These results indicate that all of the mutagenic activity in the extracts is represented in the chromatogram and none was left on the column. The percentages of each peak of activity as well as its relative retention time, presented in Table 1, demonstrate the reproducibility of this chromatographic procedure.

Table 1
PERCENTAGES OF MUTAGENIC ACTIVITY IN CHROMATOGRAPHIC PEAKS

Peak (%)[a]	No. trials	I	II	III	IV	Wash
Relative retention time[b]	4	0.94 ± 0.018	1.20 ± 0.02	1.41 ± 0.02	1.83 ± 0.08	—
Beef extract	4	21 ± 2	13 ± 2	50 ± 4	—	4 ± 1
Ground beef extract	5	26 ± 7	—	—	35 ± 4	18 ± 8
Ground beef extract (proline)	3	—	14 ± 5	—	52 ± 10	10 ± 5

[a] Based upon percentage of total activity recovered.
[b] Relative to absorbance peak (254 nm) at 290 mℓ of effluent volume.

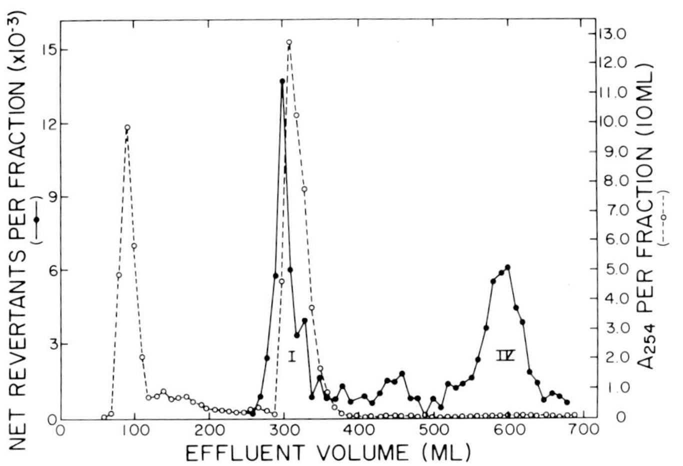

FIGURE 2. Fractionation of cooked ground beef extract on Adsorbsil-5 with 5% $MeOHCl_3$. Symbols same as for Figure 1, with the mutagenicity in peaks I and IV.

The mutagenic compounds are eluted in order of increasing polarity, with peak III the most polar. This peak has been identified as IQ based upon its MS being identical to authentic IQ. Peak I has been tentatively identified as MeIQx based on the molecular ion at m/z 213. Peak II has not as yet been identified. The major peak of activity (III) in beef extract (Figure 1) is either not present or present only in very low concentration in fried ground beef (Figure 2). It has been reported that IQ is present in fried ground beef as well as commercial beef extract,[18,19] however, this cannot be confirmed. The authors have been able to demonstrate by identical MS that peak I in beef and hamburger are the same compounds (apparently MeIQx,[20] Figure 2) and occur in approximately the same percentages (Table 1). It should be noted however that the variability in the amount of peak I in hamburger is larger than in beef extract (±27% vs. 10%). Additionally, a much more polar peak of activity is present in ground beef that is not present in commercial beef extract. This mutagen is as yet unidentified, and is present in amounts equal to the apparent MeIQx. It is perhaps not surprising that fried ground beef and beef extract, differing as they do in the manner in which they were processed, also differ in the mutagenic products formed during heating.

The addition of proline to the ground beef prior to cooking demonstrates the effects of altering the composition of the beef (Figure 3). Peak I appears not to have been a product in this sample, peak II was found and in about the same percentage as in beef extract, peak III however is still not a product, and peak IV has apparently increased. It is possible that the large variability in peak I that has been observed in various fried ground beef preparations may be due to the proline content of the beef. These data demonstrate that it is highly likely that compositional factors may play an important role, second only to temperature, in the formation of mutagenic compounds during cooking of meat products.

ACKNOWLEDGMENTS

The authors wish to thank Patricia Sharp, Jayne Storkson, Nancy Holland, and Jeff Abresch for excellent technical assistance. This research was supported in part by the College of Agricultural and Life Sciences, University of Wisconsin-Madison (CALS

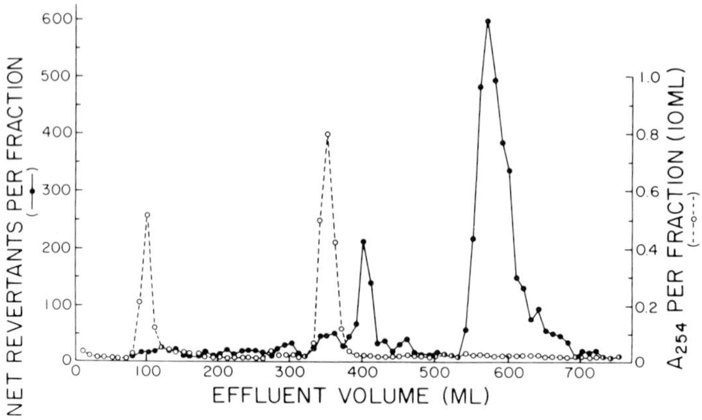

FIGURE 3. Fractionation of cooked ground beef extract with added proline (1.0 mmol/35 g ground beef) on Adsorbsil-5 with 5% $MeOHCl_3$. Symbols same as for Figure 1 with the mutagenicity in peaks II and IV.

Research Division project No. 2640); the University of Wisconsin-Madison Graduate School; the Wisconsin Agricultural Experiment Station; the U.S. Department of Agriculture; the University of Wisconsin-Madison Food Research Institute; and PHS research grant number CA29618, awarded by the National Cancer Institute, Department of Health and Human Services.

REFERENCES

1. **Vithayathil, A. J., Commoner, B., Nair, S., and Madyastha, P.,** Isolation of mutagens from bacterial nutrients containing beef extract, *J. Toxicol. Environ. Health,* 4, 189, 1978.
2. **Nagao, M., Honda, M., Seino, Y., Yahagi, T., and Sugimura, T.,** Mutagenicities of smoke condensates and the charred surface of fish and meat, *Cancer Lett.,* 2, 221, 1977.
3. **Pariza, M. W., Ashoor, S. H., Chu, F. S., and Lund, D. B.,** Effects of temperature and time on mutagen formation in pan-fried hamburger, *Cancer Lett.,* 7, 63, 1979.
4. **Pariza, M. W., Ashoor, S. H., and Chu, F. S.,** Mutagens in heated and processed meat, bakery and cereal products, *Food Cosmet. Toxicol.,* 17, 429, 1979.
5. **Spingarn, N. E. and Weisburger, J. H.,** Formation of mutagens in cooked foods. I. Beef, *Cancer Lett.,* 7, 259, 1979.
6. **Felton, J. S., Healy, D., Stuermer, D., Berry, C., Timourian, H., Hatch, F. T., Morris, M., and Bjeldanes, L. F.,** Mutagens from the cooking of food. I. Improved extraction and characterization of mutagenic fractions from cooked ground beef, *Mutat. Res.,* 88, 33, 1981.
7. **Dolara, P., Commoner, B., Vithayathil, A., Cuca, G., Tuley, E., Madyastha, P., Nair, S., and Kriebel, D.,** The effect of temperature on the formation of mutagens in heated beef stock and cooked ground beef, *Mutat. Res.,* 60, 231, 1979.
8. **Sugimura, T. and Nagao, M.,** Mutagenic factors in cooked foods, *CRC Crit. Rev. Toxicol.,* 6, 189, 1979.
9. **Matsumoto, T., Yoshida, D., Mizusaki, S., and Okamoto, H.,** Mutagenic activity of amino acid pyrolysate in *Salmonella typhimurium* TA98, *Mutat. Res.,* 48, 279, 1977.
10. **Nagao, M., Honda, M., Seino, Y., Yahagi, T., Kawachi, T., and Sugimura, T.,** Mutagenicities of protein pyrolysates, *Cancer Lett.,* 2, 335, 1977.
11. **Maillard, L. C.,** Action of amino acids on sugars, *CR Seances,* 154, 66, 1912.

12. **Hodge, E.,** Chemistry of browning reactions in model systems, *J. Agric. Food Chem.,* 1, 928, 1953.
13. **Commoner, B., Vithayathil, A. J., Dolara, P., Nair, S., Madyastha, P., and Cuca, G.,** Formation of mutagens in beef and beef extract during cooking, *Science,* 201, 913, 1978.
14. **Spingarn, N. E. and Garvie, C. T.,** Formation of mutagens in sugar-ammonia model systems, *J. Agric. Food Chem.,* 27, 1319, 1979.
15. **Shibamoto, T., Nishimura, O., and Mihara, S.,** Mutagenicity of products obtained using a maltol-ammonia browning model system, *J. Agric. Food Chem.,* 29, 643, 1981.
16. **Pariza, M. W.,** Unpublished data.
17. **Hatch, F. T., Felton, J. S., and Bjeldanes, L. F.,** Mutagens from the cooking of food: thermic mutagens in beef, in *Carcinogens and Mutagens in the Environment, Vol. 1, Food Products,* Stich, H. F., Ed., CRC Press, Boca Raton, Fla., 1983.
18. **Kasai, H., Yamaizumi, Z., Wakabayashi, K., Nagao, M., Sugimura, T., Yokoyama, S., Miyazawa, T., Spingarn, N., Weisburger, J. H., and Nishimura, S.,** Potent novel mutagens produced by broiling fish under normal conditions, *Proc. Jpn. Acad.,* 56B, 278, 1980.
19. **Spingarn, N. E., Kasai, H., Vuolo, L. L., Nishimura, S., Yamaizumi, Z., Sugimura, T., Matsushima, T., and Weisburger, J. H.,** Formation of mutagens in cooked foods. III. Isolation of a potent mutagen from beef, *Cancer Lett.,* 9, 177, 1980.
20. **Sugimura, T.,** Personal communication, 1981.
21. **Iwaoka, W. T., Krone, C. A., Sullivan, J. J., Meaker, E. H., Johnson, C. A., and Miyasata, L. S.,** A source of error in mutagen testing of foods, *Cancer Lett.,* 11, 225, 1981.
22. **Ames, B. N., McCann, J., and Yamasaki, E.,** Methods for detecting carcinogens and mutagens with the *Salmonella*/mammalian microsome mutagenicity test, *Mutat. Res.,* 31, 347, 1975.
23. **Commoner, B., Vithayathil, A., and Dolara, P.,** Mutagenic analysis as a means of detecting carcinogens in foods, *J. Food Protect.,* 41, 996, 1978.

Chapter 19

BIOCHEMICAL CONSIDERATIONS IN SPECIES AND DOSE EXTRAPOLATIONS

Ronald C. Shank

TABLE OF CONTENTS

I. INTRODUCTION

There is great pressure on the toxicologist to rapidly assess the human health risks presented by the massive number of chemicals already in the environment and chemicals about to enter the environment, including the food supply. Many remarkable advances have been made in the understanding of molecular mechanisms in chemical carcinogenesis and in developing short-term, low-cost screening tests to predict the carcinogenicity of a compound. Nevertheless, in the final analysis our knowledge in chemical carcinogenesis falls short of the mark, and we must still assay a chemical for carcinogenicity by using long-term animal testing. We know that most environmental carcinogens studied so far form adducts with target cell DNA which seems to be consistent with the somatic mutation theory of carcinogenesis, yet few scientists today are willing to label a chemical a human carcinogen on the basis of DNA adduct formation, or conclude the compound is not a carcinogen because covalent binding to the genetic material cannot be demonstrated. If a chemical gives a positive result in a mutagenicity test or in several of a battery of such tests, there are still many scientists unwilling to conclude, on such evidence alone, that that compound is carcinogenic in man. The whole-animal test is still necessary to make this prediction.

Even when a compound has been tested for carcinogenicity in a full-scale animal study, the toxicologist is still confronted with the formidable problem of extrapolating quantitatively from the animal data to the expected risk to a human population usually exposed to levels of the compound orders of magnitude lower than those used in that animal study. A bioassay cannot appraise the threat a chemical may pose to human health. It is the interpretation of the results of that bioassay and the manner in which the extrapolation from those results is made that allows one to attempt to estimate human health effects of a compound, and that interpretation and extrapolation can be at fault. This presentation will try to focus on some of the biochemical inconsistencies that can enter and confuse extrapolations from animal testing data to assessments of risk to human health.

II. METABOLIC ACTIVATION OF CARCINOGENS

A. The Electrophile as the Ultimate Carcinogen

The toxicologist and pharmacologist have long held that the biological activity of a compound is determined by the chemical activity of that compound. Chemical activity is, of course, determined by chemical structure, and hence the study of structure-activity relationships in toxicology and pharmacology is of central importance to research on mechanism of action. One of the first major contributions to chemical carcinogenesis was made by the Pullmans[1,2] in their pioneering work on PAH. In those studies a high correlation was demonstrated between the carcinogenic activities of a series of compounds and the electrical properties (i.e., chemical reactivity) of specific bonds within the individual polycyclic hydrocarbons, the so-called K and L regions of the molecules.

A number of investigations began to look for similar correlations between chemical reactivity and carcinogenicity in compounds as varied as the bifunctional alkylating agents, nitrosamines and nitrosamides, aromatic amines, the aflatoxins, pyrrolizidine alkaloids, etc. The Millers[3] brought together the large body of evidence for structure-activity relationships in carcinogenesis, to which they have been major contributors, and proposed the generalization that most, if not all, chemical carcinogens are ultimately electron-deficient reactants. In most cases examined so far, and there have been many, the carcinogens have been shown to form highly reactive electrophilic reagents that bind covalently to electron-rich (strongly nucleophilic) sites in proteins and nucleic acids. Since most environmental carcinogens, and certainly those associated with the human food supply, exist in the environment in relatively unreactive forms, there is an apparent need for the conversion of the carcinogen from its unreactive environmental form to its highly reactive electrophilic form. This points to a major biochemical consideration in species and dose extrapolations in cancer research. It is well-known that there are important quantitative and qualitative differences in metabolic pathways between animal species, and hence, these differences offer a potential for important differences in the metabolism of environmental carcinogens to the ultimate reactive forms.

B. Tissue Specificity

Boveri[4] proposed that carcinogenesis may be linked to damage of the genetic material in a cell and may be analogous to the mutation process in germ cells except that it occurs in somatic cells. This hypothesis has come in and gone out of popularity several times in the 69 years since it was introduced, yet today there is considerable biochemical evidence to support it.

The first major blow to the hypothesis came when it was shown that there was little correlation between carcinogenicity and mutagenicity for a variety of compounds. At that time it was not understood that most compounds had to be metabolically activated to exert their carcinogenicity or mutagenicity. Cancer testing was done in mammals and the mutagenicity tests were, for the most part, microbiological. Of course, the two life systems have widely varying metabolic capabilities. When the microbiological mutagenicity test systems were corrected for their deficiencies in metabolic activity by adding mammalian enzyme systems, the correlation between mutagenicity and carcinogenicity improved considerably.[5,6]

Another major blow to the somatic mutation theory was dealt by the finding that tissues that suffered extensive DNA alkylation or arylation did not necessarily become cancerous, while other tissues that experienced far less DNA adduct formation, and hence less DNA damage, readily developed malignant cancers. It has been shown subsequently that such damage to DNA can be repaired, that the adducts can be removed by a variety of mechanisms, and the genetic template can be fully restored.[7,8] If this repair of damaged DNA occurs before replication takes place, there may be no reflection in the daughter cell of the original damage to the parent cell and presumably no resultant cell transformation or cancer. Not only do various tissues differ in their abilities to repair DNA, but so do various animal species.

C. Species Differences and Similarities

Several examples of the importance of metabolic activation in interspecies differences in carcinogenesis can be discussed. DMNA and DENA are two carcinogens that have been shown to be present in certain foods, including beer and nitrite-cured meats. The active form of the dialkylnitrosamines seems to be an electron-deficient alkyl group called a carbonium ion, methonium ion, CH_3^+ , in the case of DMNA, and ethonium ion, $CH_3CH_2^+$, in the case of DENA; the electrophiles covalently bind to nucleophilic sites in DNA and other macromolecules. Bartsch and co-workers[9] compared the metabolic activation of DMNA and DENA in rat and hamster. Both compounds are carcinogenic to the liver in both species: DENA produces tumors in hamster lung but not in rat lung, and DMNA is inactive in lung in both species. Metabolic activation, measured indirectly as the in vitro conversion of ^{14}C-DMNA/DENA to $^{14}CO_2$ was high in target organs and low in nontarget organs (Table 1).

Other examples of major species differences in susceptibility to carcinogens can be

Table 1
ASSOCIATION BETWEEN NITROSAMINE METABOLISM AND TUMOR PRODUCTION IN VARIOUS ORGANS

		$^{14}CO_2$ from ^{14}C-DMN/DEN (% ^{14}C converted/g/hr)		
Compound	Animal	Liver	Lung	Sm. intestine
DMN	Rat	22	0.12	0.11
		(+)	(−)	(−)
	Hamster	36	0.2	0.05
		(+)	(−)	(−)
DEN	Rat	7.9	1.3	0.6
		(+)	(−)	(−)
	Hamster	8.9	12.4	0.9
		(+)	(+)	(−)

Note: +, − = tumor response.

given. β-Naphthylamine was first discovered to produce urinary bladder tumors in the human[10] but early attempts to demonstrate the carcinogenicity in laboratory animals failed because the rat, the usual first choice in such studies, does not metabolize β-naphthylamine in the same manner as does the human.[11] The dog does respond to the carcinogenicity of β-naphthylamine and presumably metabolizes the compound in a way similar to humans.[12] Had the compound been screened for carcinogenicity only in the rat in the absence of human data, β-naphthylamine would have been labeled erroneously as a noncarcinogen. The model carcinogen AAF is metabolically activated via N-hydroxylation by many species,[13] however, the guinea pig can only weakly N-hydroxylate AAF, and AAF does not induce cancer in this species.[13,14]

In other instances there can be remarkable consistency in the carcinogenicity of compounds across species. Schmähl and Habs[15] have shown that DENA induces liver cancer in 17 different species of rodents, primates, birds, fish, frogs, cats, dogs, and pigs. Montesano and Magee[16] demonstrated that DNA in human liver slices incubated with DMNA contained 7-methylguanine, the result of metabolic activation of the carcinogen. The author[17] demonstrated the presence of 7-methylguanine and O^6-methylguanine in the liver DNA of a 23-year-old male homicide victim poisoned with DMNA, confirming the finding that the human liver also is capable of metabolically activating DMNA. Thus there is no toxicological reason to doubt that the human should also be susceptible to the carcinogenicity of this environmental compound.

D. Dose Considerations

1. Dose and Metabolic Pathways

In addition to species considerations in cancer bioassays, dose considerations must also be made. Watanabe and co-workers[18] showed that vinyl chloride exposures that were ineffective in producing hemangiosarcomas in rat liver did not depress levels of glutathione, an endogenous biochemical intermediate that presumably reacts with activated forms of vinyl chloride to detoxify them. However, as the vinyl chloride exposure levels increased to those effective in inducing the cancers (according to the carcinogenicity bioassay of Maltoni and Lefemine[19]), the glutathione levels became severely depressed (Figure 1), implying that the reactive vinyl chloride metabolite escaped the

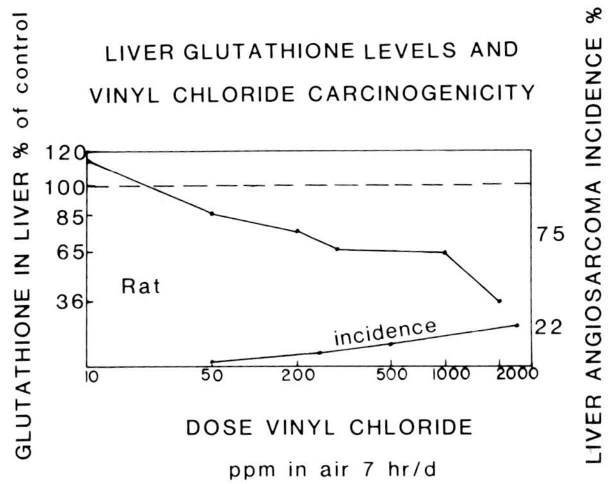

FIGURE 1. Correlation between depression of hepatic glutathione levels and incidence of liver angiosarcomas in rats exposed to vinyl chloride.

detoxication pathway via glutathione conjugation and was free to react with other nucleophilic reactants such as proteins and nucleic acids. The significance of such experiments is in the implication that exposure levels that do not deplete glutathione levels could fail to induce cancer, and that one cannot make a linear extrapolation from high-dose data to low-dose risks because there is a noncontinuous biochemical change, namely the depletion of the intracellular glutathione below a level necessary to adequately detoxify the vinyl chloride metabolites.

2. Single vs. Multiple Doses

Such discontinuities can be looked at from another point of view. In the early studies on the association between DNA alkylation and carcinogenesis, it was discovered that the 7-position of guanine was the most nucleophilic of all the available sites in double stranded DNA,[20] and with the methylating carcinogens, the most frequently occurring aberrant base in DNA was 7-methylguanine. Swann and Magee[21] later showed a poor correlation between the amount of 7-methylguanine formed in target organ DNA after a single administration of a methylating carcinogen and the potencies of those compounds as carcinogens, and investigators began to look more closely at other sites of DNA alkylation, especially at those sites involved in base-pairing. Recently the author's laboratory developed an analytical technique that permits the rapid, quantitative determination of a variety of alkylated purines in DNA without the need for radiolabeled agents.[22] A chronic study[23] was done in which rats were given 21 mg 1,2-dimethylhydrazine per kilogram of body weight s.c. weekly for 14 weeks, a regimen shown to produce cancer of the colon, liver, and kidney.[24] The levels of 7-methylguanine and O^6-methylguanine were determined at regular intervals over a 25 week period during which tumors appeared in the target organs. The methylguanine levels were measured one week after 1,2-dimethylhydrazine administration to determine the extent of accumulation of each aberrant base in target organ DNA. The results suggest that the rate at which O^6-methylguanine was removed from liver DNA increased with successive administrations of 1,2-dimethylhydrazine. They clearly showed that 7-methylguanine accumulated in DNA, and that by the end of the treatment period the amount of 7-methylguanine in liver DNA was equivalent to the amount that would be seen after

Table 2
ACCUMULATION OF 7-METHYLGUANINE IN LIVER DNA OF RATS TREATED WITH SDMH

Weeks on test	Amount 7-methylguanine in liver DNA (μmol/mol guanine)	
	Determined 1 wk after last injection[a]	Estimated for 12 hr after last injection[b]
1	190 ± 109	1600
5	761 ± 156	6300
10	804 ± 142	6700
15	1087 ± 156	9000

[a] ± S.D.; N = 6 except for 1-wk value where N = 3.
[b] It was assumed that the maximum level of 7-methylguanine in liver DNA was reached 12 hr after the last SDMH administration and that 7-methylguanine was lost from DNA at a rate of 50%/48 hr, the half-life of 7-methylguanine in DNA.[22]

a near-lethal single dose of DMNA. This indicated that over the treatment period a large amount of DNA methylation occurred that presumably resulted in apurinic sites that in turn led to strand breakage or repair. Table 2 stresses this extensive formation and loss of 7-methylguanine that potentially represents DNA damage. Thus, while we often study mechanisms of carcinogenesis by looking at biochemical changes in target cells after a single administration of the test compound, we may be studying a system that may not represent what is happening in the long induction period of cancer.

3. Dose and DNA Damage

The picture is further complicated by the fact that the dose of the carcinogen can affect the repair of damaged DNA within the same species. When a single dose of 1,2-dimethylhydrazine results in alkylation levels in excess of 400 μmol O^6-methylguanine per mole guanine in liver DNA, the rate of removal of this adduct from the DNA is 3 to 4 times slower than in instances where the alkylation levels are below 300 μmol O^6-methylguanine per mole guanine.[25] Such high doses of 1,2-dimethylhydrazine also cause extensive cytotoxicity that is followed by restorative hyperplasia, thus a forced DNA replication. The slowed removal of the methylated bases from DNA combined with the accelerated replication of that DNA, would seem to greatly increase the opportunity for somatic mutations to occur. This is demonstrated in Figure 2. At low doses of 1,2-dimethylhydrazine where little cytotoxicity occurs, the rates of DNA replication (normal rat liver hepatocytes turnover every 190 to 450 days)[26] and O^6-methylguanine removal (shortly after formation the methylated base is removed at a rate of 50% in 13 hr) are normal. As the dose of 1,2-dimethylhydrazine increases, however, DNA alkylation levels increase, removal of O^6-methylguanine slows, and restorative hyperplasia begins.

On the other hand, it has also been shown that repeated administration of low doses of 1,2-dimethylhydrazine[23] or DMNA[27] induces the repair of O^6-methylguanine-damaged DNA, so that in low-dose chronic studies, O^6-methylguanine is removed from DNA at a rate even faster than would be predicted from an experiment using a single low dose.

From this scheme it is clear that quantitative extrapolations in describing dose-re-

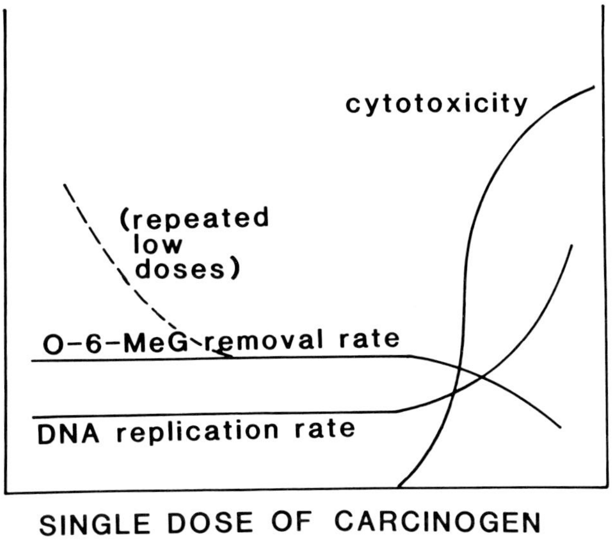

FIGURE 2. Relationships between cytotoxocity and rates of DNA repair and replication.

lated effects in liver cells are extremely difficult unless the full dose-response relationships are understood. If cancers are seen in animals treated chronically at high doses, how does one predict the cancer incidence in small animal populations exposed to much lower doses, or in large human populations exposed occasionally to still lower doses? Neither the biologist nor the mathematician has a satisfactory answer to that question.

III. TOXICITY AND CARCINOGENICITY

Most of the material covered so far is well-known and has been the subject of many papers. Perhaps the biochemical data base to support all that has been claimed is not as strong as we would wish, but at least the concepts have been discussed thoroughly for several years. The author would like to conclude on a note that, conceptually, is quite new, with the hope that the toxicologist and biochemist in the years ahead will prove its significance.

We have been interested in obtaining biochemical evidence to explore the mechanism(s), if any, that may link cytotoxicity with carcinogenicity. We have been able to demonstrate that administration of a variety of hepatotoxins at toxic levels results in the methylation of liver DNA at the 7- and O^6-positions of guanine.[28] 7-Methylguanine and O^6-methylguanine have been demonstrated in liver DNA of rats treated with hydrazine, thioacetamide, carbon tetrachloride, ethanol, yellow phosphorus, bromobenzene, and puromycin. At least in the cases of hydrazine and thioacetamide, the two cases examined so far, the source of the methyl moiety in the methylpurines appears to be *S*-adenosylmethionine. The formation of these purines is rapid. The half-maximum level of 7-methylguanine in DNA is reached only 30 min after oral administration of hydrazine, and the half-maximum level of O^6-methylguanine is reached in 45 min.[29] Maximum alkylation levels are reached approximately 6 hr after hydrazine administration and remain constant during the subsequent 18 hr. 7-Methylguanine and O^6-methylguanine form concurrently at a ratio of 10 to 20 7-methylguanine residues for every

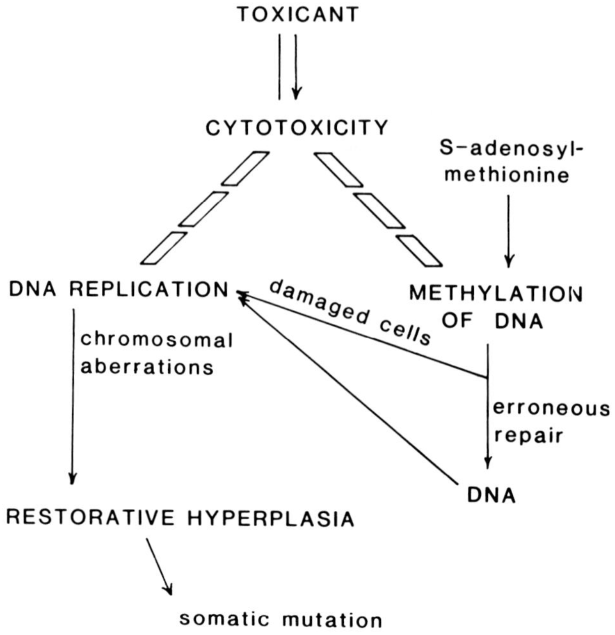

FIGURE 3. Suggested relationship between cytotoxicity and methylation of DNA leading to somatic mutation.

O^6-methylguanine molecule. The elimination rates for the two methylguanines from liver DNA are consistent with published levels for the removal of these bases from DNA in animals treated with such methylating intermediates as DMNA,[30] methylnitrosourea,[31] and 1,2-dimethylhydrazine.[25] The half-life of 7-methylguanine in liver DNA in hydrazine-treated rats is approximately 47 hr, while 0^6-methylguanine levels decrease more rapidly with an apparent half-life of 13 hr.[29] These results indicate that target organ DNA can have qualitatively the same DNA methylation pattern following a cytotoxic dose of a hepatotoxin as is seen following a less toxic dose of a strong methylating carcinogen, such as DMNA or 1,2-dimethylhydrazine. This finding may provide evidence to support a biochemical mechanism to associate cytotoxicity with carcinogenicity.[28,29,32]

A nonalkylating toxicant produces cell killing and is followed by a restorative hyperplasia. This hyperplasia takes place in an environment that may be abnormal and may be deficient in the normal regulatory feedback controls involved in homeostasis.[33] Roberts[33] suggested that erroneous DNA repair and chromosomal aberrations during the cellular regeneration may explain induction of cancers by chemicals that do not appear to directly modify the genetic material. Our recent finding of the presence of aberrant methylation in DNA following administration of a high dose of the inorganic toxicant, hydrazine, suggests that cytotoxicity may induce DNA methylation and therefore may be qualitatively equivalent to administration of a strong alkylating agent. This may add another consideration to Roberts' hypothesis, but ultimately the effect is the same: replication of damaged DNA presumably resulting in a somatic mutation (Figure 3). If this effect is a result of administration of a cytotoxic dose of the nonalkylating toxicant, rather than due to some intrinsic property of the toxicant itself, then it may be that such compounds induce somatic mutations only following cytotoxicity. Lesser exposures that do not lead to DNA damage, cytotoxicity, and tissue regeneration would not be expected to lead to somatic mutation and cancer.

As we learn more about the mechanisms of cell injury and tissue response to that injury, it becomes disappointingly clear that we know little to allow us, *on scientific bases,* to extrapolate from in vivo and in vitro carcinogenicity assays to assessing human risks. Any models designed to make such extrapolations, that fail to account for the biochemical considerations of species and dose, can only have limited application.

REFERENCES

1. **Pullman, A.,** A relation between the distribution of electronic charges and the carcinogenic potency of a certain class of hydrocarbons, *C.R. Seances Soc. Biol.,* 221, 140, 1945.
2. **Pullman, A. and Pullman, B.,** Electronic structure and carcinogenic activity of aromatic molecules, *Adv. Cancer Res.,* 3, 117, 1955.
3. **Miller, J. A.,** Carcinogenesis by chemicals: an overview. G. H. A. Clowes Memorial Lecture, *Cancer Res.,* 30, 559, 1970.
4. **Boveri, T.,** *Origin of Malignant Tumors,* (transl.), Williams & Wilkins, Baltimore, 1929, 111.
5. **Gabridge, M. G. and Legator, M. S.,** A host-mediated microbial assay for the detection of mutagenic compounds, *Proc. Soc. Exp. Biol. Med.,* 130, 831, 1969.
6. **Ames, B. N., McCann, J., and Yamasaki, E.,** Methods for detecting carcinogens and mutagens with the *Salmonella*/mammalian-microsome mutagenicity test, *Mutat. Res.,* 31, 347, 1975.
7. **Setlow, R. and Carrier, W.,** The disappearance of thymine dimers from DNA: an error-correcting mechanism, *Proc. Natl. Acad. Sci. U.S.A.,* 51, 226, 1964.
8. **Boyce, R. and Howard-Flanders, P.,** Release of ultraviolet light-induced thymine dimers from DNA in *E. coli* K-12, *Proc. Natl. Acad. Sci. U.S.A.,* 51, 293, 1964.
9. **Bartsch, H., Malaveille, C., and Montesano, R.,** In vitro metabolism and microsome-mediated mutagenicity of dialkylnitrosamines in rat, hamster, and mouse tissues, *Cancer Res.,* 35, 644, 1975.
10. **Hueper, W. C.,** *Occupational Tumors and Allied Diseases,* Charles C Thomas, Springfield, Ill., 1942.
11. **Boyland, E., Harris, J., and Horning, E. S.,** The induction of carcinoma of the bladder in rats with acetamidofluorene, *Br. J. Cancer,* 8, 647, 1954.
12. **Hueper, W. C., Wiley, F. H., Wolfe, H. D., Ranta, K. E., Leming, M. F., and Blood, F. R.,** Experimental production of bladder tumors in dogs by administration of β-naphthylamine, *J. Ind. Hyg. Toxicol.,* 20, 46, 1938.
13. **Miller, J. A. and Miller, E. C.,** The metabolic activation of carcinogenic aromatic amines and amides, *Prog. Exp. Tumor Res.,* 11, 273, 1969.
14. **Gutmann, H. R. and Bell, P.,** N-Hydroxylation of arylamides by the rat and guinea pig. Evidence for substrate specificity and participation of cytochrome P-450, *Biochim. Biophys. Acta,* 498, 229, 1977.
15. **Schmähl, D. and Habs, M.,** Carcinogenicity of *N*-nitroso compounds. Species and route differences in regard to organotropism, *Oncology,* 37, 237, 1980.
16. **Montesano, R. and Magee, P. N.,** Metabolism of dimethylnitrosamine by human liver slices *in vitro, Nature (London),* 228, 173, 1970.
17. **Herron, D. C. and Shank, R. C.,** Methylated purines in human liver DNA after probable dimethylnitrosamine poisoning, *Cancer Res.,* 40, 3116, 1980.
18. **Watanabe, P. G., Hefner, R. E., Jr., and Gehring, P. J.,** Vinyl chloride-induced depression of hepatic non-protein sulfhydryl content and effects on bromosulphalein (BSP) clearance in rats, *Toxicology,* 6, 1, 1976.
19. **Maltoni, C. and Lefemine, G.,** Carcinogenicity bioassays of vinyl chloride. I. Research plan and early results, *Environ. Res.,* 7, 387, 1974.
20. **Pullman, B.,** Aspects of the electronic structure of the nucleic acids in relation to the theories of mutagenesis and carcinogenesis, *Biopolym. Symp.,* 1, 141, 1964.
21. **Swann, P. F. and Magee, P. N.,** Nitrosamine-induced carcinogenesis. The alkylation of nucleic acids of the rat by *N*-methyl-*N*-nitrosourea, dimethylnitrosamine, dimethylsulphate, and methyl methanesulphonate, *Biochem. J.,* 110, 39, 1968.
22. **Herron, D. C. and Shank, R. C.,** Quantitative high-pressure liquid chromatographic analysis of methylated purines in DNA of rats treated with chemical carcinogens, *Anal. Biochem.,* 100, 58, 1979.
23. **Herron, D. C. and Shank, R. C.,** DNA methylation during the course of cancer induction by 1,2-dimethylhydrazine, *Carcinogenesis,* in press.

24. **Druckery, H.**, Production of colonic carcinomas by 1,2-dialkylhydrazines and azoxyalkanes, in *Carcinoma of the Colon and Antecedent Epithelium*, Burdette, W. J., Ed., Charles C Thomas, Springfield, Ill., 1970, 267.
25. **Herron, D. C. and Shank, R. C.**, In vivo kinetics of O^6-methylguanine and 7-methylguanine formation and persistence in DNA of rats treated with symmetrical dimethylhydrazine, *Cancer Res.*, 41, 3967, 1981.
26. **MacDonald, A. R.**, "Lifespan" of liver cells. Autoradiographic study using tritiated thymidine in normal, cirrhotic, and partially hepatectomized rats, *Arch. Intern. Med.*, 107, 335, 1961.
27. **Pegg, A. E. and Hui, G.**, Removal of methylated purines from rat liver DNA after administration of dimethylnitrosamine, *Cancer Res.*, 38, 2011, 1978.
28. **Barrows, L. R. and Shank, R. C.**, Aberrant methylation of liver DNA in rats during hepatotoxicity, *Toxicol. Appl. Pharmacol.*, 60, 334, 1981.
29. **Becker, R. A., Barrows, L. R., and Shank, R. C.**, Methylation of liver DNA during hydrazine hepatotoxicity: dose-response and kinetic characteristics of 7-methylguanine and O^6-methylguanine formation and persistence in rats, *Carcinogenesis*, 2, 1181, 1981.
30. **Craddock, V. M.**, Stability of deoxyribonucleic acid methylated in the intact animal by administration of dimethylnitrosamine, *Biochem. J.*, 111, 497, 1969.
31. **Kleihues, P. and Margison, G. P.**, Carcinogenicity of *N*-methyl-*N*-nitrosourea: possible role of repair excision of O^6-methylguanine guanine from DNA, *J. Natl. Cancer Inst.*, 53, 1839, 1974.
32. **Shank, R. C. and Barrows, L. R.**, Toxicity-dependent DNA methylation: significance to risk assessment, in *Health Risk Analysis*, Richmond, C. R., Walsh, P. J., and Copenhaver, E. D., Eds., Franklin Institute Press, Philadelphia, 1981.
33. **Roberts, J. J.**, Cellular responses to carcinogen-induced DNA damage and the role of DNA repair, *Br. Med. Bull.*, 36, 25, 1980.

Chapter 20

DETECTION OF SOMATIC MUTATIONS IN MAN

Richard J. Albertini, David L. Sylwester, Elizabeth F. Allen, and Bruce D. Dannenberg

TABLE OF CONTENTS

I. INTRODUCTION

Gene mutations occur spontaneously in vitro in cultured mammalian cells. The resultant mutants differ from their wild type precursors by phenotypic changes that are stable, heritable, and often associated with altered gene products. Some *de novo* mutations in mammalian cells have been localized to specific chromosomal sites.[1,2]

Because mutant frequencies generally increase following mutagen exposure in vitro, cultured cells present certain advantages over prokaryotic organisms for use in mutagenicity screening tests.[3] For example, because of the similarity of eukaryotic genetic targets, extrapolation of test results to man seems more realistic with cultured cells. When human cells are used in mutagenicity tests, extrapolation is direct. Since cytogenetic studies are also possible, multiple levels of genetic damage can be tested in the same material. Several human and subhuman cultured cell systems have been developed for mutagenicity screening purposes.[4]

There are also limitations to testing in vitro. Pharmacokinetic and metabolic factors of importance for chemical mutagenesis cannot be assessed adequately.[5,6] There is the problem of interspecies extrapolation when results obtained with subhuman systems are used to assess human effects. Even with human cell systems, the cells used arise from a single individual — often from a single clone — and no allowance is usually made for heterogeneity of susceptibility. Homeostatic mechanisms that may regulate the expression of mutation in vivo, are not present in tissue culture. Perhaps of most importance, a quantitative assessment of human health risk, which is the ultimate purpose of mutagenicity testing, cannot be made from demonstrations of mutagenicity in vitro.

II. DIRECT MUTAGENICITY TESTING

A. Definition and Advantages

Cultured mammalian cells do provide a bridge between in vitro and in vivo test systems.[3] The term "direct mutagenicity testing" is used for test systems that monitor genetic effects occurring in vivo in cells obtained directly from an intact animal or man.[7] Direct tests are capable of measuring genetic damage occurring in vivo in human cells. Such testing achieves "metabolic realism", "pharmacokinetic realism", and "target realism". It can assess in man the genetic effects of mixtures of environmental agents. The focus of direct testing is not on potentially mutagenic agents, but rather, on their genotoxic effects. Direct mutagenicity testing is able to detect differences in susceptibility among individuals to the genotoxicity of the environment. Finally, results of direct testing can be compared with data obtained from clinical and epidemiological studies of individuals in the monitored populations. This should allow linkage between the results of laboratory tests and health outcomes of the individuals being tested. Direct mutagenicity tests may thus be validated in terms of their performance as health risk predictors. Once validated, such tests may allow quantitative genetic health-risk assessments to be made in human populations.

B. Difficulties

There are at least two sorts of difficulties inherent in attempting quantitative determinations of somatic cell mutation occurring in vivo. The first has to do with the nature of the phenotypic change being monitored. As with all mutagenicity testing, it must be demonstrated that the variation under study truly has a genetic basis. Phenocopies, which are nongenetic variants that mimic cells with specific genetic damage, are almost certain to occur for any phenotypic variation. Since breeding tests and segregation analysis cannot be performed, criteria have been established for somatic cells in vitro by which a phenotypic change is judged as having a genetic basis.[8] To satisfy most of these, the variant cells, or some portions of their genome, must be propagatable in vitro. Although this is easily accomplished with cultured cells, it has not been possible with direct test material since the test cells are not usually grown in vitro. Technologies are only now developing to allow demonstrations of this kind.

The second sort of difficulty with direct mutagenicity testing lies in quantitation of mutation, which is the event of interest. At best, mutant cells resulting from mutation are being counted. With direct mutagenicity testing, the size of the cell population at risk and the number of cell generations upon which to base calculations are unknown. Assumptions must be made regarding cell kinetics and the way specific cell populations undergo expansion in vivo. Selection for or against mutant phenotypes in vivo, and the result of this on mutant frequencies, are largely unknown. When assessing the effects of known mutagens, the average dosage as well as cell killing are also unknown. The methods developed for quantitating mutagenesis in vitro in mammalian cells are not strictly applicable.

Finally, it is likely that any in vivo genetic effects that are measurable will be small in magnitude. It seems unreasonable to expect that doses of environmental agents sufficient to give several orders of magnitude increases in mutant frequencies will be delivered to cell populations in vivo. The resulting cellular cytotoxicity would probably be incompatible with life of the host. Even if small populations of cells receive such doses, the survivors are likely to be diluted with unexposed cells prior to assay. Direct mutagenicity testing systems should be capable of detecting small effects.

C. Direct Mutagenicity Tests

Because of their advantages, several direct mutagenicity tests have been proposed. At least three, using human cells, have been discarded because of an inability to distinguish between phenocopies and true genetic mutants.[9,10]

Three tests are currently in use or are being developed for direct mutagenicity testing in animals. These are the mammalian spot test,[11,12] the rat granuloma pouch assay,[13,14] and the lactic dehydrogenase-X (LDH-X) system in murine sperm.[15]

In addition to those discarded, a variety of direct mutagenicity tests remain available for use in humans. Cytogenetic methods measuring chromosomal aberrations have long been used in this regard.[16,17] Detection of sister chromatid exchanges occurring in vivo is a newer, and possibly more sensitive, test system.[18,19,20] The micronucleus test may be used in humans as well as in animals.[21] Morphologic abnormalities of sperm heads have been described for both human and animal sperm.[22] The frequencies with which abnormal sperm occur may reflect genetic damage in precursor cells. Finally, the measurement of double Y bodies in human sperm also qualifies as a direct mutagenicity test.[23]

At present, two methods are under development to quantify specific locus mutations occurring in vivo in human cells. The first of these determines primary structural changes in human hemoglobin (Hb), and uses mature RBCs as test cells.[10] This test is designed to enumerate in normal individuals of Hb type AA the frequency of mature RBCs in the peripheral blood that contain large amounts of a specific mutant Hb such as Hb-S, Hb-C, etc. Such variant RBCs, carrying structurally mutant Hbs, can arise only by mutation in the structural Hb genes of RBC processor cells. Rare variant RBCs, occurring in normal individuals at frequencies of approximately 1.1×10^{-7} for single amino acid substitutions, are recognized by immunological methods.[10]

The second method is that which detects 6-thioguanine resistant (TG^r) peripheral blood lymphocytes (PBLs) that arise in vivo in man. Such cells are presumed to arise by somatic cell mutation. This system evolved from the in vitro mutagenicity test that uses diploid human fibroblasts.

III. THE TG^r VARIANT LYMPHOCYTE ASSAY

A. Rationale

TG^r variants arising in vitro in populations of cultured diploid human fibroblasts increase in frequency following exposures of these cells to a variety of mutagens.[24-29] Many of the TG^r fibroblasts appearing in vitro also acquire a deficiency of the activating enzyme hypoxanthine-guanine phosphoribosyl transferase (HPRT). Furthermore, cells from individuals with the DNA repair disorder Xeroderma pigmentosum, that are exquisitely sensitive in vitro to the cytotoxicity of UV light, are also much more sensitive than normal cells to UV induction of TG^r variants.[30] For several reasons, TG^r diploid human fibroblasts are felt to arise in vitro by a process of somatic cell mutation.[25]

Since TG^r variants of human somatic cells arise *de novo* in vitro by a process of somatic cell mutation, it seems likely that similar cells arise also by a like mechanism in vivo. These variants should be useful for direct mutagenicity testing if a suitable cell type (i. e., allows representative sampling) and a method for recognizing rare TG^r cells can be developed. The problem is one of detecting a very low order mosaicism arising in vivo in cell samples obtained directly from intact humans.

The naturally occurring Lesch-Nyhan (LN) mutation is the human prototype for mutations resulting in TG^r HPRT-deficient somatic cells.[7,24,31] LN cells lack HPRT activity, which is necessary to activate purine analogs to cytotoxic agents.[32,33] LN cells aided greatly in the development of the human fibroblast system[24] that has been used widely as an in vitro mutagenicity screening test.

The LN phenotype is expressed also in PBLs.[34] PBLs are easily obtained, and since they are polyclonal in origin, have the potential for reflecting multiple independent genetic events. Methods are available for stimulating quiescent populations of PBLs to DNA synthesis in vitro. In summary, the PBL appears to be a suitable cell for direct mutagenicity testing.

Because of the X-chromosomal location of the HPRT gene[32,33] and the single active X-chromosome principle in mammalian cells,[35,36] females who are heterozygous for the LN mutation have two populations of somatic cells — those that express the wild type allele for HPRT and those that express the LN mutant allele. As expected, LN heterozygotes have, on average, approximately equal frequencies of normal and LN fibroblasts recovered from skin biopsies.[37-43] However, heterozygous females show a low-order mosaicism for LN PBLs, with mutant cells usually comprising 1% or less of the total.[34,44] Presumably this results from negative selection in vivo against PBLs with the LN phenotype.

LN heterozygous females provide an excellent model for detecting infrequent TG^r PBLs arising in vivo in humans. Because of apparent selection in vivo against mutant PBLs, these cells constitute a minority population in the peripheral blood of heterozygotes, and methods for their detection approximate those necessary to detect rare TG^r PBLs in non-LN individuals.

Negative selection also serves to keep TG^r PBLs from accumulating in vivo with time in non-LN individuals, where it is assumed that they arise by somatic cell mutation. Increases in their frequencies should therefore be related temporarily to causative events — an important consideration if the system is to be useful for direct mutagenicity testing.

B. Experience to Date

TG^r in PBLs is defined as resistance to the inhibition of phytohemagglutinin (PHA) stimulated tritiated thymidine (^{3}HTdr) incorporation in vitro by this agent. LN PBLs are resistant to inhibition because they lack HPRT activity. Early studies, using 8-azaguanine (AG) as the inhibiting purine analog, quantitated ^{3}HTdr incorporation by scintillation spectrometry.[34] With this method, approximately 1% azaguanine-resistant (AG^r) LN PBLs could be detected in artificial mixtures with normal cells, and LN heterozygous females were diagnosed on the basis of lymphocyte mosaicism. However, this method showed a lower limit of approximately 1% for detection of resistant LN cells in mixtures, which made it too insensitive for detecting rare AG^r or TG^r PBLs in non-LN individuals.

Autoradiography allows direct enumeration of cells incorporating ^{3}HTdr. It is sufficiently sensitive to allow detection of rare TG^r PBLs occurring at frequencies of the order of 10^{-6}. Autoradiographic studies of LN heterozygotes showed them to be mosaics in the peripheral blood, with LN TG^r PBL frequencies ranging from 10^{-3} to 5×10^{-2}.[44] Blood samples from LN males showed from 23 to 100% of PBLs to be resistant to TG at the concentration used.[44-47] Since all cells from such individuals are HPRT-deficient, this reflects either the error of the method or partial sensitivity of prototype LN PBLs to high concentrations of TG. TG^r LN PBLs were easily detected and quantified by autoradiography in artificial mixtures with normal cells.[46,47]

Several studies were then performed in order to estimate TG^r PBL variant frequencies (V_fs) in normal non-LN individuals and in individuals undergoing a variety of known mutagen exposures. TG^r PBLs were found in normals at a median frequency of 1.1×10^{-4} when determined at 2×10^{-4} *M* TG.[46,47] By contrast, the median TG^r PBL V_f in a group of cancer patients being treated with cytotoxic agents was 8.5×10^{-4}, with most individual values being higher than the highest of the normal group.[46,47] Similarly, psoriatic patients receiving photochemotherapy (PUVA) with 8-methoxypsoralen (P) and long-range UV light (UV-A), and individuals receiving X-irradiation therapy, were shown to have elevated TG^r PBL V_fs when compared to normals.[47,48]

Although these early studies showed the expected elevations of V_fs in humans following exposures to known mutagens, the values were too high, even in control groups, to be easily ascribable to somatic cell mutation.[47] Furthermore, some psoriatic or cancer patients prior to treatment had V_f values in the treated range.[47,48] There was also a great deal of day-to-day variability, even in control individuals.[47] These all suggested that a large component of phenocopies was being detected in these studies.

The presence of phenocopies was clearly suggested by further experiments.[49] In the TG^r PBL assay, phenocopies are TG-sensitive cells that mimic TG^r cells by incorporating, under the conditions of assay, ^{3}HTdr in the presence of TG. With the lymphocyte method, PBLs are exposed to TG for only 42 hr in vitro. This interval was chosen because a longer in vitro culture interval might allow for somatic cell mutation in vitro as well as for multiplication of preexisting mutant cells.[46] The vast majority of PBLs are in the G_0 phase of the cell cycle when put into culture and enter the cell cycle only following PHA stimulation in vitro. These cells must proceed through G_1 before beginning DNA synthesis, and should be inhibited from synthesis if sensitive to TG. However, a small but unknown fraction of PBLs is in cycle or is otherwise committed to DNA synthesis when put into culture. Such cells seemingly cannot be inhibited by TG from entering their first in vitro S-phase (DNA synthesis) and thus, incorporate ^{3}HTdr and become labeled. Such cells are almost certainly not mutant somatic cells.

Phenocopies caused errors in early estimates of TG^r PBL V_fs. It was then found, quite by accident, that cryopreserving the mononuclear cell (MNC) fractions of human peripheral blood prior to test serves to eliminate or greatly reduce the scoring of phenocopies in the variant lymphocyte assay.[49] Closer study of this suggested that cells committed to DNA synthesis are probably not removed per se by cryopreservation, but rather, are induced by this to initiate DNA synthesis immediately upon being put into culture. In this regard, the committed cells act as though they have been synchronized. Cooling of MNC samples to 4°C has a similar effect. In either case, the result is that committed cells seemingly complete DNA synthesis before the interval of labeling in the variant lymphocyte assay. Although not physically removed, the effects of phenocopies seem to be eliminated. Furthermore, cryopreservation does not cause prototype TG^r LN PBLs to become TG-sensitive.[49] The authors feel that cryopreservation of MNCs allows a better estimation of TG^r PBL V_fs resulting from somatic cell mutation.

All determinations of TG^r PBL V_fs in normal and mutagen-exposed non-LN individuals have been repeated with cryopreserved MNC samples. Normal non-LN individuals without known mutagen exposures now show TG^r PBL V_f values of the order of 10^{-5} to 10^{-6}. Cancer patients receiving known mutagens still show TG^r PBL V_f elevations when compared to normals. These are modest, with frequencies in the range of 10^{-5} to 10^{-4}.

C. Method

The method of assay for TG^r PBLs has been described elsewhere in detail.[50] Venous blood is obtained in heparinized syringes. Whole blood is fractionated by the Ficoll-Hypaque method and the MNC fraction removed and washed with PBS. MNCs are resuspended in medium RPMI 1640 (GIBCO) with additives (25 m*M* HEPES, 2 m*M* glutamine, 100 units of penicillin, and 100 μg streptomycin per milliliter medium) supplemented with 10% human AB serum and 7.5% DMSO for freezing. MNCs are suspended at 5×10^6 to 10^7/mℓ and aliquoted at approximately 1 mℓ per Nunc tube. Freezing is in a Union Carbide® biological freezing unit for controlled freezing. Frozen cells are stored in the vapor phase of liquid nitrogen.

For testing, frozen MNCs are rapidly thawed and diluted with warm (37°C) RPMI 1640, centrifuged and resuspended in medium RPMI 1640 containing additives, and supplemented with 20% human AB serum at a final density of approximately 10^6 cells per milliliter.

Two methods are currently used for culture. For the standard method, MNC suspensions are aliquoted as 1 mℓ samples into point bottom glass tubes to which are added 0.2 mℓ PHA-P (Burroughs, Wellcome®) to achieve a final PHA concentration of 5 μg/mℓ. For test cultures, TG (0.1 mℓ) solution in RPMI 1640 and 0.1 *N* NaOH is added to give a final TG concentration of 2×10^{-4} *M*. For control cultures, 0.1 mℓ of pH-adjusted RPMI 1640 is added. Cultures are incubated at 37°C in a 5% CO_2 humidified atmosphere for 30 hr, after which 5 μC:^{3}HTdr (New England Nuclear) is added and the incubation continued for another 12 hr. Cultures are then terminated.

In order to study larger numbers of cells, mass cultures in plastic tissue culture flasks are now employed. MNC suspensions are prepared to contain 10^6 cells per milliliter as described. Cell suspensions are inoculated into flasks in volumes ranging from 25 to 100 mℓ per flask, depending on size. PHA and either TG- or pH-adjusted medium are added to achieve the same concentrations as for point bottom tube cultures. Culture conditions and duration are the same as for point bottom tube cultures. Tissue culture flasks are placed on end, and the MNCs settle to the bottom in a loose layer. After 30 hr, the supernatant medium is gently aspirated without disturbing the loose MNC layer at the bottom of the upright flask, leaving 2 to 5 mℓ of medium above the layer. ^{3}HTdr is added at 5 μCi/mℓ of remaining culture medium and the cultures are incubated for another 12 hr before termination.

Tube cultures are terminated by adding 4 mℓ of 0.1 *M* citric acid at 4°C to each culture tube. Tubes are centrifuged at 600 × G for 10 min, washed in 4 mℓ of methanol-acetic acid fixative (3:1), resuspended in 0.2 mℓ of fixative, and refrigerated for several hours. Flask cultures are agitated so as to resuspend MNCs in the remaining medium. The suspended cells are then transferred to point bottom tubes and 4 vol 0.1 *M* citric acid at 4°C is added per tube. The remaining steps are identical for the two culture methods.

A 0.025 mℓ sample of each cell suspension in fixative is counted with an electronic particle counter (COULTER® Counter ZBI model). The cell suspensions *in toto* are then added in measured volumes to 18 × 18 mm coverslips affixed with permount to microscope slides. A single slide is made from the contents of each tube. The slides are dried, stained with aceto-orcein for 20 sec, and rinsed with water. More recently, slides of cells from TG containing cultures have been autoradiographed without staining.

For autoradiography, slides are dipped into NTB-2 (Eastman Kodak®) emulsion, exposed at 4°C for 24 hr, and developed according to Kodak instructions.

Autoradiographed slides are thus available for control (containing pH-adjusted medium only without TG) and test (containing 2×10^{-4} *M* TG) cultures from each individual.

The TG^r PBL V_f for each individual is calculated from the labeling index (LI_t) of the test cultures divided by the labeling index (LI_c) of the control cultures: LI_t (test cultures)/LI_c (control cultures) = V_f.

The LI_t of test cultures is determined by counting all labeled nuclei on all slides made from test cultures, and dividing by the total number of cells added to slides, determined by COULTER® count: Number of labeled nuclei on all slides/total number of cells on all slides = M/T = LI_t. Test slides are scanned at low power light microscopy (160X), and the presence of rare labeled nuclei is confirmed with oil immersion (1000X).

The LI_c of control cultures is determined from a random differential count of 5000 nuclei on slides to which have been added nuclei from control, non-TG containing cultures: Number of labeled nuclei per 5000 nuclei/5000 = LI_c. The differential counts are usually made under oil immersion microscopy (1000X) on control slides. Therefore, $V_f = LI_t/LI_c = M/LI_c \cdot T$, where $LI_c \cdot T = N$ = number of cells in the test that are evaluatable.

D. Statistical Concerns

In order to determine the precision of the PBL assay and the optimum size of experiments, confidence intervals (CIs) for the true V_f values were determined. This is done by assuming that M is a Poisson variable and LI_c is a binomial fraction. Exact CIs for the true Poisson mean of labeled nuclei are available from tables, using the observed count M. Also, since N is a binomial random variable with parameters T and p (the probability that a nucleus from a control culture will be labeled), it can be shown that the approximate standard deviation of N is $s_N = T\sqrt{pq/c}$ where q = 1−p and c = number of nuclei evaluated in control cultures = 5000.

Two methods can be used to establish CIs for true V_f.[50] A method that provides the narrower CI treats N as a constant and divides N into the endpoints (a_M and b_M obtained from tables) of the CI for the true Poisson mean value. Thus the CI is a_M/N, b_M/N. A method that provides a wider CI is obtained by dividing the lower CI bound (a_M) by the upper CI bound for N (i.e., b_N), and the upper CI bound, b_M, by the lower CI bound for N (i.e., a_N). Thus, the CI is a_M/b_N, b_M/a_N.

In comparing two V_fs (V_{f1}, V_{f2}) from, say, two individuals, or two points in time we ask if the ratio V_{f1}/V_{f2} is significantly different ($p < 0.05$) from 1. This is true if the approximate 95% CI for the ratio, i.e., the lower 95% CI bound for V_{f1}, divided by the upper 95% CI bound for V_{f2}, and vice versa, excludes the special value 1.

Elsewhere the authors[50] present a method for using the transformation $Y = \ln(X)$ (where ln denotes the natural logarithm) to find approximate CIs. X may represent a Poisson variable such as M or a binomial fraction such as LI_c. The method allows close approximation of the CI for the true value measured by V_f for moderate values of M, i.e., 10 or more. Using the transformation, an estimate of the approximate variance is $s^2 = 1/M + 1 - LI_c/C \cdot LI_c$ and the approximate 95% CI for the true V_f is $(V_f e^{-1.96s}, V_f e^{+1.96s})$.

The precision of the estimate, V_f, is governed mostly by the size of M. Ideally, TG^r PBL V_f determinations should involve experiments of sufficient size so that two-fold differences in V_fs are detectable. Elsewhere it is shown that a V_f determination must be based on a minimum of 14 labeled nuclei in test cultures in order to detect a doubling of the V_f as statistically different.[50] Therefore, if the median normal TG^r PBL V_f is approximately 5×10^{-6}, experiments must be of sufficient size so that at least 3×10^6 evaluatable nuclei are recovered from test cultures. The authors have seldom achieved this using the tube cultures and thus have begun using the flask cultures as described.

E. Sample Results

1. The Appearance of Labeled Nuclei

Control culture slides are scanned at low (X160) power and oil immersion (X1000) light microscopy. The differential count is usually made at the latter power. Test slides are scanned at low power microscopy, and "positive"-labeled nuclei are confirmed under oil immersion.

Figure 1A is a photomicrograph of a portion of a slide from which the LI_c was determined while Figure 1B is a photomicrograph of a portion of a test slide (TG culture) for the same individual, from which the LI_t was determined. Both photomicrographs are shown at X160. There is a single-labeled nucleus shown in Figure 1B. Figure 2A and 2B are photomicrographs of the same portions of slides as seen under oil immersion (X1000). A labeled nucleus on a test slide is scored if it has the same appearance as labeled nuclei on control slides.

2. TG^r PBL V_f Determinations

The results of 39 recent TG^r PBL V_f determinations performed on cryopreserved MNC samples from 37 individuals are shown graphically in Figure 3, where the abscissa gives the ages of the individual tested and the ordinate depicts the V_fs. Both scales are linear and the scale on the ordinate is broken. Ten of the individuals tested are normal adults, two of whom have psoriasis. Eight of the individuals have a history of multiple skin cancers and none is receiving chemotherapy. Three individuals are renal transplant recipients, all of whom are receiving azathioprine continuously and 11 are patients with various visceral malignancies, all of whom receive intermittent X-irradiation and/or cytotoxic chemotherapy. Six samples were from umbilical cord blood samples (five individuals), but mutagen exposure histories are not available.

It is apparent from these results that V_f values do not vary simply as a result of age. This is consistent with earlier observations using fresh blood samples.[46] The oldest individual tested (85 years) has a V_f of 5.4×10^{-6}, comparable to the values of younger individuals.

TG^r PBL V_f values for normal, nonmutagen exposed adults range from less than 1.1 to 26×10^{-6}, with only two greater than 10×10^{-6}. One of these two is employed as a chemist, while the other is under dermatological care for psoriasis. By contrast, 9 of the 11 cancer patients receiving X-irradiation or cytotoxic chemotherapy have TG^r PBL V_f values greater than 10×10^{-6}. Patients with multiple skin cancers were studied because of the possibility that these multiple malignancies result from unusual mutagenic influences in vivo.[51] Half of the 8 patients tested had V_f values greater than 10×10^{-6}.

Renal transplant recipients receiving azathioprine are individuals in whom there is pharmacological selection in vivo for TG^r cells. TG^r PBL V_fs might rise in such persons as a result of selection,[46] although the degree of increase may be a function of time. Two of the three transplant recipients showed seemingly elevated TG^r PBL V_fs, one at 51.9 and the other at 217.5×10^{-6}. It is difficult to recover large numbers of PBLs from such patients because of the lymphopenia associated with therapy. These values therefore are based on small numbers of cells evaluated (N) and small numbers of labeled nuclei scored (M) (7 and 8 nuclei, respectively).

TG^r PBL V_f values for the 6 umbilical cord blood samples range from 6.9 to 489.0 $\times 10^{-6}$ and are usually based on large numbers of cells scored. Replicate determinations on a single frozen sample gave V_f values of 6.9 and 11.2×10^{-6}, which provides some measure of technical variability. Because there is no information regarding possible mutagen exposures of mothers and fetuses, it is difficult to speculate on possible interpretations of this range of values. Umbilical cord lymphocytes may not be comparable to adult PBLs in several respects, some of which may be important in the lymphocyte assay. (See Note Added in Proof.)

Table 1 gives results in tabular form for a sample of these individuals and lists some of the factors that may be important in determining TG^r PBL V_f values in vivo. Listed also are 95% confidence intervals for the V_f values. It may be noted that many of the V_f determinations are based on small numbers of labeled nuclei so that confidence intervals are wide.

3. *Relationship of V_fs to LI_cs*

It has been previously reported that when fresh samples are tested there is no relationship between the LI_c of a control PBL culture and the V_f determined on its companion test cultures.[46] Thus, the elevated TG^r PBL V_f values of cancer patients receiving chemotherapy are not simply a reflection of lower LI_cs in this group. Is this true also for cryopreserved MNC samples? Figure 4 shows graphically the V_f values (ordinate) presented in Figure 3 and the LI_cs of the respective control cultures (abscissa). Both scales are linear. The LI_cs, range from 0.01 to 0.55, with most being in the range of 0.02 to 0.25. Cultures from umbilical cord blood samples have the highest LI_cs, with all being in excess of 0.40. Again, the results show that there is no correlation between LI_cs and V_fs. With the exception of the umbilical cord samples, the several groups of individuals tested are not distinguishable on the basis of LI_cs. Thus, for cryopreserved MNC samples also, an increase in TG^r PBL V_fs in treated cancer patients is not correlated with a decrease in LI_cs as compared to healthy controls.

IV. SUMMARY AND DISCUSSION

Traditional epidemiological studies focus on diseased individuals as the most realistic endpoints for identifying human health hazards while human direct mutagenicity tests propoi t to recognize somatic cell mutation occurring in vivo. It is assumed that mutation is an early event that may end in "genetic" disease. If so, mutation and disease may be linked by demonstrating that quantitation of the former allows prediction of the latter. By coupling epidemiological studies with human direct mutagenicity testing, the latter may be validated as to usefulness in assessing human health risks. This is the goal of direct mutagenicity testing.

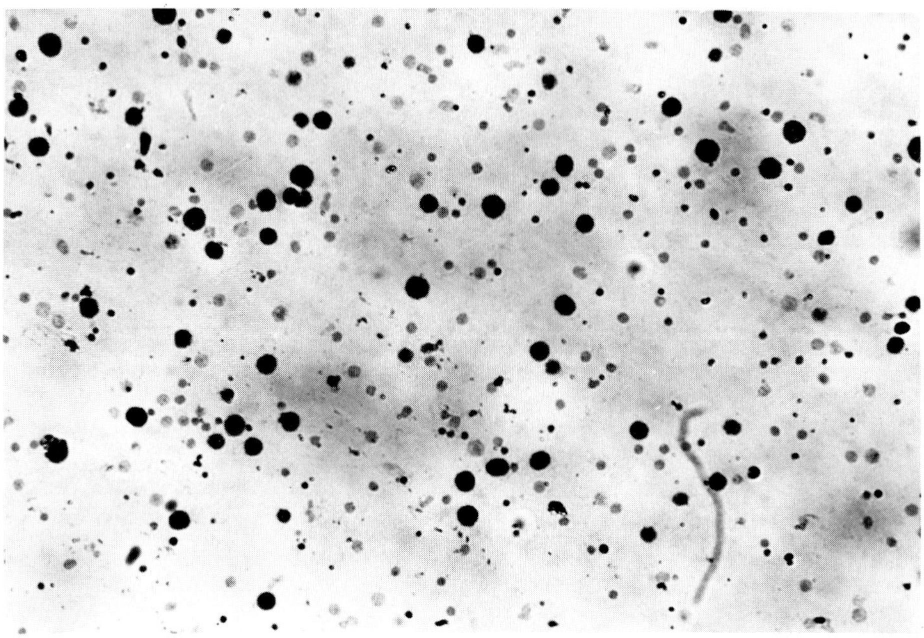

FIGURE 1A. Appearance of nuclei from a control culture with a labeling index (LI_c) of 21%. (Magnification X160).

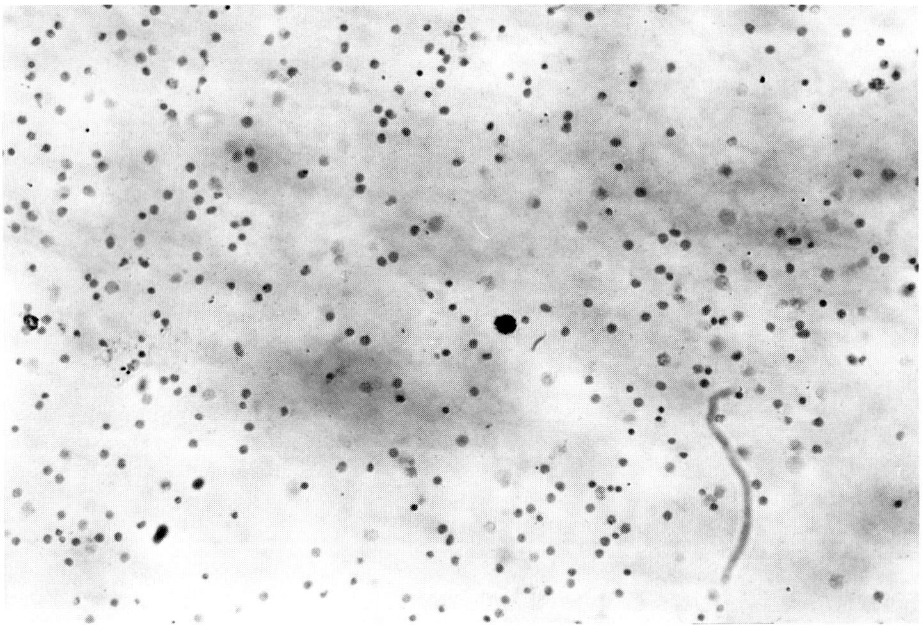

FIGURE 1B. Appearance of nuclei from a test culture from the same person as 1A showing one labeled nucleus. (Magnification X160).

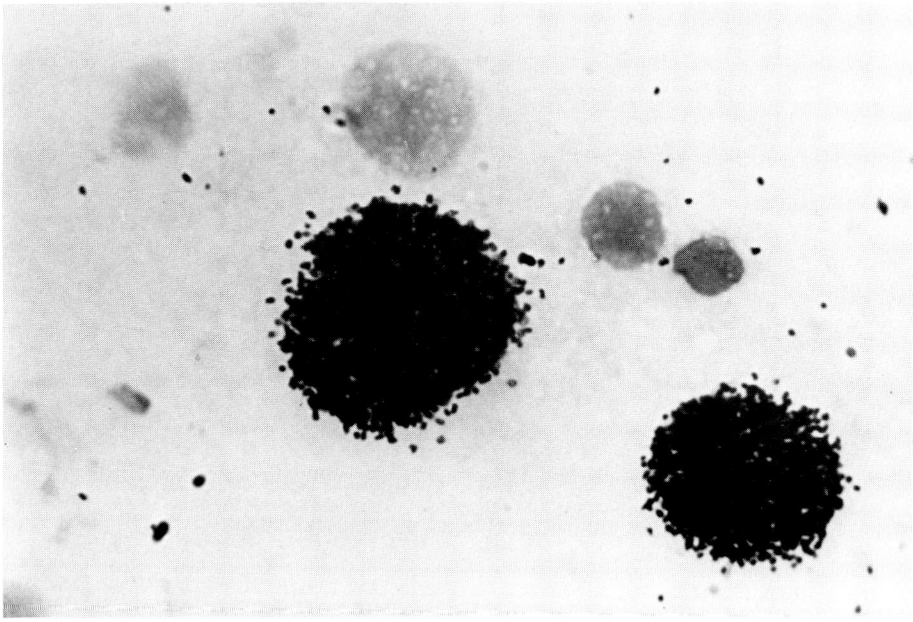

FIGURE 2A. Appearance of nuclei from control culture shown in 1A with two labeled nuclei. (Magnification X1000).

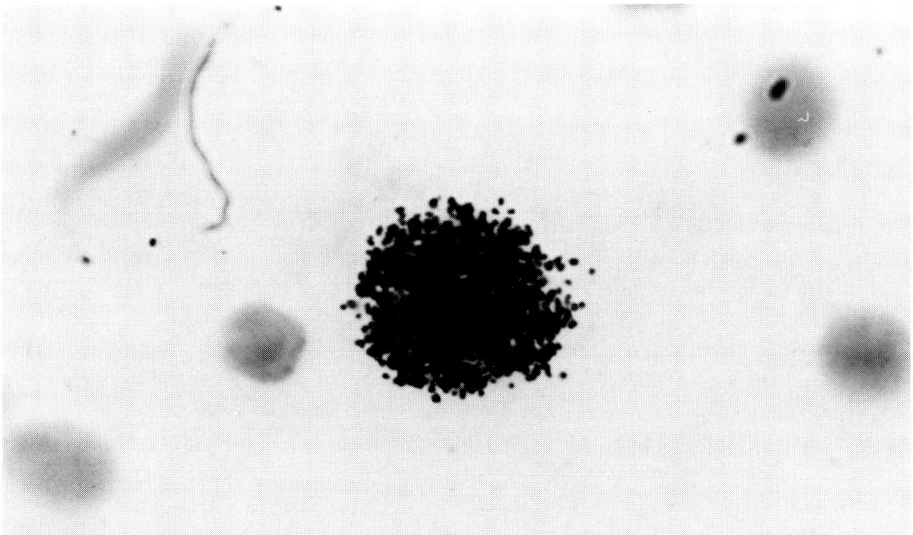

FIGURE 2B. Appearance of nuclei from test culture shown in 1B with one labeled nucleus. (Magnification X1000).

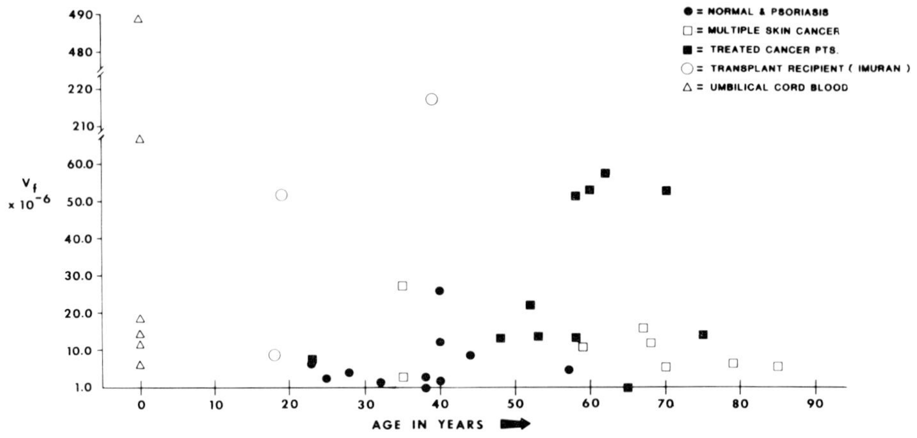

FIGURE 3. Frequencies of TG^r PBLs (V_fs) shown by age of the individual. The ordinate shows V_f on a linear scale. The abscissa shows age in years.

The variant TG^r PBL assay has undergone considerable evolution since its original description. Phenocopies were unrecognized initially, causing the authors[46,47] to report TG^r PBL V_fs of 10^{-4} and higher in normal, nonmutagen-exposed individuals. It can now be recognized that at least one source of phenocopies exist giving greater frequencies than due to the cells that may be true genetic mutants. Fortunately, the effects of phenocopies are eliminated or reduced greatly by freezing.[49] As stated, although this may not per se remove the cells committed to DNA synthesis from test populations, it does remove their effects.

The authors now find that normal, nonmutagen-exposed individuals have TG^r PBL V_fs of the order of 10^{-6} to 10^{-5} with occasional higher values that might reflect occult, perhaps occupational, mutagen exposures. Some individuals receiving mutagenic chemotherapeutic agents have elevated TG^r PBL V_fs as compared to normals. These elevations are modest and are not seen in all mutagen-exposed individuals. Similarly, some persons with multiple skin cancers that may reflect somatic mutagenic influences in vivo, also show modest V_f elevations. As reported earlier, patients treated with azathioprine, converted to 6-mercaptoprine (6MP) to give selection in vivo for TG^r PBLs, show elevated TG^r PBL V_fs.[46] This is consistent with replication in vivo of the variant cells.

Reported here for the first time are TG^r PBL V_fs determined on umbilical cord blood samples of newborn infants. These few early determinations show a wide range of values. The reason for this is not currently known. There may have been large differences in somatic cell mutation rates during the different pregnancies, perhaps reflecting different mutagen exposures. Alternatively, these differences in variants may result from only slight differences in mutation rates. Lymphocyte kinetics during gestation may permit mutations to undergo greater amplification than they do in adults, or, selection against the mutant phenotype may be less *in utero.* Also, phenocopies still may be confounding the results. Since these were pilot studies, there is no information on mutagen exposures. Ongoing monitoring of umbilical cord blood samples and established baseline V_f values for the newborn will provide this. (See Note Added in Proof.)

At present, evidence that variant TG^r PBLs are true genetic mutants is indirect. LN PBLs, which are known mutants, behave as do these cells.[46] LN PBLs can be detected quantitatively in artificial mixtures with normal cells under the conditions used for the

Table 1
TG[r] PBL V_fs (CRYOPRESERVED MNCs)

Individuals tested							Test results				
Subject nc.	Sex	Age	Status	Mutagen	Duration[a]	Time since exposure[b]	LI_c	Evaluatable nuclei = N	Labeled nuclei = M	$V_f \times 10^{-6}$	C.I.
1	F	38	Normal	—	—	—	0.37	2,048,000	6	2.9	1.1,6.4
5	F	44	Normal	—	—	—	0.21	2,830,000	24	8.5	5.7,12.7
6	F	57	Normal	—	—	—	0.03	555,000	3	5.4	1.1,15.6
14	M	85	Skin Ca[c]	—	—	—	0.06	547,000	3	5.4	1.1,16.1
18	M	67	Skin Ca	—	—	—	0.12	552,000	9	16.3	7.2,31.0
25	—	O[d]	Newborn	—	—	—	0.51	852,000	57	66.9	51.5,86.9
29	F	52	Cancer	L-PAM[e]	12 mo	1.5 mo	0.31	508,000	11	21.6	12.0,39.1
33	F	53	Cancer	CTX[f]	1 week	1 week	0.17	1,373,000	19	13.8	8.8,21.8
35	F	60	Cancer	CTX[f]	month	(?) 1—2 wk	0.17	562,000	30	53.4	37.1,76.7

[a] Estimated duration of treatment.
[b] Estimated interval since last treatment.
[c] Multiple skin cancer; no chemotherapy.
[d] Umbilical cord blood sample.
[e] L-phenylalanine mustard (alkylating agent).
[f] Cyclophosphamide (alkylating agent).

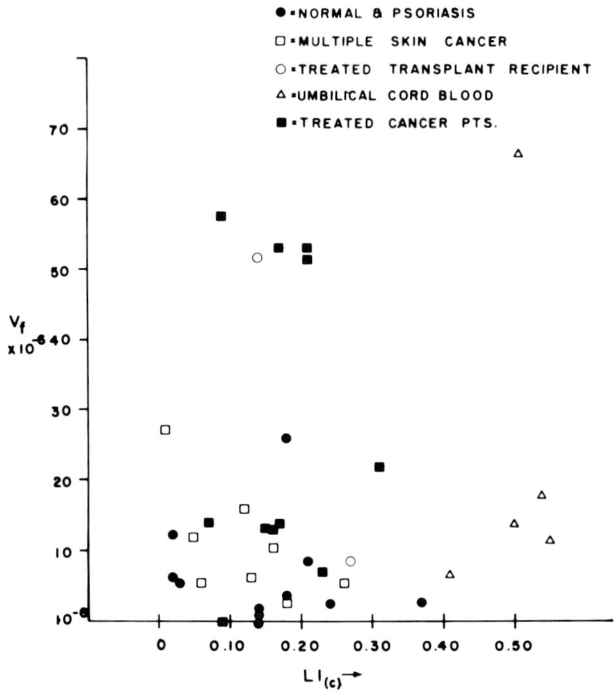

FIGURE 4. Frequencies of TG^r PBLs (V_f) shown by labeling index of control cultures (LI_c). The ordinate shows V_f on a linear scale. The abscissa shows labeling index as a proportion of total cells present.

variant lymphocyte assay.[46,47] Also, infrequent mutant LN PBLs are clearly detectable in LN heterozygous females using the variant lymphocyte assay. These cells certainly are mutants and are detected in mixed in vivo populations.

Twice the authors have propagated in vitro, directly in TG, PBLs obtained from azathioprine-treated renal transplant recipients with high TG^r PBL V_fs.[47] In neither case, however, were the cells removed from TG so that maintenance of the resistant phenotype without selection could be demonstrated. Both cultures were lost before HPRT activity was determined.

The variant lymphocyte assay has been developed to the point where V_f values are reasonable and consistent with a mutational origin of TG^r cells. If the variants are mutants, frequencies such as shown in Figure 3 probably reflect different mutagen exposures or other mutagenic influences in vivo and the test is valuable for monitoring unknown environments. However, V_f values such as shown in Figure 3 cannot themselves be used to determine that variants are mutants. This determination remains to be made before the lymphocyte system is used for unrestricted mutagenicity monitoring.

The lymphocyte is a propagatable cell. Methods are now available for cloning human PBLs in vitro using TCGF.[52-55] In principle, PBL cultures can be initiated in the presence and absence of TG, and clonal populations can be derived and characterized. A demonstration of the genetic basis of TG-resistance in human PBLs is now a major effort in this laboratory. This demonstration has now been made.[56]

Even if all the variant cells are eventually shown to be mutants, the relationship between them and mutation may not be direct. Although selection in vivo against the mutant phenotype should help to maintain temporal relationships between mutations and mutants, other factors such as the time during embryogenesis of mutation,[10] clonal expansion of mutants, and cell kinetics may be important. Mutation is the event of interest and is inferred through the enumeration of mutants. However, nonmutagenic factors may influence mutant frequencies. These complications must be considered for any direct mutagenicity test.

The human variant lymphocyte assay is described in its present state of development. A major effort is now directed toward documenting the genetic basis of the variant PBLs. The test is presented as a candidate for direct mutagenicity testing in man.

NOTE ADDED IN PROOF

Recently determined TG^r PBL V_f values for umbilical cord blood samples are in the normal adult range.

ACKNOWLEDGMENTS

This work was supported by National Institutes of Health Grant No. PHS RO1 25036-03, Cancer Center Support Grant CA 22435, Genetics Center Support Grant MCH 46-1001-02-0, March of Dimes Grant 52633-4.

REFERENCES

1. **Kahan, B. and DeMars, R.**, Localized derepression on the human intact X-chromosome in mouse-human cell hybrids, *Proc. Natl. Acad. Sci. U.S.A.*, 72, 1510, 1975.
2. **Kavathas, P., DeMars, R., and Bach, F. H.**, Hemizygous HLA mutants of lymphoblastoid cells: a new type reagent for histocompatibility testing, *Hum. Immunol.*, 1, 317, 1980.
3. **Abbondandolo, A.**, Prospects for evaluating genetic damage in mammalian cells in culture, *Mutat. Res.*, 42, 279, 1977.
4. **Hsie, A. W., O'Neill, J. P., and McElheny, V. K.**, Eds., Banbury Rep. 2: *Mammalian Cell Mutagenesis: the Maturation of Test Systems*, Cold Spring Harbor Laboratory, Cold Spring Harbor, N.Y., 1979.
5. **Magee, P.**, Activation and inactivation of chemical carcinogens and mutagens in the mammal, *Essays Biochem.*, 10, 105, 1974.
6. **Matter, B. E.**, Problems of testing drugs for potential mutagenicity, *Mutat. Res.*, 38, 243, 1976.
7. **DeMars, R. and Jackson, J. L.**, Mutagenicity detection with human cells, *J. Environ. Pathol. Toxicol.*, 1, 35, 1977.
8. **Chu, E. H. Y. and Powell, S. S.**, Selective systems in somatic cell genetics, in *Advances in Human Genetics*, Vol. 7, Harris, H. and Hirschorn, K., Eds., Plenum Press, New York, 1976, chap. 5.
9. **Atwood, K. C. and Petter, F. J.**, Erythrocyte automosaicism in some persons of known genotype, *Science*, 134, 210, 1961.
10. **Stamatoyannopoulos, G.**, Possibilities for demonstrating point mutations in somatic cells, as illustrated by studies of mutant hemoglobins, in *Genetic Damage in Man Caused By Environmental Agents*, Berg, K., Ed., Academic Press, New York, 1979, 49.
11. **Russell, L. B. and Major, M. H.**, Radiation induced presumed somatic mutations in the house mouse, *Genetics*, 42, 161, 1957.
12. **Fahrig, R.**, The mammalian spot test: a sensitive *in vivo* method for the detection of genetic alterations in somatic cells of mice, in *Chemical Mutagens: Principles and Methods for Their Detection*, Vol. 5, Hollander, A. and de Serres, F. J., Eds., Plenum Press, New York, 1978, chap. 7.
13. **Maier, P., Manser, P., and Zbinden, G.**, Granuloma pouch assay. I. Induction of ouabain resistance by MNNG in vivo, *Mutat. Res.*, 54, 159, 1978.
14. **Lee, I. P. and Zbinden, G.**, Differential DNA damage induced by chemical mutagens in cells growing in a modified Selye's granular pouch, *Exp. Cell Biol.*, 47, 92, 1979.

15. **Ansari, A. A., Baig, M. A., and Malling, H. V.**, An approach to *in vivo* germinal mutation detection with a "mono"-specific antibody against lactic dehydrogenase-X, *Proc. Natl. Acad. Sci. U.S.A.*, 77, 7352, 1980.
16. **Evans, H. J. and O'Riordan, M. L.**, Human peripheral blood lymphocytes for the analysis of chromosome observations in mutagen tests, in *Handbook of Mutagenicity Test Procedures*, Kilbey, B. J., Legator, M., Nichols, W., and Ramel, C., Eds., Elsevier/North-Holland, Amsterdam, 1977, 261.
17. **Brogger, A.**, Chromosome damage in human mitotic cells after *in vivo* and *in vitro* exposure to mutagens, in *Genetic Damage in Man Caused by Environmental Agents*, Berg, K., Ed., Academic Press, New York, 1979, 87.
18. **Perry, P. and Evans, H. J.**, Cytological detection of mutagen-carcinogen exposure by sister chromatid exchange, *Nature (London)*, 258, 121, 1975.
19. **Wolff, S.**, Sister chromatic exchanges, *Ann. Rev. Genet.*, 11, 183, 1977.
20. **Latt, S., Allen, J. W., Rogers, W. E., and Juerglus, L. A.**, In vitro and in vivo analysis of sister chromatid exchange formation, in *Handbook of Mutagenicity Test Procedures*, Kilbey, B. J., Legator, M., Nichols, W., and Ramel, C., Eds.,Elsevier/North-Holland, New York, 1977, 275.
21. **Schmid, W.**, The micronucleus test, in *Handbook of Mutagenicity Test Procedures*, Kilbey, B. J., Legator, M., Nichols, W., and Ramel, C., Eds., Elsevier/North-Holland, Amsterdam, 1977, 235.
22. **Wyrobek, A. J. and Bruce, W. R.**, The induction of sperm-shape abnormalities in mice and humans, in *Chemical Mutagens: Principles and Methods for Their Detection*, Vol. 5, Hollander, A. and de Serres, F. J., Eds., Plenum Press, New York, 1978, 257.
23. **Kapp, R. W., Jr., Picciano, D. J., and Jackson, C. B.**, Y-chromosomal nondisjunction in dibromochloropropane-exposed workmen, *Mutat. Res.*, 64, 47, 1979.
24. **Albertini, R. J. and DeMars, R.**, Detection and quantification of x-ray induced mutation in cultured, diploid human fibroblasts, *Mutat. Res.*, 18, 199, 1973.
25. **DeMars, R.**, Resistance of cultured human fibroblasts and other cells to purine and pyrimidine analogues in relation to mutagenesis detection, *Mutat. Res.*, 24, 335, 1974.
26. **Jacobs, L. and DeMars, R.**, Quantification of chemical mutagenesis in diploid human fibroblasts. Induction of azaguanine-resistant mutants by *N*-methyl-*N*-nitrosoguanidine, *Mutat. Res.*, 53, 29, 1978.
27. **Jacobs, L. and DeMars, R.**, Chemical mutagenesis with diploid human fibroblasts, in *Handbook of Mutagenicity Test Procedures*, Kilbey, B. J., Legator, M., Nichols, W., and Ramel, C., Eds., Elsevier/North-Holland, Amsterdam, 1977, 193.
28. **Burger, P. H. and Simons, J. W. I. M.**, Mutagenicity of 8-methoxypsoralen and long-wave ultraviolet irradiation in diploid human skin fibroblasts: an improved risk estimate of photochemotherapy, *Mutat. Res.*, 63, 371, 1979.
29. **Maher, V. M. and Wessel, J. E.**, Mutations to azaguanine resistance in cultured diploid human fibroblasts by the carcinogen *N*-acetoxy-2-acetylamino fluorine, *Mutat. Res.*, 28, 277, 1975.
30. **Maher, V. M., Ouelette, L. M., Curren, R. D., and McCormick, J. J.**, Frequency of ultraviolet light-induced mutations is higher in Xeroderma pigmentosum variant cells than in normal human cells, *Nature (London)*, 271, 593, 1976.
31. **DeMars, R.**, Genetic studies of HG-PRT deficiency and the Lesch-Nyhan syndrome with cultured human cells, *Fed. Proc. Fed. Am. Soc. Exp. Biol.*, 30, 944, 1971.
32. **Seegmiller, J. E., Rosenbloom, F. M., and Kelley, W. N.**, Enzyme defect associated with a sex-linked human neurological disorder and excessive purine synthesis, *Science*, 155, 1682, 1967.
33. **Caskey, C. T. and Kruh, G. D.**, The HPRT locus, *Cell*, 6, 1, 1979.
34. **Albertini, R. J. and DeMars, R.**, Mosaicism of peripheral blood lymphocyte populations in females heterozygous for the Lesch-Nyhan mutation, *Biochem. Genet.*, 11, 397, 1974.
35. **Lyon, M. G.**, Gene action in the X-chromosome of the mouse (*Mus muscalas* L.), *Nature (London)*, 190, 372, 1961.
36. **Russell, L. B.**, Mammalian X-chromosome action: inactivation limited in spread and in region of origin, *Science*, 40, 976, 1963.
37. **Dancis, J., Cox, R. P., Berman, P. H., Jansen, V., and Balis, M. E.**, Cell population density and phenotypic expression of tissue culture fibroblasts from heterozygotes of Lesch-Nyhan's disease (inosinate-pyrophosphorylase deficiency), *Biochem. Genet.*, 3, 609, 1969.
38. **Felix, J. S. and DeMars, R.**, Detection of females heterozygous for the Lesch-Nyhan mutation by 8-azaguanine-resistant growth of cultured fibroblasts, *J. Lab. Clin. Med.*, 27, 596, 1971.
39. **Fujimoto, W. Y. and Seegmiller, J. E.**, Hypoxanthine-guanine phosphoribosyltransferase deficiency: activity in normal, mutant and heterozygote-cultured human skin fibroblasts, *Proc. Natl. Acad. Sci., U.S.A.*, 65, 577, 1970.
40. **Fujimoto, W. Y., Subak-Sharpe, J. H., and Seegmiller, J. E.**, HG-PRT deficiency: clinical agents selective for mutant or normal cultured fibroblasts in mixed and heterozygous cultures, *Proc. Natl. Acad. Sci. U.S.A.*, 68, 1516, 1971.

41. **Migeon, B. R., Der Kaloustian, V. M., Nyhan, W. L., Young, W. J., and Childs, B.,** Hypoxanthine-guanine phosphoribosyltransferase deficiency: heterozygote has two clonal populations, *Science*, 16, 425, 1968.
42. **Rosenbloom, F. M., Kelley, W. N., Henderson, J. F., and Seegmiller, J. E.,** Lyon hypothesis and X linked disease, *Lancet*, 2, 305, 1967.
43. **Salzmann, J., DeMars, R., and Benke, L.,** Single allele expression at an X linked hyperuricemia locus in heterozygous human cells, *Proc. Natl. Acad. Sci., U.S.A.*, 60, 545, 1968.
44. **Strauss, G. H., Allen, E. F., and Albertini, R. J.,** An enumerative assay of purine analogue resistant lymphocytes in women heterozygous for the Lesch-Nyhan mutation, *Biochem. Genet.*, 18, 529, 1980.
45. **Strauss, G. H. and Albertini, R. J.,** 6-Thioguanine resistant lymphocytes in human peripheral blood, in *Progress in Genetic Toxicology*, Scott, D., Bridges, B. A., and Sobels, F. H., Eds., Elsevier/North-Holland, Amsterdam, 1977, 327.
46. **Strauss, G. H. and Albertini, R. J.,** Enumeration of 6-thioguanine resistant peripheral blood lymphocytes in man as a potential test for somatic cell mutation arising, *in vivo*, *Mutat. Res.*, 62, 353, 1979.
47. **Albertini, R. J.,** Drug resistant lymphocytes in man as indicators of somatic cell mutation, *Teratogen. Carcinogen. Mutagen.*, 1, 25, 1980.
48. **Strauss, G. H., Albertini, R. J., Krusinski, P. A., and Baughman, R. D.,** 6-Thioguanine resistant peripheral blood lymphocytes in humans following psoralen long-wave UV light therapy, *J. Invest. Dermatol.*, 73, 211, 1979.
49. **Albertini, R. J., Allen, E. F., Quinn, A. S., and Albertini, M. R.,** Human somatic cell mutation: *in vivo* variant lymphocyte frequencies as determined by 6-thioguanine resistance, in *Population and Biological Aspects of Human Mutation: Birth Defects Institute Symposium XI*, Hook, E. B. and Porter, I. H., Eds., Academic Press, New York, 1981, 235.
50. **Albertini, R. J., Sylwester, D. L., and Allen, E. F.,** The 6-thioguanine resistant peripheral blood lymphocyte assay for direct mutagenicity testing in man, in *Mutagenicity: New Horizons in Genetic Toxicology*, Heddle, J. A., Ed., Academic Press, New York, 1982.
51. **Albertini, R. J. and Allen, E. F.,** Direct mutagenicity testing in man, in *Human Health Risk Analysis, Proc. 3rd Life Sci. Symp.*, Walsh, P. J., Richmond, C. R., and Copenhaver, E. D., Eds., Franklin Institute Press, Philadelphia, 1981, 131.
52. **Gillis, S., Ferm, M. M., Ou, W., and Smith, K. A.,** T-cell growth factor: parameters of production and a quantitative micro assay for activity, *J. Immunol.*, 120, 2027, 1978.
53. **Smith, K. A.,** T-cell growth factor, *Immunol. Rev.*, 51, 337, 1980.
54. **Inouye, H., Hauk, J. A., Chardonnens, X., Segall, M., Alter, B. J., and Bach, F. H.,** Cloned primed lymphocyte test reagents in the dissection of HLA-D, *J. Exp. Med.*, 152, 143, 1980.
55. **Schendel, D. J. and Wank, R.,** Production of human T-cell growth factor, *Hum. Immunol.*, 2, 325, 1981.
56. **Albertini, R. J., Castle, K. L., and Borchnering, W. R.,** The human 6-thioguanine resistant lymphocytes that arise *in vivo* are mutant somatic cells, submitted.

Chapter 21

INHIBITION OF GENOTOXIC ACTIVITIES OF COMPLEX MIXTURES BY NATURALLY OCCURRING AGENTS

Miriam P. Rosin

TABLE OF CONTENTS

I. INTRODUCTION

Attention has recently shifted from the identification of the mutagenic, clastogenic, and recombinogenic activities of isolated single compounds to the assessment of the potential genetic hazard of complex mixtures.[1] The reasons for this altered emphasis are manifold. Man is seldom exposed to a single mutagen/carcinogen, but rather to complex mixtures of several chemical and/or physical agents. Such mixtures may consist of one or more mutagens/carcinogens in combination with a variety of possible modulating agents: "antimutagens/anticarcinogens", repair inhibitors, promoters, antipromoters, mitosis-stimulating agents, mitosis-inhibiting agents, toxic agents, and sensitizers. The overall carcinogenic/mutagenic activity of a mixture will be dependent on the balance of the large number of interactions present between its components and the target cell(s). Since the number of possible interactions among the numerous components of a complex mixture, such as a food product, are astronomical, an in-depth assessment of such interactions in the classical animal assay for tumorigenicity would be prohibitive. Thus, short-term assays for carcinogen activity may prove to be invaluable tools in unraveling the effects of the exposure of man to mixtures.

It is the purpose of this monograph to describe some of the naturally occurring inhibitors of the genotoxic activities of single carcinogens/mutagens as well as complex mixtures. Such studies may act as a basis for understanding the variety of interactions possible in a complex mixture, some of which will lead to a suppression of the genotoxic, and by implication, the carcinogenic activity of the mixture.

Table 1 lists several in vitro assays that have been used to identify inhibitors of carcinogen-induced genotoxicity. These assays employ a variety of endpoints: DNA damage, DNA repair, mutation, and cell transformation. The importance of using more than one of these assays at a time to study mixtures is indicated by the following examples. Gebhart[2] and Gebhart and Kappauf[3] have shown that cysteine suppresses the clastogenic activity of trenimon and 8-hydroxyquinoline sulfate in human lymphocyte cultures without having any effect on the frequency of sister chromatid exchanges induced by these chemicals. These authors suggest that cysteine is affecting the frequency of break formation (chromosome aberration) rather than repair processes (SCE formation). Miller et al.[4] studied the influence of two retinoids, the trimethylmethoxyphenyl analog of *N*-ethyl retinamide, and β-all-*trans*-retinoic acid, on the induction of sister chromatid exchanges and radiation-induced transformation in mouse C3H/10T½ and Syrian hamster embryo cells. The retinoids reduced the in vitro transformation induced by ionizing radiation as well as completely eliminating the promotional effects of the phorbol ester, TPA. In contrast, the retinoids had no effect on the SCE frequency induced by radiation. These two examples suggest that modulators may show some specificity in their effects on carcinogen-induced genotoxicity by acting at only certain stages of carcinogenesis, e.g., prior to DNA damage, during DNA repair, or during the "promotional" phase of carcinogenicity as opposed to the "initiation" phase.

It would appear logical to look for inhibitors of carcinogen-induced genotoxicity among naturally occurring and prevalent components. Table 2 lists the limited number of such inhibitors that have been identified. Only a few of these compounds have also been assayed for the ability to suppress tumorigenesis in animals (Table 3). Furthermore, several compounds in groups such as the indoles, coumarins, lactones, and isothiocyanates have been shown to be potent inhibitors of tumorigenicity in animals,[5] but have yet to be assayed for their effect on carcinogen-induced genotoxicity. Several of these plant components, together with the flavones,[6] have been shown to affect the enzyme activation of carcinogens to their active metabolic forms.

Table 1
IN VITRO ASSAYS FOR INHIBITORS OF CARCINOGEN-INDUCED GENOTOXICITY

Endpoint	Assay	Ref.
DNA damage	Filter binding assay for "nicked DNA"	45
	DNA fragmentation in cultured fibroblasts	20, 21
	DNA fragmentation of stomach DNA in treated animals	17
DNA repair	Induction of DNA repair synthesis in tissue cultures	20, 21
Mutation	Reversion to tryptophan independence in *Salmonella* cultures	22—24, 27, 37, 38, 46—50
	Host-mediated assay for *Salmonella* mutagenesis	51
	Induction of point mutation in mammalian cell cultures	52
	Induction of gene conversion in *Saccharomyces* cultures	72
	Induction of chromosome aberrations	2, 20, 21, 25
	Micronuclei formation	53
	Induction of sister chromatid exchanges	26
Cell transformation	Frequency of cell transformation in cell lines (C3H/10T½) or primary cultures (e.g., hamster lung cultures)	13—16, 54

The remainder of this chapter will be concerned with two aspects. First, a brief summary will be presented on the activity of several of the more well-studied naturally occurring agents (ascorbate, cysteine, selenium, Cu^{++}, Mn^{++} and Fe^{+++}) on the genotoxicity of single mutagens/carcinogens. For a comprehensive discussion of the role that plant phenols may play in modulating the genotoxic activity of carcinogens, the reader is referred to the review of Stich and Powrie in this volume. Secondly, data will be presented that illustrate how these same inhibitors can be assayed for activity in more complex model systems.

II. NATURALLY OCCURRING INHIBITORS OF GENOTOXIC ACTIONS OF CARCINOGENS

A. Antioxidants

1. Ascorbate

One of the more extensively studied antioxidants is ascorbate. Not only has this compound been shown to inhibit carcinogen-induced neoplasia in animals,[8-10] but it has been suggested that it be used to reduce the incidence of malignant tumors in humans[11] as well as to control the progression of cancer in patients.[12] Table 4 summarizes the capacity of ascorbate to reduce carcinogen-induced genotoxic activity. This vitamin appears to be able to suppress the DNA-damaging as well as the mutagenic and cell-transforming capacities of several carcinogens, including complex mixtures.

An interesting series of papers has been published on the ability of ascorbate to affect the process of transformation in carcinogen-treated cell cultures. Leuchtenberger and Leuchtenberger[13] have reported that the malignant transformation of hamster lung cultures by either tobacco or marijuana cigarette smoke could be suppressed by the addition of ascorbate to the culture media. Furthermore, cultures that had already become transformed due to exposure to the cigarette smoke could be "reversed" to normal growth when grown in the presence of this vitamin. Ascorbate has also been reported to inhibit the "promotion" phase of 3MC-initiated transformation of C3H/

Table 2
NATURALLY OCCURRING INHIBITORS OF CARCINOGEN-INDUCED GENOTOXICITY

- Vitamins
  - Ascorbate[13-15,17,20-22,38,50,53]
  - α-Tocopherol[16,38,53]
  - Retinoids[4,16,52]
  - Vitamin K[74]
- Thiols
  - Cysteine[3,13,20,21,23-27]
  - Cysteamine[2,21,24,53,54]
  - Glutathione[20,24]
- Trace metals
  - Selenium[22,37,38]
  - Fe^{+++}[7,29-32]
  - Fe^{++}[30,32]
  - Cu^{++}
- Tannins[7]
  - Catechin
  - Tannic acid
- Phenolics[7]
  - Caffeic acid
  - Gallic acid
  - Catechol
  - Chlorogenic acid
  - Pyrogallol[55]
  - 4-Methyl catechol
  - Resorcinol
- Unsaturated fatty acids[56]
  - Oleic acid
  - Linoleic acid
- Pyrrole pigments
  - Hemin[57,58]
  - Biliverdin[58]
  - Chlorophyllin[47]
- Vegetable factors[46-48]
- Browning reaction products
  - Caramelization
  - Maillard

10T½ mouse embryo fibroblast cells.[14,15] This effect of ascorbate appears to be "irreversible", since after several subcultures in ascorbate the vitamin can be removed without reexpression of the morphological transformation. In this respect, the effect of ascorbate appears to differ from that of vitamin A. Derivatives of vitamin A also suppress 3MC-induced transformation in the C3H/10T½ cultures but, in contrast to ascorbate, the subculturing of retinoid-blocked, 3MC-initiated cells for 4 to 5 weeks in the absence of the retinoid leads to a reversion to a transformed state.[16]

The overall effect of ascorbate on carcinogen activity is a very complicated issue. This fact is exemplified by apparently conflicting reports of inhibitory or enhancing activities of ascorbate on the genotoxic[15,17] and tumorigenic[8,18] actions of several carcinogens. It appears that a variety of variables, such as the relative concentration of carcinogens/mutagens and ascorbate used in an assay, and the procedure employed for exposing the target cell or animal to these test mixtures, have a large role in determining the overall mutagenic/carcinogenic response of a test system.[8,15,17]

2. *Thiols: Cysteine, Glutathione, Cysteamine*

Thiol-containing antioxidants, such as cysteine, glutathione, and cysteamine, are

Table 3
NATURALLY OCCURRING INHIBITORS OF TUMORIGENESIS IN CARCINOGEN-TREATED ANIMALS

- Vitamins
  - Ascorbate[8-10]
  - α-Tocopherol[59]
  - Retinoids[60]
- Thiols
  - Cysteamine[54]
- Trace metals
  - Selenium[33,34]
  - Cu^{++}[61]
- Isothiocyanates[5]
  - Benzyl isothiocyanate
  - Phenethyl isothiocyanate
- Indoles[5]
  - Indole
  - Indole-3-carbinol
  - 3,3′-Diindolylmethane
  - Indole-3-acetonitrile
  - DL-Tryptophan[62]
- Coumarins and lactones[5]
  - Coumarin
  - Limettin
  - Scopoletin
  - α-Angelicalactone
- Phenols
  - Gallic acid[63]

Table 4
INHIBITORY EFFECT OF ASCORBATE ON CARCINOGEN-INDUCED GENOTOXICITY

Endpoint	Genotoxic activity reduced in presence of ascorbate	Carcinogen/mutagen	Ref.
DNA damage	DNA fragmentation in human fibroblasts	DMN	21
	In vivo DNA fragmentation of mouse gastric mucosal cell DNA	MNNG	17
	DNA repair in human fibroblasts	Nitrosation products of methylguanidine, DMN	20, 21
Mutation	Reverse mutation to his^+ independence in *Salmonella*	MNNG, malonaldehyde, β-propiolactone, DMN, nitrosated fish	22, 38, 49, 50, 64
	Chromosome aberration in human fibroblasts	DMN	21
	Formation of micronucleated polychromatic RBC in BM cells of mouse	Dipyrone	53
Transformation	Induction of transformed foci in C3H/10T½ cell cultures	3MC	14, 15
	Transformation of primary hamster lung cultures	Tobacco or marijuana smoke	13

Note: DMN: dimethylnitrosamine; MNNG: *N*-methyl-*N*′-nitro-*N*-nitrosoguanidine; and 3MC: 3-methylcholanthrene.

very potent inhibitors of carcinogen-induced genotoxicity. Glutathione and glutathione S-transferases are important components of liver S9 microsomal fractions.[19] This thiol-containing antioxidant and its enzyme represent a detoxification process in mammalian cells that acts via the conjugation of the active electrophiles of mutagens/carcinogens to glutathione. The efficiency of thiol-containing antioxidants in suppressing carcinogen-induced genotoxicity is indicated by the wide range of carcinogens that are affected by their presence. For example, cysteine has been shown to suppress the genotoxic activity of alkylating agents (MNNG, MMS, and DMN),[20-23]aromatic amines (*N*-acetoxy-2-acetylaminofluorene and *N*-hydroxy-2-acetyl-aminofluorene),[22-24] a nitro aromatic (4-nitroquinoline-1-oxide),[23] two nitrofurans (5-nitro-2-furaldehyde semicarbazone, 2-(2-furyl)-3-(5-nitro-2-furyl)acrylamide,[23] a fungal toxin (aflatoxin B_1),[23] nitrosation products of methylurea and methylguanidine,[23] two pyrrolizidine alkaloids (monocrotaline and lasiocarpine),[25] hydrazines (hydrazine and isoniazid),[26] and several pesticides (captan, captafol, folpet, and NBT).[27]

B. Transition Metals

Table 5 summarizes the modulating effects of the transition metals Mn^{++}, Cu^{++}, and Fe^{+++} on the clastogenic activities of a variety of compounds on mammalian fibroblast cells. The concentration of the transition metals that is employed is similar to that observed in various body tissues as well as in food sources. This modulating activity of the transition metals has been observed in other in vitro systems that employ different test organisms and genotoxic endpoints. Thus the presence of Cu^{++}, Mn^{++}, or Fe^{+++} in a treatment mix can influence a compound's ability to induce DNA damage as measured by unscheduled DNA synthesis in mammalian fibroblasts and to induce sister chromatid exchanges in CHO cells, gene conversion in *Saccharomyces cerevisiae* D7 cultures, and reverse mutation in cultures of *Salmonella typhimurium* (see footnotes, Table 5). The mechanism underlying the effect of these metals on genotoxicity is not known. However, the importance of studying this effect is indicated by two examples. First, compounds such as tannic acid, caffeic acid, and chlorogenic acid have been reported to produce a negative response in the *Salmonella* mutagenesis assay. However, if Mn^{++} is added to the incubation mix, there is a significant induction of his^+ revertants by each of these chemicals.[28] Secondly, the reverse situation may occur when complex mixtures containing transition metals are tested. The presence of Fe^{+++} in a treatment mixture abolishes the genotoxic activities of betanin, catechin, tannins isolated from various sources, isoniazid, chlorogenic acid, and quercetin, as well as the clastogenic activity of several complex mixtures, including animal and human fecal extracts, caramel powder, and cigarette smoke.[7,29-32] These results suggest that transition metals may play an important role in determining the genotoxicity and possibly the carcinogenicity of the mutagens/carcinogens to which man is exposed.

C. Trace Metals: Selenium

1. Effect on Carcinogen-Induced Genotoxicity

The effect of selenium on carcinogen activity has been an area of intensive study in recent years. This interest is based upon the observation that selenium reduces neoplasia in animals exposed to a range of carcinogens.[33,34] Indeed, epidemiological studies have been reported that show an inverse correlation between the concentration of selenium in the environment and the incidence of various malignancies in the human population.[35,36] In vitro studies performed with selenium-containing derivatives are supportive of a role for this trace metal in suppressing the genotoxicity of several carcinogens.[22,37,38] The mechanism for these protective actions of selenium is unknown.

Table 5
MODULATING ACTION OF TRANSITION METALS[a] ON THE CLASTOGENIC ACTIVITY OF VARIOUS COMPOUNDS AND MIXTURES

Group	Compounds tested	Activity in presence of metal (10^{-4} *M*)[b]			Ref.
		Fe^{+++}	Cu^{++}	Mn^{++}	
Catecholamines	1-Epinephrine		+		31
	L-Dopa			+	31
Vitamins	Ascorbate	+	+[c,d]	+	65—68
Reducing agents	Cysteamine, glutathione, cysteine		+[c,d]		67, 68
Alkaloids	Arecoline	N.E.	N.E.	+[e]	Footnote g
Betanins	Betanin	−		+	Footnote g
Tannins	Tannic acid (commercial)	−	N.E.	+[e,f]	7
	Tannins isolated from grape juice, apple juice, wine, and betel nut	−	+	+	Footnote g
	Catechin	−			Footnote g
Hydrazines	Isoniazid	−	+[d]	+[c,d,e]	29, 68—70
Phenolics	Catechol, 4-methyl catchol, gallic acid		+	+	71
	Resorcinol, *p*-hydroxybenzoic acid, salicylic acid, *p*-coumaric acid[e], ferulic acid, eugenol		N.E.[e]	N.E.[e]	71, Footnote g
	Phloroglucinol, pyrogallol, protocatechuic acid, vanillic acid, caffeic acid,[e,f] 3,4-dihydroxyphenylacetic acid, syringic acid		N.E.[e,f]	+[e,f]	28, 71
	Chlorogenic acid	−	+[e]	+[e,f]	28
	Quercetin	−	N.E.	+	Footnote g
Complex mixtures	Fecal extracts, animal and human	−	+	+	30, 31
	Caramel powder #1	−	N.E.	N.E.	32
	Cigarette smoke	−	+[e]	N.E.	Footnote g
	Sewage outfall			N.E.	75
	Human sewage sludge (commercial fertilizer)			−	75

[a] Exposures were to chelated metals: cupric sulfate and manganese chloride chelated with glycine (molar ratio 1:10) and ferric chloride with EDTA (1:2).

[b] Enhancing effect, +; suppression, −; N.E., no effect.

[c−f] Indicated response observed in one of the following assays: SCE induction in CHO cells[c]; unscheduled DNA synthesis in mammalian fibroblast cultures[d]; gene conversion in *Saccharomyces cerevisiae* D7[e]; *Salmonella* mutagenesis assay[f].

[g] Unpublished data.

2. Effect on Spontaneous Mutation Frequencies

Recent studies indicate that selenium-containing compounds may also affect the rate at which a cell culture undergoes spontaneous mutagenesis.[39] Figure 1 illustrates the effect of sodium selenite on the frequency of spontaneous mutation to histidine independence in ten strains of *Saccharomyces cerevisiae.* The strains that were used include a wild-type (nonmutator) strain and nine isogenic mutator strains. Two other inorganic selenium derivatives, sodium selenide and sodium selenate, were also shown to possess

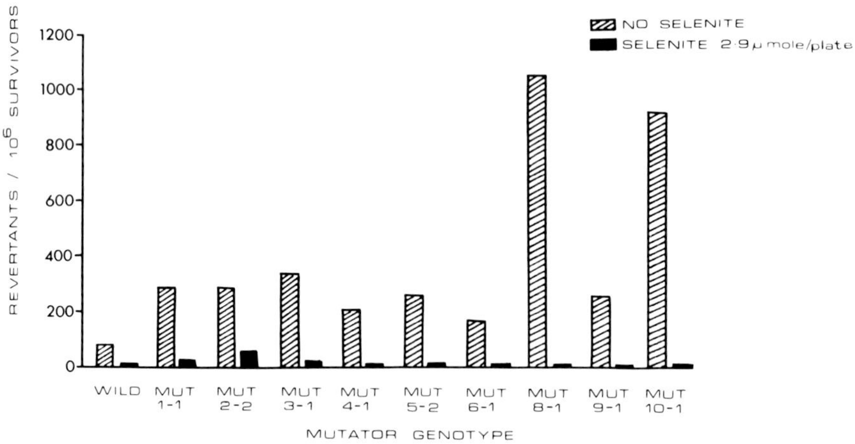

FIGURE 1. The effect of sodium selenite on the frequency of spontaneous mutation to histidine prototrophy of several yeast mutator strains. For procedure, see Reference 39.

an inhibiting capacity against spontaneous mutagenesis in these yeast strains. There is some support in the literature for a similar activity of selenium in animals.[40] Whether or not the ability of selenium-containing compounds to suppress mutagenesis in vitro is linked to an inhibition by selenium of spontaneous tumorigenesis in C3H/St mice is, at this point in time, unknown.

Selenium is not the only trace metal that has been shown to affect spontaneous mutation frequencies. Cobaltous chloride has also been reported to suppress spontaneous mutation induction in *Bacillus subtilis*[41] and *Escherichia coli.*[42] At present there are no available data that show comparable effects of either selenium or cobalt on mammalian mutation frequencies. However, Raj and Heddle[43] have reported that the thiol-containing antioxidant, cysteine, suppresses chromosomal breakage in both Fanconi's anemia and normal human fibroblasts. Thus, naturally occurring agents may have a significant effect on the genetically controlled predisposition of an organism to mutagenesis.

III. INHIBITORY EFFECTS OF NATURALLY OCCURRING AGENTS ON GENOTOXIC ACTIVITY OF COMPLEX MIXTURES

The interactions already described in this paper are, with a few exceptions, those that occur between two compounds. Even with this approach, the number of possible combinations to be tested among the many components in the mixtures to which man is exposed is beyond the scope of a realistic study. An intermediary step is necessary; one that will lie somewhere between the study of the genotoxicities of single compounds and the analysis of very complicated mixtures containing hundreds of components. This laboratory employs two model systems for complex mixtures as such a step.

A. Cigarette Smoke

The first model system involves the exposure of mammalian fibroblast cultures to cigarette smoke. Cultures of either CHO or human fibroblast cells are grown on glass coverslips. These coverslip cultures are placed into the test chamber of a cigarette inhalation machine where they are exposed to a puff of cigarette smoke once a minute. The volume and duration of the cigarette puff, the dilution of the cigarette smoke in

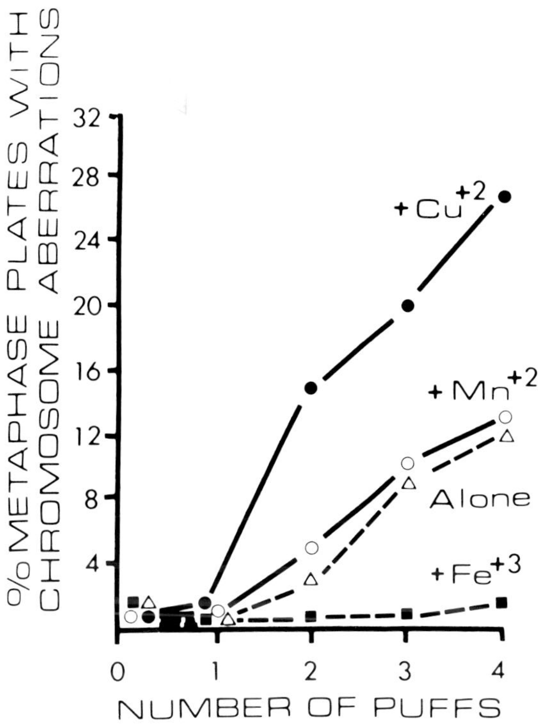

FIGURE 2. The effect of transition metals on the clastogenic activity of cigarette smoke in CHO cells. Chromosome aberrations scored included both chromatid breaks and exchanges. Concentration of metals used was 10^{-4} *M*.

the test chamber with fresh air, and the rate of removal of the smoke from the chamber are all processes that can be controlled so as to produce laboratory conditions that simulate human exposure patterns. Three puffs of cigarette smoke are sufficient to induce a significant number of chromatid breaks and exchanges in the treated cells. The effect of various modulating agents on the genotoxic activity of this cigarette smoke can be assayed by adding the test agents to the culture media for 1 hr prior to exposure of the cells to the cigarette smoke. Figure 2 shows the effect of the three transition metals, Fe^{+++}, Cu^{++}, and Mn^{++}, on the clastogenic activity of the cigarette smoke. Cultures that were pretreated with Cu^{++} displayed an increase in the frequency of metaphase plates with chromosome aberrations. In contrast, Fe^{+++} abolished the chromosome-damaging action of the cigarette smoke, while Mn^{++} had no effect. Figure 3 illustrates the effect of ascorbate, cysteine, and glutathione on the frequency of metaphase plates with chromosome aberrations in treated CHO cells. Each of these antioxidants suppressed the aberration frequency observed after smoke treatment.

Recently, the yeast cells *Saccharomyces cerevisiae* D7 have also been shown to be responsive to treatment with cigarette smoke. Logarithmically growing cultures of this yeast that were exposed on a filter paper to 5 to 7 puffs of cigarette smoke displayed up to a ten-fold increase above spontaneous levels of the frequency of gene conversion to tryptophan independence. This cigarette smoke-induced gene conversion was elevated in cells pretreated for 1 hr with Cu^{++}. The search for modulating agents of the convertogenic and clastogenic activities of cigarette smoke should be highly relevant due to the strong correlation observed between cigarette smoking and lung cancer in man.

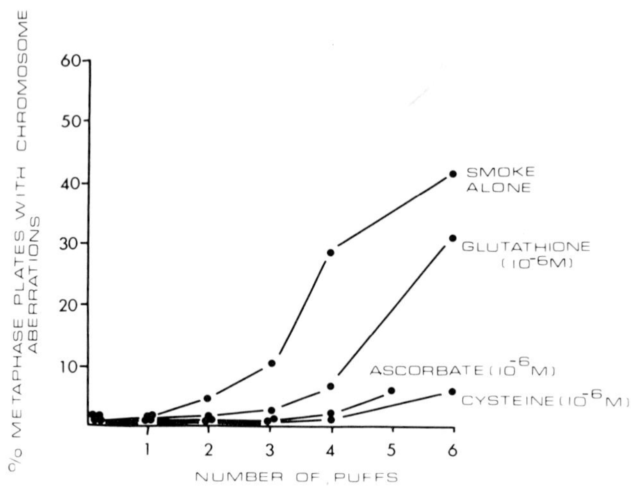

FIGURE 3. The effect of several antioxidants on the clastogenic activity of cigarette smoke in CHO cells.

B. Nonenzymatic Browning Reaction Products

The second model that was used was a nonenzymatic browning reaction system created by heating glucose and lysine to 121°C for 1 hr. The browning reaction products formed during this process induce chromosome aberrations in CHO cells, gene conversion in *S. cerevisiae* and reverse mutation in *Salmonella typhimurium*.[44] There are several advantages to using these browning reaction products as a model system for complex mixtures. First, such studies would be relevant to the human situation, since nonenzymatic browning reactions occur in all heated sugar- and protein-containing foods. Secondly, the production of this complex mixture can be controlled, and its contents reproduced under rigid laboratory conditions. Finally, the number of products formed in the reaction is small enough to allow it to be fractionated and the components identified to see which chemicals possess genotoxic, antigenotoxic, or promoting activities. Figure 4A illustrates the convertogenic activity of a heated glucose-lysine mixture in cultures of *Saccharomyces cerevisiae* D7. The identification of possible modulating agents that affect this convertogenic activity is easily performed by including a concurrent exposure of the test agent with the heated glucose-lysine products.[72] Figure 4B shows the effect of the metals Cu^{++}, Mn^{++} Fe^{+++}, and selenite on the convertogenic activity induced by these browning reaction products. Fe^{+++}, and to a lesser degree Cu^{++}, both suppressed the convertogenic activity of this model system, while selenite and Mn^{++} had no effect. Figure 4C shows the inhibitory effects of three enzyme systems in this test system. Rat liver S9 fraction, horseradish peroxidase, and beef liver catalase all suppressed the convertogenic activity of the browning reaction products. These results indicate a few of the protective mechanisms that cells have towards the products of browning reactions.

Recent work on browning reaction model systems indicates that one or more of the components of this reaction must also possess antigenotoxic properties. Figure 5 shows the effect of these reaction products on the convertogenic activity of a fungal toxin, sterigmatocystin, in the *S. cerevisiae* test system. The concurrent addition of heated glucose-lysine and the fungal toxin to the yeast cells results in a suppression of the sterigmatocystin-induced gene conversion. This anticonvertogenic activity is also observed with treatments containing caramel powder. Caramel powder is a commercially

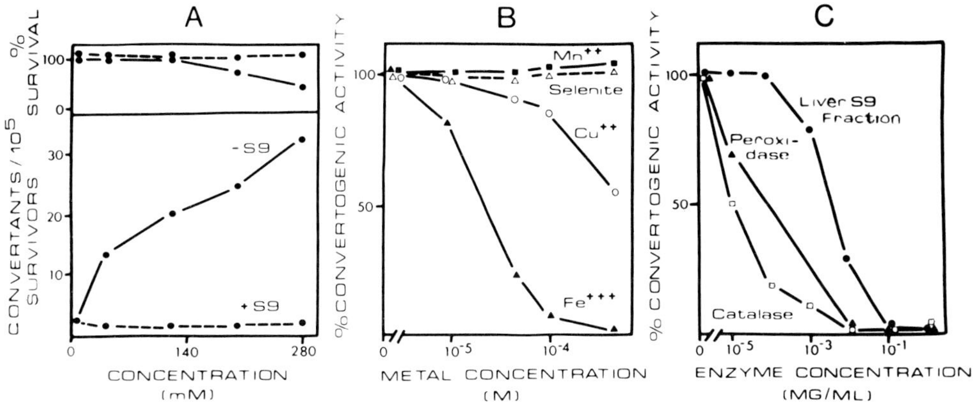

FIGURE 4. The influence of several modulating agents on the convertogenic activity of a heated glucose-lysine solution in *Saccharomyces cerevisiae* strain D7. (A) Frequency of gene conversion to tryptophan prototrophy in the presence and absence of an S9 activation mix. (B) Effect of trace metals on convertogenic activity of heated glucose-lysine reaction products (70 m*M*). (C) Effect of enzymes on the convertogenic activity of heated glucose-lysine (70 m*M*). Enzyme systems included horseradish peroxidase, beef liver catalase, and rat liver S9 fraction (−NADP, −G6P). Concentration of rat liver S9 fraction expressed as milligrams of protein per milliliters of treatment mix.

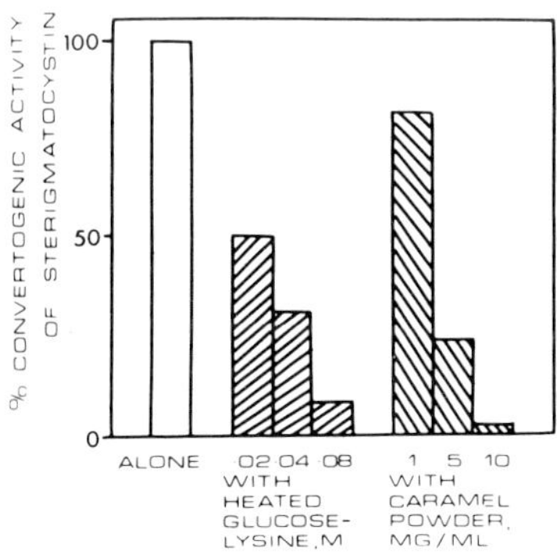

FIGURE 5. Inhibition of sterigmatocystin-induced gene conversion in *S. cerevisiae* D7 by browning reaction products. Concentration of sterigmatocystin used was 2×10^{-5} *M*.

prepared mixture containing browning reaction products that is added as a colorant to a variety of foods and beverages.[32] Browning reaction products have been shown to suppress the mutagenic activity of the S9-activated carcinogen aflatoxin B_1, the direct-acting alkylating agent MNNG, and the nitrosation products of methylurea (Figure 6), as measured in the *Salmonella* mutagenesis assay.[73] The identification of the components of this model system that are responsible for its antigenotoxic properties is an area of current research in this laboratory.

The data presented in this paper indicate the way that "standardized" model systems can be used to explore the host of interactions possible in complex mixtures. Realistic assessments of the genetic hazard of mixtures containing mutagenic/carcinogenic components will rely on our ability to identify modifying agents present either in the mixture itself or in the target cell.

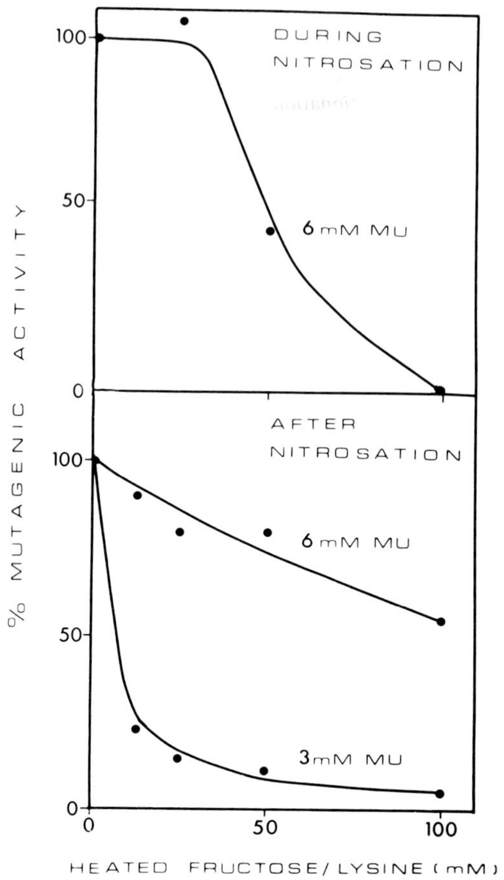

FIGURE 6. Inhibition of mutagenic activity of nitrosation products of methylurea (MU) of *Salmonella typhimurium* strain TA1535 by heated fructose-lysine reaction products. Heated fructose-lysine products were added either (A) during the nitrosation reaction, or (B) after the nitrosation reaction but during the 20 min exposure of the bacteria to the nitrosation products. Concentrations of nitrosated MU assume complete nitrosation (starting concentration).

REFERENCES

1. **Waters, M. D., Nesnow, S., Huisingh, J. L., Sandhu, S. S., and Claxton, L.**, Eds., *Application of Short-Term Bioassays in the Fractionation and Analysis of Complex Environmental Mixtures,* Plenum Press, New York, 1979.
2. **Gebhart, E.**, The anticlastogenic effect of various combinations of cysteamine, AET, HCT, and amino acids on chromosome damage by Trenimon and bleomycin in human lymphocytes *in vitro, Hum. Genet.,* 43, 185, 1978.
3. **Gebhart, E. and Kappauf, H.**, The action of three anticlastogens on the induction of sister chromatid exchange by Trenimon and 8-hydroxyquinoline sulfate in human lymphocyte cultures, *Environ. Mutagen.,* 2, 191, 1980.
4. **Miller, R. C., Geard, C. R., Osmak, R. S., Rutledge-Freeman, M., Ong, A., Mason, H., Napholz, A., Perey, N., Harisiadis, L., and Borek, C.**, Modification of sister chromatid exchanges and radiation-induced transformation in rodent cells by the tumor promoter 12-*O*-tetradecanoylphorbol-13-acetate and two retinoids, *Cancer Res.,* 41, 655, 1981.
5. **Wattenberg, L. W.**, Naturally occurring inhibitors of chemical carcinogenesis, in *Naturally Occurring Carcinogens-Mutagens and Modulators of Carcinogenesis,* Miller, E. C., Miller, J. A., Sugimura, T., Takayama, S., and Hirono, I., Eds., University Park Press, Baltimore, 1979, 315.
6. **Buening, M. K., Chang, R. L., Huang, M-T., Fortner, J. G., Wood, A. W., and Conney, A. H.**, Activation and inhibition of benzo(a)pyrene and aflatoxin B_1 in human liver microsomes by naturally occurring flavonoids, *Cancer Res.,* 41, 67, 1981.
7. **Stich, H. F. and Powrie, W. D.**, Plant phenolics as genotoxic agents and as modulators for the mutagenicity of other food components, in *Carcinogens and Mutagens in the Environment, Vol. 1, Food Products,* Stich, H. F., Ed., CRC Press, Boca Raton, Fla., 1983.
8. **Mirvish, S. S., Pelfrene, A. F., Garcia, H., and Shubik, P.**, Effect of sodium ascorbate on tumor induction in rats treated with morpholine and sodium nitrite, and with nitrosomorpholine, *Cancer Lett.,* 2, 109, 1976.
9. **Schlegel, J. U.**, Proposed uses of ascorbic acid in prevention of bladder carcinomas, *Ann. N.Y. Acad. Sci.,* 258, 432, 1975.
10. **Slaga, T. J. and Bracken, W. M.**, The effects of antioxidants on skin tumor initiation and aryl hydrocarbon hydroxylase, *Cancer Res.,* 37, 1631, 1977.
11. **Cameron, E., Pauling, L., and Leibowitz, B.**, Ascorbic acid and cancer: a review, *Cancer Res.,* 39, 663, 1979.
12. **Cameron, E. and Pauling, L.**, Ascorbic acid as a therapeutic agent in cancer, *J. Int. Acad. Prev. Med.,* 5, 8, 1979.
13. **Leuchtenberger, C. and Leuchtenberger, R.**, Protection of hamster lung cultures by L-cysteine or vitamin C against carcinogenic effects of fresh smoke from tobacco or marihuana cigarettes, *Br. J. Exp. Pathol.,* 58, 625, 1977.
14. **Benedict, W. F., Wheatley, W. L., and Jones, P. A.**, Inhibition of chemically induced morphological transformation phenotype by ascorbic acid in C3H/10T½ cells, *Cancer Res.,* 40, 2796, 1980.
15. **Rosin, M. P., Peterson, A. R., and Stich, H. F.**, The effect of ascorbate on 3-methylcholanthrene-induced cell transformation in C3H/10T½ mouse-embryo fibroblast cell cultures, *Mutat. Res.,* 72, 533, 1980.
16. **Merriman, R. L. and Bertram, J. S.**, Reversible inhibition by retinoids of 3-methylcholanthrene-induced neoplastic transformation in C3H/10T½ clone 8 cells, *Cancer Res,* 39, 1661, 1979.
17. **Koropatnick, D. J. and Stich, H. F.**, The modifying effect of sodium ascorbate on DNA damage and repair after *N*-methyl-*N*′-nitro-*N*-nitroso-guanidine treatment *in vivo, Biochem. Biophys. Res. Commun.,* 92, 292, 1980.
18. **Banič, S.**, Vitamin C acts as a cocarcinogen to methylcholanthrene in guinea-pigs, *Cancer Lett.,* 11, 239, 1981.
19. **Summer, K-H., Göggelmann, W., and Greim, H.**, Glutathione and glutathione S-transferases in the *Salmonella* mammalian-microsome mutagenicity test, *Mutat. Res.,* 70, 269, 1980.
20. **Wei, L., Whiting, R. F., and Stich, H. F.**, Inhibition of chemical mutagenesis: an application of chromosome aberration and DNA synthesis assays using cultured mammalian cells, in *Short-Term Tests for Chemical Carcinogens,* Stich, H. F. and San, R. H. C., Eds., Springer-Verlag, New York, 1981, 428.
21. **Lo, L. W. and Stich, H. F.**, The use of short-term tests to measure the preventive action of reducing agents on formation and activation of carcinogenic nitroso compounds, *Mutat. Res.,* 57, 57, 1978.
22. **Rosin, M. P. and Stich, H. F.**, Assessment of the use of the *Salmonella* mutagenesis assay to determine the influence of antioxidants on carcinogen-induced mutagenesis, *Int. J. Cancer,* 23, 722, 1979.
23. **Rosin, M. P. and Stich, H. F.**, The inhibitory effect of cysteine on the mutagenic activities of several carcinogens, *Mutat. Res.,* 54, 73, 1978.

24. **Rosin, M. P. and Stich, H. F.,** Inhibitory effect of reducing agents on *N*-acetoxy- and *N*-hydroxy-2-acetylaminofluorene- induced mutagenesis, *Cancer Res.,* 38, 1307, 1978.
25. **Avanti, S.,** Chromosome breakage by pyrrolizidine alkaloids and modification of the effect by cysteine, *Caryologia,* 14, 231, 1961.
26. **Speit, G., Wick, C., and Wolf, M.,** Induction of sister chromatid exchanges by hydroxylamine, hydrazine, and isoniazid and their inhibition by cysteine, *Hum. Genet.,* 54, 155, 1980.
27. **Moriya, M., Kato, K., and Shirasu, Y.,** Effects of cysteine and a liver metabolic activation system on the activities of mutagenic pesticides, *Mutat. Res.,* 57, 259, 1978.
28. **Stich, H. F., Rosin, M. P., Wu, C. H., and Powrie, W. D.,** A comparative genotoxicity study of chlorogenic acid (3-*O*-caffeoylquinic acid), *Mutat. Res.,* 90, 201, 1981.
29. **Whiting, R. F., Wei, L., and Stich, H. F.,** Enhancement by transition metals of chromosome aberrations induced by isoniazid, *Biochem. Pharmacol.,* 29, 842, 1980.
30. **Stich, H. F. and Kuhnlein, U.,** Chromosome breaking activity of human feces and its enhancement by transition metals, *Int. J. Cancer,* 24, 284, 1979.
31. **Stich, H. F., Rosin, M. P., San, R. H. C., Wu, C. H., and Powrie, W. D.,** Intake, formation and release of mutagens by man, in *Gastrointestinal Cancer: Endogenous Factors,* Bruce, W. R., Correa, P., Lipkin, M., Tannenbaum, S. R., and Wilkins, T. D., Eds., Banbury Rep. 7, Cold Spring Harbor Laboratory, Cold Spring Harbor, N.Y., 1981, 247.
32. **Stich, H. F., Stich, W., Rosin, M. P., and Powrie, W. D.,** Clastogenic activity of caramel and caramelized sugars, *Mutat. Res.,* 91, 129, 1981.
33. **Shamberger, R. J.,** Relationship of selenium to cancer. V. Inhibitory effect of selenium on carcinogenesis, *J. Natl. Cancer Inst.,* 44, 931, 1970.
34. **Jacobs, M. M., Jansson, B., and Griffin, A. C.,** Inhibitory effects of selenium on 1,2-dimethylhydrazine and methylazoxymethanol acetate induction of colon tumors, *Cancer Lett.,* 2, 133, 1977.
35. **Schrauzer, G. N., White, D. A., and Schneider, C. J.,** Cancer mortality correlation studies. III. Statistical associations with dietary selenium intakes, *Bioinorg. Chem.,* 7, 23, 1977.
36. **Shamberger, R. J., Tytko, S. A., and Willis, C. E.,** Antioxidants and cancer. VI. Selenium and age-adjusted human cancer mortality, *Arch. Environ. Health,* 31, 231, 1976.
37. **Jacobs, M. M., Matney, T. S., and Griffin, A. C.,** Inhibitory effects of selenium on the mutagenicity of 2-acetylaminofluorene (AAF) and AAF derivatives, *Cancer Lett.,* 2, 319, 1977.
38. **Shamberger, R. J., Corlett, C. L., Beaman, K. D., and Kasten, B. L.,** Antioxidants reduce the mutagenic effect of malonaldehyde and β-propiolactone. IX. Antioxidants and cancer, *Mutat. Res.,* 66, 349, 1979.
39. **Rosin, M. P.,** Inhibition of spontaneous mutagenesis in yeast cultures by selenite, selenate, and selenide, *Cancer Lett.,* 13, 7, 1981.
40. **Schrauzer, G. N. and Ishmael, D.,** Effects of selenium and of arsenic on the genesis of spontaneous mammary tumors in inbred C3H mice, *Ann. Clin. Lab. Sci.,* 4, 441, 1974.
41. **Inoue, T., Ohta, Y., Sadaie, Y., and Kada, T.,** Effect of cobaltous chloride on spontaneous mutation induction in *Bacillus subtilis* mutator strain, *Mutat. Res.,* 91, 41, 1981.
42. **Kada, T. and Kanematsu, N.,** Reduction of *N*-methyl-*N'*-nitro-*N*-nitrosoguanidine-induced mutations by cobalt chloride in *Escherichia coli, Proc. Jpn. Acad.,* 54B, 234, 1978.
43. **Raj, A. S. and Heddle, J. A.,** The effect of superoxide dismutase, catalase, and L-cysteine on spontaneous and on mitomycin C-induced chromosomal breakage in Fanconi's anaemia and normal fibroblasts as measured by the micronucleus method, *Mutat. Res.,* 78, 59, 1980.
44. **Powrie, W. D., Wu, C. H., Rosin, M. P., and Stich, H. F.,** Clastogenic and mutagenic activities of Maillard reaction model systems, *J. Food Sci.,* 46, 1433, 1981.
45. **Kuhnlein, U.,** Disulfiram inhibits DNA breakage by hydroxyradical-producing agents, *Biochim. Biophys. Acta,* 609, 75, 1980.
46. **Kada, T., Morita, K., and Inoue, T.,** Antimutagenic action of vegetable factor(s) on the mutagenic principle of tryptophan pyrolysate, *Mutat. Res.,* 53, 351, 1978.
47. **Morita, K., Hara, M., and Kada, T.,** Studies on natural desmutagens: screening for vegetable and fruit factors active in inactivation of mutagenic pyrolysis products from amino acids, *Agric. Biol. Chem.,* 42, 1235, 1978.
48. **Lai, C.-N., Butler, M. A., and Matney, T. S.,** Antimutagenic activities of common vegetables and their chlorophyll content, *Mutat. Res.,* 77, 245, 1980.
49. **Guttenplan, J. B.,** Inhibition by L-ascorbate of bacterial mutagenesis induced by two *N*-nitroso compounds, *Nature (London),* 268, 368, 1977.
50. **Guttenplan, J. B.,** Mechanisms of inhibition by ascorbate of microbial mutagenesis induced by *N*-nitroso compounds, *Cancer Res.,* 38, 2018, 1978.
51. **Moriya, M., Ohta, T., Watanabe, K., Watanabe, Y., Sugiyama, F., Miyazawa, T., and Shirasu, Y.,** Inhibitors for the mutagenicities of colon carcinogens, 1,2-dimethylhydrazine and azoxymethane, in the host-mediated assay, *Cancer Lett.,* 7, 325, 1979.

52. **Trosko, J. B. and Chang, C. C.**, Chemical modification of radiation-induced mutagenesis and its role in carcinogenesis, *J. Supramol. Struct.*, Suppl. 2, 42, 1978.
53. **Sylianco, C. Y.**, Some interactions affecting the mutagenicity potential of dipyrone, hexachlorophene, thiodan, and malathion (meet. abstr.), *Mutat. Res.*, 53, 271, 1978.
54. **Marquardt, H., Sapozenk, M. D., and Zedeck, M. S.**, Inhibition by cysteamine-HCl of oncogenesis induced by 7,12-dimethylbenz(a)anthracene without affecting toxicity, *Cancer Res.*, 34, 3387, 1974.
55. **Rahimtula, A. D., Zachariah, P. K., and O'Brien, P. J.**, The effects of antioxidants on the metabolism and mutagenicity of benzo(a)pyrene *in vitro*, *Biochem. J.*, 164, 473, 1977.
56. **Hayatsu, H., Arimoto, S., Iogawa, K., and Makita, M.**, Inhibitory effect of the ether extract of human feces on activities of mutagens: inhibition by oleic and linoleic acids, *Mutat. Res.*, 81, 287, 1981.
57. **Arimoto, S., Negishi, T., and Hayatsu, H.**, Inhibitory effect of hemin on the mutagenic activities of carcinogens, *Cancer Lett.*, 11, 29, 1980.
58. **Arimoto, S., Ohara, Y., Namba, T., Negishi, T., and Hayatsu, H.**, Inhibition of the mutagenicity of amino acid pyrolysis products by hemin and other biological pyrrole pigments, *Biochem. Biophys. Res. Commun.*, 92, 662, 1980.
59. **Haber, S. L. and Wissler, R. W.**, Effect of vitamin E on carcinogenicity of methylcholanthrene, *Proc. Soc. Exp. Biol. Med.*, 111, 774, 1962.
60. **Moon, R. C., Grubbs, C. J., and Sporn, M. B.**, Inhibition of 7,12-dimethylbenz(a)anthracene-induced mammary carcinogenesis by retinyl acetate, *Cancer Res.*, 36, 2626, 1976.
61. **Kamamoto, Y., Makiura, S., Sugihara, S., Hiasa, Y., Arai, M., and Ito, N.**, Inhibitory effect of copper on DL-ethionine carcinogenesis in rats, *Cancer Res.*, 33, 1129, 1973.
62. **Miyakawa, M. and Yoshida, O.**, Protective effect of DL-tryptophan on benzidine-induced hepatic tumor in mice, *Gann*, 71, 265, 1980.
63. **Mirvish, S. S., Cardesa, A., Wallcave, L., and Shubik, P.**, Induction of mouse lung adenomas by amines or ureas plus nitrite and by *N*-nitroso compounds: effect of ascorbate, gallic acid, thiocyanate, and caffeine, *J. Natl. Cancer Inst.*, 55, 633, 1975.
64. **Marquardt, H., Rufino, F., and Weisburger, J. H.**, On the aetiology of gastric cancer: mutagenicity of food extracts after incubation with nitrite, *Food Cosmet. Toxicol.*, 15, 97, 1977.
65. **Stich, H. F., Wei, L., and Whiting, R. F.**, Enhancement of the chromosome-damaging action of ascorbate by transition metals, *Cancer Res.*, 39, 4145, 1979.
66. **MacRae, W. D. and Stich, H. F.**, Induction of sister chromatid exchanges in Chinese hamster cells by the reducing agents bisulfite and ascorbic acid, *Toxicology*, 13, 167, 1979.
67. **Stich, H. F., Wei, L., and Lam, P.**, The need for a mammalian test system for mutagens: action of some reducing agents, *Cancer Lett.*, 5, 199, 1978.
68. **MacRae, W. D. and Stich, H. F.**, Induction of sister-chromatid exchanges in Chinese hamster ovary cells by thiol and hydrazine compounds, *Mutat. Res.*, 68, 351, 1979.
69. **Zetterberg, G. and Böstrom, J.**, Mitotic gene conversion induced in yeast by isoniazid. Influence of a transition metal and of the physiological conditions of the cells, *Mutat. Res.*, 91, 215, 1981.
70. **Whiting, R. F., Wei, L., and Stich, H. F.**, Enhancement by transition metals of unscheduled DNA synthesis induced by isoniazid and related hydrazines in cultured normal and Xeroderma pigmentosum human cells, *Mutat. Res.*, 62, 505, 1979.
71. **Stich, H. F., Rosin, M. P., Wu, C. H., and Powrie, W. D.**, The action of transition metals on the genotoxicity of simple phenols, phenolic acids, and cinnamic acids, *Cancer Lett.*, 14, 251, 1981.
72. **Rosin, M. P., Stich, H. F., Powrie, W. D., and Wu, C. H.**, Induction of mitotic gene conversion by browning reaction products and its modulation by naturally occurring agents, *Mutat. Res.*, in press.
73. **Chan, R. I. M., Stich, H. F., Rosin, M. P., and Powrie, W. D.**, Antimutagenic activity of browning reaction products, *Cancer Lett.*, 615, 27, 1982.
74. **Israels, L. G.**, Personal communication.
75. **San, R.**, Personal communication.

Chapter 22

U.S. APPROACHES TO REGULATING CARCINOGENS AND MUTAGENS IN FOOD

W. Gary Flamm

TABLE OF CONTENTS

I. INTRODUCTION

The views reflected in this presentation are the author's, though they reflect much of the current thinking of the FDA. This chapter, out of necessity, is devoted primarily to the subject of the U.S. policy on carcinogens in food and is virtually silent on policy regarding mutagens. The reason for this is simply because policy on the latter has not yet been promulgated. Nevertheless, two points should be made in passing. First, the Environmental Protection Agency (EPA) has proposed, in the U.S. *Federal Register,*[1] guidelines that deal with the assessment of germ cell risk from mutagenically active substances, and second, the FDA uses mutagenicity test data in some cases as a means of deciding whether carcinogenicity testing is required for a particular chemical substance.

It is, of course, not possible to discuss U.S. food safety policy without centering the discussion on the anticancer provisions of the U.S. food law known as the Delaney Clause. This is true despite the surprising fact that the Delaney Clause has been invoked only 5 times in the past 23 years. It is argued, however, that the clause has had much greater influence than is apparent from its infrequent use. Furthermore, it stands as evidence in the minds of many that the Congress of the U.S. has intended a "no risk" policy for food additive substances. Perhaps at no time since the passage of the food additive amendments in 1958 has there been so much intense discussion about this clause and so many proffered solutions to the perceived dilemmas in the food law. It is believed that the current dialogue is useful and is helping all of us to develop a better understanding of the issues, what the problems truly are, and what can and cannot be done to solve them.

II. OVERVIEW OF CURRENT POLICY

It should be recognized that law and policy often are dichotomous with respect to the presence of chemical carcinogens in food. Food additives, defined under Section 201(s) as "any substance the intended use of which results or may reasonably be expected to result . . . in its becoming a component . . . of any food . . .," are subject to the Delaney Clause under Section 409. On the other hand, unavoidable contaminants, pesticides, and natural toxicants are not. Indeed, even certain food ingredients, such as those considered "prior sanctioned" by FDA or the U.S. Department of Agriculture before 1958, and GRAS (generally recognized as safe) compounds are not, in a strict sense, covered by the Delaney Clause either. Insofar as unavoidable contaminants and natural toxicants are concerned, the law states under Section 402 that "A food shall be deemed to be adultered . . . if it bears or contains any poisonous or deleterious substance which may render it injurious to health; but in the case the substance is not an added substance such food shall not be considered adultered under this clause if the quantity of such substance in such food does not ordinarily render it injurious to health." This is often referred to as the general food safety provision of the law. Rather than ban a food found to contain a carcinogen that is either an unavoidable contaminant or natural toxicant, such as PCB, dioxin, or aflatoxin, the law allows that their levels in food be regulated as described in Section 406. This section states that, "Any poisonous or deleterious substance added to food, *except where such substance is required in the production thereof or cannot be avoided by good manufacturing practice* shall be deemed to be unsafe . . ." This exception within Section 406 of the law allows FDA to exercise judgment. Thus, it has the authority to regulate carcinogens that are either unavoidable contaminants or natural toxicants in food at levels that balance public health considerations against the consequences of banning the food and thus affecting the nation's food supply.

It might be argued, therefore, that Section 406 provides for risk/benefit decision making and that risk/benefit balancing for certain substances in food was the clear and certain intent of the U.S. Congress. Just as clearly, however, the Congress has prescribed a no-risk policy for those carcinogenic substances that are legally defined as food additives. Or has it? For instance, the legislative history of Section 409 indicates that the Congress considered, and rejected, the concept that safety could be proven to an absolute certainty, and instead asserted a standard of reasonable certainty that no harm shall occur. Additionally, food law authorities such as former FDA chief counsel, Peter B. Hutt,[2] and consumer advocate James Turner,[3] agree that Congress intended to incorporate into the Delaney Clause substantial scientific discretion in determining whether a substance has been shown to induce cancer. Once scientists have made the determination that a food additive is a carcinogen, Congress has established the policy that it must be banned. But considerable leeway exists in deciding, on scientific grounds, whether a substance is carcinogenic.

It could, however, be argued that for scientific and "political" reasons little of this leeway has been exercised to date and it begs the question whether, by the exercise of this leeway, the food safety dilemmas would be solved and criticism of the Delaney Clause and the FDA would be silenced. This question will be examined again in Section IV, but first it should prove instructive to describe what the so-called Delaney Clause is in legal terms.

The Federal Food, Drug and Cosmetic Act actually contains the following three anticancer clauses: Section 409 (c) (3) (A) of the Act dealing with food additives states that "no additive shall be deemed to be safe if it is found, after tests which are appropriate for the evaluation of the safety of food additives, to induce cancer in man or animals . . ." Section 706 (b) (5) (B) which is included in the Color Additive Amendment of 1960 states that "A color additive . . . shall be deemed unsafe, and shall not be listed for any use which will or may result in ingestion of all or part of such additive if the additive is found by the Secretary to induce cancer when ingested by man or animal . . ." Section 512 (d) (1) (H) of the Animal Drug Amendments of 1968 states that a new animal drug application shall not be approved if "such drug induces cancer when ingested by man or animals after tests which are appropriate for the evaluation of the safety of such drug . . ." It should be noted that these three sections allow the addition of carcinogenic food additives, colorants, and drugs to food-producing animals on the condition that these substances do no harm to the animal and that they leave no detectable residue in the tissues of the animal as determined by a method approved by the Secretary of Health and Human Services. This is the famous DES (diethylstilbestrol) proviso, and, as will be described later, has led to years of effort by the FDA to develop a policy that interprets the meaning of ". . . leave no detectable residue in the tissues . . ." The three clauses, Sections 409(c)(3)(A), 706(b)(5)(B), and 512(d)(1)(H), including the so-called DES proviso, constitute what is known generally as the "Delaney Clause."

While widely referred to and hotly debated, the Delaney Clause has been invoked only five times by the Agency. The first time was in 1967 for Flectol, an adhesive used in packing material.The second occasion was with chloroaniline, an ingredient used in a packaging adhesive. Other attempts to invoke the Delaney Clause have involved saccharin and DES. It has been argued by the Agency and others, that enforcing the clause was not necessary in any of these instances since the same action could have been taken under the general food safety provisions of the law. Given this history, why all the fuss over the Delaney Clause? The following section attempts to describe and explain what some of the concern is about.

III. PROBLEMS AND POLICY FORCING ISSUES

To those who oppose it, the Delaney Clause represents a symbol of what is wrong with current food safety policy in the U.S. and is often held responsible for having created a "no-risk" regulatory climate that has inhibited the development of new food additives. The experience with aspartame, an artificial sweetener composed of two naturally occurring amino acids, that has been under review for many years, has been cited as an illustration of the type of inhibitory effect the Delaney Clause has had, and is continuing to have, on the approval process for food additives. Actions or attempted actions on cyclamate, saccharin, and nitrite, that were in part or wholly driven by the Delaney Clause, have prompted additional criticism of FDA from many quarters and probably have given further substance to the ongoing debate.

Hutt[2] recently suggested that three major factors had combined to create what was referred to as "our current food safety crisis." First, it was indicated that as more and more food ingredients are tested for potential carcinogenicity in animals, more and more are found to be carcinogenic in at least one sex of one strain of one species of test animals. The experience of the National Cancer Institute's Carcinogen Bioassay Program was cited as supporting this contention. Hutt further alluded to the fact that many natural substances found in food have been shown to be carcinogenic. Given the thousands of chemicals added to food either directly or indirectly, it can be expected that eventually many will be found to be carcinogenic in animal tests.

In addition, greatly enhanced sensitivity of analytical methods has enabled chemists to detect carcinogenic substances in food at parts per billion and even parts per trillion. At these low levels it is expected that such ubiquitous carcinogens as lead, PAH, halogenated hydrocarbons, nitrosamines, and aromatic amines will be found in virtually all chemical substances, including regulated food and color additives. Does the presence of trace amounts of carcinogens in regulated additives *a priori* make them illegal under Sections 409 and 706 because of the Delaney Clause? Likewise, does the exquisite sensitivity of present analytical methods render the DES proviso useless as a means of regulating carcinogenic substances administered to food producing animals? These are questions that are still being addressed by the Agency, and will undoubtedly be the subject of public debate for some time to come.

Finally, Hutt[2] asserts that epidemiological studies that have implicated a number of important components of the American diet as carcinogenic also contribute to "the food safety crisis." Coffee and protein were cited as having been associated with increased rates of human cancer. Indeed, Hutt points out that the ubiquity and frequency of carcinogens in food makes it impossible to survive on a diet consisting only of substances that have not been found to be carcinogenic by some types of scientific study. In support of this view Hutt quotes a Federal Register Document published by FDA in October 1979 that states: "A requirement for warnings on all foods that contain an inherent carcinogenic ingredient or a carcinogenic contaminant (in contrast to deliberately added carcinogenic substance) would apply to many, perhaps most, foods in a supermarket. Such warnings would be so numerous they would confuse the public, would not promote informed consumer decision making and would not advance the public health."

Hutt[2] contends that FDA has lost control over the food safety issue now that the inevitability of carcinogens in food has been recognized. Clearly, such recognition represents a significant challenge to the longstanding FDA food safety policy of "no-risk." Hutt[2] claims the country will not tolerate an ever-increasing number of public confrontations on food safety, such as those experienced with saccharin and nitrite. Basic changes in the law may take years to achieve, leaving to FDA the critical task of exerting its customary leadership role to establish new policy in this area.

Indeed, the National Academy of Sciences, in a report entitled "Food Safety Policy",[4] expressed some of the same thinking as regards the need for FDA to assume leadership in the food safety area. The recent Monsanto court decision on acrylonitrile[5] ruled that the Agency had the authority to declare the level of a carcinogen migrating from packaging material into food *de minimus*. This ruling may indicate that there exists sufficient legal latitude to enable significant policy changes without necessarily having to change the current food law. We can expect this to be a subject of hot debate over the coming months and perhaps years. The next section of this paper describes some of the current activities on the part of FDA to address the concerns listed above.

IV. NEWLY PROPOSED APPROACHES

Many critics of the Delaney Clause and of the FDAs administration of the U.S. food laws would agree that virtually all food safety problems today can be resolved through a change in interpretation of the existing statutory provisions. Whether or not this is true remains to be seen, but the main thrust of the argument is the proposition that FDA can, possibly through rule making, reinterpret the Delaney Clause to apply only to significant risk of cancer and not to insignificant or *de minimus* risk. Actually, the FDA has been pursuing a related approach since 1973, when it began work on a Federal Register proposal to regulate carcinogenic animal drugs under the DES proviso that permits the use of such drugs in food-producing animals on the condition that no detectable residues appear in the edible portions of the animal at the time of slaughter.

The task before the Agency was and remains that of determining through appropriate scientific procedures the levels of residue or residues that are considered safe within the meaning of the law. Were the Agency to declare through rule-making or litigation that the only level of residue considered safe is zero, it would in effect, thwart the intent of the Congress. Instead, the Agency has embarked on a long and, at times, agonizing effort to formulate a scientifically defensible and well-balanced approach to regulating carcinogenic residues of animal drugs and feed additives in food-producing animals.

After much debate and public comment, a final order, known as the SOM regulation, was published in 1977.[6] It was soon remanded back to the Agency by the courts, citing possible violations of the Administrative Procedures Act. The court contended that the Agency had made major changes in the proposal when publishing its final regulation and that reproposal would have been appropriate.

In any event, the Agency went back to the drawing board, and after considering the court's opinion and additional public comment made two significant changes in the reproposal of the regulation, which was published again in 1979.[7] The first of these was a so-called threshold assessment component that was added to the preamble of the proposal. The purpose of threshold assessment is to establish whether a sponsor who is seeking new animal drug approval must subject the drug to the extensive testing requirements of the SOM regulation.

Threshold assessment combines structure activity relationships with mutagenicity testing and other short-term tests in laboratory animals, as a means of deciding whether extensive carcinogenicity testing of the proposed new animal drug and its metabolites is required. The Agency is now considering treating threshold assessment as a guideline rather than as part of the SOM regulation. The reason for this is the belief that the entire science of structure activity relationship and the use of mutagenicity tests for predicting results in animal carcinogenesis bioassay is still developing at a rapid rate. A final regulation would tend to set in concrete the prescribed approach, while a guideline, on the other hand, would be easier to change and will allow more flexibility and latitude.

The Agency currently is applying a threshold assessment approach to regulating new animal drugs and has identified the Ames test, mutagenicity tests employing mammalian cells in tissue culture, DNA repair tests, and the sex-linked recessive lethal tests in *Drosophila* as a possible means of demonstrating whether the substance or substances in question are genotoxic. Since the usefulness of these tests vary, depending upon the chemical class under consideration, case-by-case decisions with respect to which and how many tests must be conducted will be made by FDA scientists in discussion, and consultation with sponsors of new animal drug applications.

Once it has been decided that a proposed new animal drug must be tested for carcinogenicity under the specifications of the SOM regulation, extensive metabolic studies and residue identification work is required. But the critical issue with regard to the U.S. food safety policy for carcinogens is the fact that, if the animal drug or its metabolites is found carcinogenic, it may still be approvable as long as the risk from residues of the drug and its metabolites is determined to be acceptable. The level of risk that has been proposed as acceptable is a lifetime risk of one cancer in a population of a million people exposed to the drug at the residue tolerance level for an entire lifetime.

The SOM regulation, published in 1977[6] proposed to use the Mantel-Byran log-probit approach to the extrapolation of animal carcinogenicity data to humans. This was changed when the Agency reproposed the SOM regulation in 1979.[7] The most recent proposal specifies the use of linear extrapolation as the most appropriate way to estimate the upper limit or upper bound of cancer risk to humans from lifetime exposures to residues of carcinogenic animal drugs or their metabolites in the edible portions of food-producing animals. To date, the Agency has taken the position that linear extrapolation provides a conservative estimate of risk, and that combined with a series of other conservative assumptions, it offers substantial assurance that the estimated risk is an overestimate of the real risk. Indeed, this is the intent of the approach, but care must be exercised in the application of these conservative assumptions if a return to a policy of "no-risk" is to be avoided. The tendency to layer one conservative assumption on top of another is always present. The effect of such practice, however, is that for all intents and purposes, one returns to a "no-risk" policy. To avoid this trap, strict adherence to stated goals must be rigorously exercised and combined with a logic that avoids "worst case assessments of risk" in favor of a "worst feasible case assessment of risk." To proceed differently is unlikely to provide the degree of discrimination that is recognized to be necessary.

In the final analysis, to adequately protect the public health from the effects of chemical carcinogens in food, reasonable approaches must be devised that provide the level of discrimination necessary to preclude any significant risk of cancer in exposed human populations. The challenge to the Agency is to define in a scientifically meaningful sense the word "significant", and to develop the rules and tools necessary to discriminate between those carcinogens and their levels in food that are "significant" from those that are not.

Toward this end, scientists in the Bureau of Foods of the FDA have been engaged in reviewing carcinogenicity data from more than 100 substances, many of which were initially tested by the National Cancer Institute as part of its carcinogen bioassay program. These case studies together with analysis of the so-called ED_{01} study of 2-acetylaminofluorene conducted at the FDAs National Center for Toxicological Research, promise to offer new tools, and even new rules, for the evaluation of carcinogenicity data from laboratory animals. Questions that relate to both the qualitative and the quantitative aspects of risk of cancer are being addressed and are expected to result in a more scientifically rigorous means of evaluating the significance of cancer risk, thus providing the improved discrimination desired.

The Agency is also considering treating adventitious contaminants as well as other carcinogenic constituents in regulated food and color additives under the same risk-assessment procedures as proposed for animal drugs under the SOM regulation. The legal reasoning is that unless the "whole additive" is found to be carcinogenic the Delaney Clause does not apply. If, for instance, the "whole additive" is found, in tests appropriate to evaluate the carcinogenicity of food additives, to be noncarcinogenic but is nevertheless known to contain a constituent that has been demonstrated to induce cancer in animals, the constituent will be subjected to the general food safety provisions of the law.

Under these provisions, the Agency is considering subjecting constituents of regulated food and color additives to a carcinogen risk assessment policy, analogous to that proposed for animal drugs and feed additives. At issue with respect to this policy, and that pertaining to animal drugs, is the appropriateness of the scientific approach, that in essence uses experimental animals to determine cancer risk to humans through linear extrapolation of animal data. The challenge to FDA seems to be the development of those discriminators (rules and tools) that will separate those levels of carcinogens that pose significant cancer risk to exposed human populations from those levels of carcinogens that do not.

Without question, the FDA is responding to the food safety challenges of the 1980s. However, it cannot address all of these issues simultaneously. To attempt to do so would result in failure. Priorities must be carefully set to insure that resources are sufficient to tackle the worst or most appropriate problem first, as good management in government always demands.

REFERENCES

1. U.S. Environmental Protection Agency, Mutagenicity Risk Assessment — Proposed Guidelines, Fed. Regist. Publ. FRL 1563-2, 74984, U.S. Government Printing Office, Washington, D.C., 1980.
2. **Hutt, P. B.,** FDA can handle food safety issues most effectively, *Legal Time of Washington,* p. 37, April 27, 1981.
3. **Turner, J.,** The Delaney anticancer clause: a model environmental protection law, *Vand. L. Rev.,* 24, 889, 1971.
4. National Academy of Sciences, Food Safety Policy, Scientific and Societal Considerations, Inst. of Med. Rep., National Academy of Sciences, Washington, D.C., 1979.
5. Monsanto vs. Kennedy, 613 F 2d 947, D.C. Cir., 1979.
6. U.S. Food and Drug Administration, Chemical compounds in food-producing animals, *Fed. Regist.,* 42, 10412, 1977.
7. U.S. Food and Drug Administration, Chemical compounds in food-producing animals, *Fed. Regist.,* 44, 17070, 1979.

Chapter 23

RECOMMENDATIONS CONCERNING MUTAGENS AND CARCINOGENS IN FOOD GIVEN BY SCIENTIFIC COMMITTEES WITHIN THE EUROPEAN ECONOMIC COMMUNITY (EEC)

Ib Knudsen and Emil Poulsen

TABLE OF CONTENTS

I. INTRODUCTION

At present the European Economic Community (EEC) has ten member states: Belgium, Denmark, Federal Republic of Germany, France, Greece, Ireland, Italy, Luxembourg, The Netherlands, and the U.K.

The regulatory base of the EEC is the Treaty of Rome. One of its articles, Article 100, forms the basis for community action on national rules for foodstuffs, when such rules give rise to trade barriers between the member states.

The legal instrument most commonly used for the approximation of food laws is the Directive. Such directives are binding on the member states to which they are addressed, but they do not have to be implemented literally in the national legislation, though the general sense of the directive has to be implemented.

The EEC Commission and the EEC Council provide the main day-to-day impetus in the decision-making processes of the Community. It is the Council that makes the final decision, but without the proposals of the Commission no progress can be made. During the preparation of the proposals the Commission consults with the national authorities in the member states, with professional groups within the EEC, and with its scientific committees.

The scientific committees are established by the Commission by Commission Decisions. Relating to food, the following scientific committees are involved: the Scientific Committee for Food, the Scientific Committee for Animal Nutrition, and the Scientific Committee for Pesticides. The members of the committees are nominated by the Commission from highly qualified scientists who have competence in the fields relating to the committees.

II. SCIENTIFIC COMMITTEE FOR FOOD

In questions concerning the toxicological evaluation of food and food ingredients the Commission may consult the Scientific Committee for Food (SCF), established in 1974. In general, the acceptance at Community level of a process or manufacture of a foodstuff or the composition of a food or food ingredient is dependent upon the assurance that there is no hazard to the health or safety of the consumer of such food.

The advice of the SCF is normally published. It is indicative of the practical nature of this contribution that the Commission has accepted the advice of the SCF, in principle, on every occasion. So far, ten series of reports have been published and one more is in press.

A. Testing Guidelines

The SCF issued guidelines in 1980 for the safety assessment of food additives.[1] For the purpose of these guidelines, the SCF considers as a food additive any substance that is added to food at any stage during manufacture with the purpose of changing its characteristics. The SCF stresses that no fixed protocol for conducting the toxicological examinations should be laid down, to be followed rigidly in every case. Rather, a framework is proposed for planning a sequence of steps enabling the safety-in-use of a food additive to be established.

The guidelines deal, among other things, with genetic toxicity studies. It is recommended that any new food additive should be investigated for mutagenic potential by a testing procedure that covers both gene and chromosome damage, both in vitro and in vivo. The SCF recognizes the evidence that somatic cell mutations, as opposed to germ cell mutations, may be associated with the development of certain types of cancer. Therefore the suggested battery is supposed to cover both carcinogenicity and mutagenicity by the selection of screening tests investigating the following endpoints:

1. Detection of gene mutations in bacteria, both with and without the use of mammalian metabolic activation systems
2. Determination of the ability to produce damage to the chromosomes of mammalian cells grown in vitro (metaphase analysis)
3. Testing of the ability to penetrate to and interact with genes in the complex arrangements of DNA in eukaryotic cells in vitro (fungi, yeasts or mammalian cells) or in vivo (*Drosophila melanogaster* or mice)
4. Determination of mutagenic activity in in vivo tests in a mammal, e.g., a metaphase analysis of BM in the rodent, a micronucleus test in the rodent, a dominant lethal test in the rodent
5. Following specific circumstances, assays for primary nonspecific damage to DNA or stimulation of unscheduled DNA synthesis, and for cell transformation using appropriate in vitro cultured mammalian or human cell lines may be desirable, as well as tests for induction of heritable translocations and tests for somatic point mutations in rodents.

The guidelines for carcinogenicity studies of food additives stress the need for studies in both mice and rats unless specific considerations dictate otherwise. The highest dose level for food additives should generally be set at 5% of the diet, because as the SCF says, dose levels in excess of 5% of the diet may cause nonspecific chronic effects and these may, in turn, influence tumor incidence nonspecifically. The results of a carcinogenicity test in mice or rats would not normally be regarded as negative unless there is also a 75% to 80% survival of the animals in each test group after 18 months or 24 months, respectively, from the start of dosing.

The SCF discusses the extrapolation of results in animals to man and states that special considerations apply to substances shown to have carcinogenic and/or mutagenic effects: these may act, either directly or after metabolic activation, with appropriate macromolecular targets in the affected body cells and constitute primary genotoxic compounds for which thresholds probably do not exist. There are also other compounds with similar end effects that do not act by these direct mechanisms, and these are named epigenetic carcinogens or mutagens. The SCF consider the latter substances as having different properties from genotoxic carcinogens, and for these a threshold may exist. The SCF concludes that, in principle, substances shown by adequate bioassays to be potentially genotoxic in animals should not be added to the human food supply, although this consideration may not apply as rigidly to substances acting by epigenetic mechanisms.

For substances with purely heritable mutagenic potential that do not produce any positive results in adequately performed carcinogenicity bioassays, there is, at present, no possibility of making quantitative risk assessments in terms of the induction of increased mutation frequencies in man. Here, only broad categorization of mutagenic activity can be achieved, based on appropriate combinations of the outcome of in vitro and in vivo genetic toxicity tests for different genetic endpoints. The SCF states that interpretation of these results on their own, in terms of human mutagenic hazard, is not possible at present and any evaluation has to be made on the basis of all available toxicological data. Each compound has, therefore, to be considered on its merits.

Finally the SCF stresses that any interpretation and evaluation of test results represents a "state of the art" type of conclusion. The SCF states that because the safety of a food additive can never be proven absolutely, the word "safe" has to be interpreted in some circumstances, as meaning a socially acceptable potential risk under the existing or predicted conditions of consumption or exposure.[1]

A similar report, although not as detailed as the one on food additives, is issued by the SCF covering the toxicological evaluation of substances from materials and articles intended to come into contact with food.[2] The SCF emphasizes that the program of toxicity testing necessary for these compounds depends on their physicochemical properties, their chemical structure, a knowledge of the toxicity of related compounds, and the quantity migrating into the food. The normal demand of SCF will be that such substances should be tested for long-term toxicity and the effects on reproduction and teratogenicity. However, under certain circumstances the SCF considers it may not be necessary to test a substance for long-term toxicity, for example where the chemical nature of the substance, the results of metabolic, biochemical, or short-term studies, the degree of man's exposure to the substance, or other relevant information suggest that it is not essential to insist on these long-term studies. The minimum requirement would then be acute and 90-day tests in at least two relevant animal species using oral administration. At that point in time (October 1, 1976) the SCF limited itself to acknowledging the technical developments of mutagenicity tests, especially in vitro screening tests.[2]

B. Opinions on Specific Food Additives

Saccharin was considered by the SCF in 1977.[3] Neither the mutagenicity studies in the Ames test nor the cell transformation assay available to the SCF demonstrated any positive results for saccharin.

The SCF considered the then-available interim report of the Canadian two-generation study in rats on the impurity of saccharin *o*-toluenesulfonamide (OTS) and on 5% sodium saccharin. An increased incidence of bladder carcinoma in the male animals of the F_0 and especially the F_1 generations was reported.

The SCF was aware that in the Canadian study the examination of the bladder sections revealed little evidence of premalignant lesions such as hyperplasia and metaplasia. The tumors found had arisen *de novo* and the criterion for judging malignancy was one of invasion.

The SCF listed the other studies carried out, and the majority of these studies in all other species were deemed negative. On the basis of the animal studies, the SCF endorsed the temporary ADI of 0 to 2.5 mg/kg bw/per day established by JECFA in April 1977 for total intake of saccharin. The SCF stresses, however, that this will have to be reassessed in the light of new information as it becomes available.

After considering all the data, negative as well as positive, the SCF expressed the opinion that it is particularly important that the intake of saccharin by children and pregnant women should be limited. The SCF is quite aware that a risk/benefit assessment might be made for critical population groups such as diabetics and those suffering from diseases originating from obesity, and suggested that these individuals be guided by their physician.[3]

C. Opinions on Flavorings

The SCF expressed its opinion on an appropriate approach for controlling the health hazards associated with the use of natural flavorings containing certain pharmacologically active principles.[4]

The SCF endorsed the approach of the Council of Europe, which is the publication of a list of pharmacologically active principles present in natural flavoring materials, but that require limitation. As an interim measure the SCF has selected those active principles that, in its opinion, deserve particular attention as a risk to health, e.g., if an active principle is likely to produce irreversible toxic effects. The SCF concludes that the residues should be the smallest technologically achievable, and their presence through transfer to food or beverages should arise entirely from the use of the natural

Table 1
LIMITS FOR CERTAIN PHARMACOLOGICALLY ACTIVE PRINCIPLES PRESENT IN THE FINAL FOOD

(Arising entirely through transfer to food or beverages from the use of natural source materials as flavoring or flavorings prepared from these natural source materials)[4]

Substance	Food (mg/kg)	Beverages (mg/kg)	Exceptions (mg/kg)
β-asarone	0.1	0.1	1 in alcoholic beverages 1 in seasonings used on snack food
Coumarin	2	2	10 in certain types of caramel confectionery 10 in alcoholic beverages 50 in chewing gum
Total hydrocyanic acid	1	1	1 for each alcoholic volume % in alcoholic beverages 5 in stone fruit juices 25 in confectionery 50 in marzipan or its substitutes
Total safrole and isosafrole	1	1	5 in alcoholic beverages above 25% alcohol 15 in foods containing mace and nutmeg 2 in alcoholic beverages with less than 25% alcohol
Total thujone (α and β)	0.5	0.5	5 in alcoholic beverages with less than 25% alcohol 10 in alcoholic beverages with more than 25% alcohol 35 in bitters 50 in foods containing sage 100 in sage stuffings

source material as flavoring, or of flavorings prepared from these natural source materials. The SCF has proposed a list of limits for pharmacologically active principles in the final food (Table 1).

D. Opinions on Processing Aids

1. Asbestos

The SCF concluded that although there is convincing evidence that occupationally exposed persons are at risk from inhaled and ingested asbestos fibers, there is at present no clear scientific evidence that persons not occupationally exposed to asbestos are at risk from asbestos fibers in air, water, and food.[4] However, in the opinion of the SCF a categorical answer is not possible. The SCF makes a number of recommendations, mainly stressing the need for avoiding asbestos in the food industry through the development of suitable alternatives.[4]

2. Extraction Solvents

In the latest report the SCF deals with extraction solvents.[5] Among several other types, the halogenated hydrocarbon solvents are considered, and are judged partly on their mutagenic and/or carcinogenic properties.

Chloroform — Hepatotoxic, nephrotoxic, fetotoxic,and carcinogenic properties are considered. The opinion of the SCF is that this substance is unacceptable for use as an extraction solvent in food.

Dichloromethane — Most of the administered dose is excreted unchanged in the expired air. A few short-term experiments on foods extracted with this solvent showed no toxicity from the minute residues of the solvent, or from any potential reaction products with food components. The available short-term tests in rats and mice using drinking water as a vehicle showed hepatotoxicity at practically all levels tested. Cell transformation studies were negative, but in vitro mutagenicity tests in bacteria were positive. The available long-term studies in rats and mice were inadequate for meaningful evaluation, but new studies are either in progress or planned. The SCF considers that the available toxicological evidence is insufficient to establish an ADI, meanwhile, the use of this solvent is temporarily acceptable. In addition, use shall only result in minimum residues, and not in any case exceed 10 mg/kg food. The SCF intends to make a review not later than 1983.

Trichloroethylene — The SCF has found a wealth of data on metabolism as well as short-term, long-term, and mutagenicity studies. The substance has also been used for many years in human anesthesia. An oral carcinogenicity study in mice, in which large doses are used, produced hepatocellular carcinoma and lung tumors in both sexes. Two oral carcinogenicity studies in rats are not considered adequate by the SCF. Other, more recent, studies have identified carcinogenic/mutagenic stabilizers present in the samples tested that make the carcinogenicity findings questionable in the opinion of the SCF. Furthermore, additional inhalation studies in rats and mice, and feeding studies using low levels, have not shown any evidence of carcinogenicity. The SCF considers the use of this substance temporarily acceptable as an extraction solvent for food. The results of well-conducted oral carcinogenicity studies are needed by 1985.

1,2-Dichloroethane — The rat and mouse studies showing a carcinogenic effect were performed on technical grade material, in which any contaminants present were not determined. Another oral long-term study has not shown any carcinogenic effects, and long-term inhalation studies in mice and rats are also deemed negative with regard to carcinogenicity. The substance is mutagenic in the bacterial systems tested but a reproduction study shows no effects on the offspring. The overall evidence on carcinogenicity is considered conflicting by the SCF. Furthermore, the SCF does find some evidence of interaction with certain food constituents to form toxic compounds. The SCF considers this substance not acceptable for use as an extraction solvent for food based upon available data.

Carbon tetrachloride —The data available to the SCF does not allow the Committee to establish an ADI. The substance is rapidly absorbed and has demonstrated considerable hepatotoxicity. The oral long-term studies in rats, mice, and hamsters demonstrate the induction of benign liver tumors and hepatocellular carcinomas. The compound is nonmutagenic for bacteria. The SCF considers this substance unacceptable for use as an extraction solvent for food.

Dichlorodifluoromethane — Data from oral administration are available from short-term studies in rats and dogs, as well as from a long-term study in rats. Mutagenicity has been investigated without revealing any adverse effects. The SCF agrees with the ADI on 0 to 15 mg/kg bw for man established by JECFA in 1975 and thus considers this compound acceptable for use as a solvent for food. The SCF states that it has not considered the environmental implications that might arise from this type of compound.[5]

E. Opinions on Packaging Materials

1. Vinyl Chloride Monomer (VC)

The SCF considered in 1975 that VC is carcinogenic to man as well as to other species, and that VC has been shown to produce other adverse effects on the CNS, bone, and liver.[6] The SCF concludes that the aim should therefore be to take all possible steps to reduce all forms of exposure to VC. The SCF recommends that VC should not be detectable in food or potable water by an agreed method, and attempts should be made to develop generally applicable analytical methods for VC with a sensitivity of the order of 0.001 to 0.002 mg/kg.[6] Based upon the recommendation of the SCF the Commission has issued a directive on VC monomer, limiting the permissible level to no more than 0.01 mg VC/per kilogram of food.[7] Therefore, an official limit for a carcinogenic compound in food has been set.

2. Vinylidene Chloride Monomer (VDC)

The SCF considered VDC in 1979 and 1981.[8,9]

The carcinogenic effect in mice is considered to be specific for this species and related to the cytotoxic action on the mouse kidney as the primary target organ rather than to any direct interaction with cellular DNA. This view is based on the observation that VDC, and its metabolites with an epoxide structure, are eliminated essentially by reaction with glutathione or bind covalently with cellular macromolecules, though only to a minimal extent with DNA.

The toxicological data available from other species demonstrate, in the opinion of the SCF, the absence of any significant toxic potential of VDC for the rat, the chinese hamster, and the rabbit with regard to reproduction and teratology as well as carcinogenicity. On the other hand, the SCF considers the data insufficient to establish a tolerable daily intake (TDI). In establishing TDI values the SCF chooses a particularly cautious approach involving either the choice of a larger safety factor than usual, or the consideration of a very low concentration being present in the packaging materials.

The SCF has no objection on toxicological grounds to accepting the continued use of plastic materials and articles made with VDC and intended to come into contact with foodstuffs, but because the SCF is unable to establish a TDI for this monomer, it recommends that the levels of VDC monomer residues should be reduced as much as possible.[9]

3. Acrylonitrile Monomer (AN)

The hazards to human health of AN were considered in 1979 and 1981.[8,9]

The toxicological data demonstrate, in the opinion of the SCF, that AN administered to rats orally or by inhalation induces tumors in various organs in a dose-related manner. This increase is induced by concentrations in the diet above 1 mg/kg. AN also induces fetal malformations and early resorptions in rats. AN is mutagenic in some bacterial systems, but not in others. The mutagenicity tests in eukaryotic systems, with and without metabolic activation, yield no conclusive results.

The epidemiological studies indicate an increase in total tumor frequency, but there is no clear evidence for an increased occurrence of specific tumors among exposed workers. An increased incidence of symptoms relating to the CNS, cardiovascular system, and hematopoietic system has been reported in exposed workers.

The SCF considered AN to be carcinogenic for animals and possibly for man.[9] The aim should be, in the opinion of the SCF, to take all possible steps to reduce all forms of exposure to AN. The SCF recommends that the levels of AN monomer residue in materials and articles intended to come into contact with foodstuffs should be reduced as much as possible. The SCF stresses that in every instance AN should not be detectable in food by a method of analysis that would be applicable generally to most foods by most control laboratories. The analytical methods available today have a limit of detection of 0.005 to 0.01 mg/kg. The SCF proposes that attempts be made to develop a method with greater sensitivity.[9]

III. SCIENTIFIC COMMITTEE FOR ANIMAL NUTRITION

The Scientific Committee for Animal Nutrition (SCAN) was established in 1976 and may be consulted by the Commission on any question of scientific or technical origin relating to animal nutrition and health. Especially, SCAN may be consulted concerning the composition of animal feedingstuffs, treatments that may alter them, additives as well as substances, or products that may be considered undesirable in feedingstuffs.

SCAN has now finished its second series of reports.[10] In the second series SCAN issued its guidelines for the evaluation of feed additives. In order to clarify the biological consequences of the proposed use of a feed additive SCAN lists the toxicological examinations needed. Concerning mutagenicity it is said that examination of the potential mutagenicity, including in vitro screening tests with metabolic activation, may add to the evaluation of toxicity. Carcinogenicity studies are covered by the term chronic toxicity. Testing for chronic toxicity is, in general, performed in two species of rodents, given three dosages orally. In order to evaluate the carcinogenic potential, attention is given to time to occurrence, histological types of tumors, and their frequency.

None of the opinions expressed so far by SCAN give any clear indication how the information on carcinogenic/mutagenic potential is applied in practical terms.

IV. SCIENTIFIC COMMITTEE FOR PESTICIDES

The Scientific Committee for Pesticides (SCP) was established in 1978 because the Commission wished to submit to the Council two proposals concerning maximum levels for pesticide residues.

A. Pesticide Residues in Cereals and Foodstuffs of Animal Origin

The SCP issued a report in 1979 with the Committee's opinion on the proposal of the Commission concerning maximum levels for certain pesticide residues in cereals and foodstuffs of animal origin.[11] In general, the proposals of the Commission are supported by SCP without explaining the considerations underlying them. Only in the cases of 1,2-dibromoethane and 1,2-dichloroethane did the SCP feel that the levels proposed called for a special comment. The SCP found indications that both compounds may induce mutations, and recent studies have revealed that they are also most likely carcinogenic in both rats and mice. The SCF, considering the same data on carcinogenicity, found the evidence conflicting. In the light of this information, the SCP has considered that these compounds should be regarded as having potential carcinogenic properties in man and their use as food fumigants should therefore be discouraged. The maximum residue levels should be as low as possible, and the values recommended by the SCP have consequently been set at or about the current limits of determination.[11]

B. Ethylene Oxide as a Fumigant of Foodstuffs

The report on ethylene oxide as a fumigant of foodstuffs was released in December 1979. Both ethylene oxide and its reaction products, mainly ethylene chlorohydrin and ethylene bromohydrin are discussed. The SCP has found very variable residue levels that ranged up to 5000 mg/kg under some circumstances. Because of limited use, the actual exposure of man to its dietary residues is generally low. Only consumers of large amounts of highly spiced foodstuffs or herbs drawn from limited sources treated under poor fumigation conditions are, in the opinion of the SCP, likely to receive appreciable quantities of ethylene chlorohydrin and other as yet unidentifiable reaction products.

The SCP does note the mutagenic activity of ethylene oxide in a variety of test systems and concludes that, as various epoxides have shown an oncogenic activity in animals, ethylene oxide has to be suspected of carcinogenic properties.[12] Also, a Swedish study that reported three cases of leukemia among exposed workers is cited, as well as a negative long-term study in rats receiving food fumigated with ethylene oxide. The latter study is considered inadequate by the SCP because of technical limitations. Ethylene chlorohydrin is mutagenic in various systems, and these findings may be indicative of a possible carcinogenic effect, in the opinion of the SCP. No adequate carcinogenicity studies on ethylene chlorohydrin have been reported to the SCP.

In the light of the available information, the SCP is of the opinion that the potential hazards to the consumer associated with the continued use of ethylene oxide for fumigation of staple foods, e.g., cereals, milk powder, fruits, and vegetables are such as to indicate that a total prohibition of these uses should be envisaged. In addition, the SCP agrees that the use of ethylene oxide on herbs and spices, which are minor commodities, should be kept under review with a view to its discontinuance as such, when acceptable alternative methods of treatment become available. The SCP also agrees that the maximum levels for ethylene oxide in cereals should be extended to the metabolites ethylene bromohydrin and ethylene chlorohydrin.[13] The maximum levels should be set at or about their lower limits of detection: ethylene oxide 0.2 mg/kg, ethylene bromohydrin 0.1 mg/kg, and ethylene chlorohydrin 1 mg/kg.

V. CONCLUSIONS

The problems concerning the steady supply of healthy food, without addition or contamination with chemicals at levels that might create a risk to human health, are handled by a number of scientific committees, national experts, consumer groups, and producers before the Commission finalizes its proposals for regulation.

The guidelines according to which the toxicological examinations are performed differ from one scientific committee to another and from one part of the regulatory body to another. However, there is an effort within the Commission to harmonize the testing guidelines in order to get those guidelines developed by the various scientific committees to agree with each other and with those presently prepared by OECD.

It should be realized that, in the final development of regulatory actions and legal instruments within the EEC, the recommendations by the scientific committees are used by the Commission in its discussions with national experts and representatives of the member states.

Any regulation concerning mutagens and carcinogens in food will be based on a proposal from the EEC Commission which after consultation with the Economic and Social Committee and the European Parliament, submits this to the Council of Ministers for final acceptance, alteration, or rejection.

Changes in such regulations will either be possible through negotiations between the member states and the Commission (Committee procedure) or will have to go through the Council again.

Finally, it should not be forgotten that in areas where no EEC regulations exist (or are not proposed) each member state can have its own regulations pertaining to carcinogens and mutagens in food.

REFERENCES

1. Report of the Scientific Committee for Food, Food — science and techniques, 10th ser., Commission of the European Communities, Brussels, 1980.
2. Report of the Scientific Committee for Food, Toxicological evaluation of a substance for materials and articles intended to come into contact with foodstuffs, 3rd ser., Commission of the European Communities, Brussels, 1977.
3. Reports of the Scientific Committee for Food, 4th ser., Commission of the European Communities, Brussels, 1977.
4. Reports of the Scientific Committee for Food, 9th ser., Commission of the European Communities, Brussels, 1979.
5. Reports of the Scientific Committee for Food, 11th ser., Commission of the European Communities, Brussels, 1981.
6. Report of the Scientific Committee for Food, 1st ser., Commission of the European Communities, Brussels, 1977.
7. Council Directive of January 30, 1978, On the approximation of the laws of the member states relating to materials and articles containing vinyl chloride monomer and intended to come into contact with foodstuffs, *Off. J. Eur. Commun.*, 21, (No. L. 44), 15, 1978.
8. Minutes of the 24th meet., Scientific Committee for Food, Brussels, March 22 and 23, 1979.
9. Minutes of the 31st meet., Scientific Committee for Food, Brussels, January 15 and 16, 1981.
10. Reports of the Scientific Committee for Animal Nutrition, 2nd ser., Commission of the European Communities, Brussels. 1980.
11. Report of the Scientific Committee for Pesticides, Maximum levels for certain pesticide residues in cereals and foodstuffs of animal origin, Commission of the European Communities, Doc. 3051/VI/79-E-final, 1979.
12. Report of the Scientific Committee for Pesticides, The Use of Ethylene Oxide as a Fumigant of Foodstuffs, Doc. 3532/NI/79-E-final, Commission of the European Communities, Brussels, 1979.
13. Summary report of the 5th meet., Scientific Committee for Pesticides, Doc. 920/VI/80-EN, Commission of the European Communities, Brussels, 1980.

INDEX

A

B

C

E

F

I

Q

R

S

T